MODERN CONTROL OF DC-BASED POWER SYSTEMS

MODERN CONTROL OF DC-BASED POWER SYSTEMS

A Problem-Based Approach

MARCO CUPELLI, ANTONINO RICCOBONO,
MARKUS MIRZ, MOHSEN FERDOWSI AND
ANTONELLO MONTI

Academic Press is an imprint of Elsevier
125 London Wall, London EC2Y 5AS, United Kingdom
525 B Street, Suite 1650, San Diego, CA 92101, United States
50 Hampshire Street, 5th Floor, Cambridge, MA 02139, United States
The Boulevard, Langford Lane, Kidlington, Oxford OX5 1GB, United Kingdom

British Library Cataloguing-in-Publication Data
A catalogue record for this book is available from the British Library

Library of Congress Cataloging-in-Publication Data
A catalog record for this book is available from the Library of Congress

ISBN: 978-0-12-813220-3

For Information on all Academic Press publications
visit our website at https://www.elsevier.com/books-and-journals

Publisher: Jonathan Simpson
Acquisition Editor: Lisa Reading
Editorial Project Manager: Katie Chan
Production Project Manager: Vijayaraj Purushothaman
Cover Designer: Greg Harris

Typeset by MPS Limited, Chennai, India

CONTENTS

LIST OF FIGURES

LIST OF TABLES

LIST OF CONTRIBUTORS

Edoardo DeDin
Institute for Automation of Complex Power Systems, E.ON Energy Research Center, RWTH Aachen University, Aachen, Germany

Sriram K. Gurumurthy
Institute for Automation of Complex Power Systems, E.ON Energy Research Center, RWTH Aachen University, Aachen, Germany

LIST OF ACRONYMS

2DOF 2 degree of freedom
AC alternating current
BIBO bounded-input, bounded-output
CCM continuous conduction mode
CLF control Lyapunov function
CPL constant power load
DAB dual active bridge
DC direct current
DCM discontinuous conduction mode
DSP digital signal processor
FPGA field programmable gate array
HiL hardware in the loop
ISPSs integrated shipboard power systems
LRC line regulating converter
LQG linear quadratic Gaussian
LQR linear quadratic regulator
LSF linearization via state feedback
LSS load shedding scheme
LTI linear time invariant
MIMO multiple input multiple output
MMC modular multilevel converter
MVDC medium-voltage DC
NI national instruments
PCC point of common coupling
PD proportional derivative
PI proportional integral
PID proportional integral derivative
POL point of load converter
PWM pulse width modulated
RTDS real-time digital simulator
RTOS real-time operative system
RMS root mean square
SAB single active bridge
SISO single input single output
SMC sliding mode control
VSS variable structure system
VMC voltage mode control

SYMBOLS USED IN THIS WORK

	USA ANSI Y32	EUROPE IEC 60617	This Work
Resistor			
Inductor			
Capacitor			
Current source			
Voltage source			+

PREFACE

Control Engineering is a very fascinating topic. The whole idea of shaping the full behavior of a system thanks to a mathematical design process is the heart of the engineering philosophy and of the engineering way of thinking. Possible applications are in any field of our society and this also makes control engineering a fascinating research and application field. The requirements to be valuable in a large variety of applications made the control theory evolve more and more in the direction of applied mathematics. Such abstraction created incredibly powerful results that found innumerable applications in every field.

On the other hand, this process of abstraction made the connection to the real problems not so obvious, particularly from the point of view of young engineers still trying to grasp the basics of a specific field. The goal of this book is to try to fill this gap for the power engineering field.

This idea of the book started from the perception that many graduate level engineers in power systems, while feeling educated in the control area, did not find a good way to use their skills in a real design process. In a nutshell, many students felt that they had great analytical knowledge in the analysis phase and very little understanding of the design phase. Filling this gap in a very generic way is basically impossible: on the other hand, when the application field is fixed, it is much easier to find a way to define a design approach and to link theory to practical considerations.

The main idea of this book is to help students in the process of understanding how to use control knowledge in the practical field of power system engineering. A specific up-to-date challenge is used as a reference to depict practical considerations in the application of a variety of modern control solutions. In this sense, this book is not just another control theory book but rather a process of educating power engineers to exploit control knowledge.

From an education point of view, we use a problem-based approach. Direct Current (DC) Technology is emerging as a valid alternative to classical Alternative Current (AC) systems. This transition is also happening together with a growing penetration of power electronics in power systems. The same process of transformation has already happened in other power applications, such as ships. All in all, the transformation means also that control is becoming a more critical skill for power engineering. The book uses the challenges introduced by DC technology to review how different control solutions could be applied in modern power grids.

For this reason, we do not enter into the details of the mathematical formulation and demonstration. A lot of excellent books are already available for this purpose [1]. Here the idea is to have a quick review of the theoretical prerequisites and then to focus on the understanding of how to use a control design formulation in a given context. For this reason a significant set of simulation results is also provided.

The hope is to empower the students with a set of design tools and with the right level of understanding that enables them to select the right tool for the right problem.

DC Power Systems are deemed to be solutions for high efficiency power conversion scenarios. Examples of those include Off-shore Windfarms, Industrial Power Grids, Microgrid Applications, More Electric Aircraft, Electric Vehicles, and All Electric Ships. The last three cases have been of particular interest as a full DC distribution has been realized or planned for the future. These types of systems have in common that the generation capacity is limited and that the maximum load demand may exceed the available generation capacity. Furthermore, the installed inertia is very low. This characteristic is expected to be more evident when in 30 years our energy generation will move towards a fully renewable model.

Therefore, DC microgrids offer a compelling use case to study the challenges that the future power engineers will face in 30 years. As result, this book places an emphasis on parallel source and load DC power systems.

The main goal of this book is to capture this knowledge in a structured way. To support the migration from AC to DC Systems, the actual power engineers have to learn about their peculiarities. In the current context, DC Systems are a domain of the Power Electronics research field while classical power engineers lack the necessary background in power electronics or control theory to fully master this topic. The contributors' research activity delivered a significant amount of knowledge that the authors would like to share with this book. This book aims at:

- being used in an interdisciplinary university program to teach about future challenges of DC Grids in a problem-based context;
- practicing power engineers who will face the challenge of integrating DC grids in their existing architecture.

Stability of DC Microgrids is an interdisciplinary topic covering: Power Electronics, Power Systems, and Control Engineering aspects.

Typically, Power Electronics and Control for Power Electronics books deal with the control of a single converter, or at most they extend to the cascaded systems, while leaving multisource and load systems not covered. Power System books classically do not cover DC Power Systems at all. Applications of Control Theory are often only shown at stand-alone converter level, i.e., loaded by a simple resistive component and supplied by an ideal voltage source, without looking at the system-level aspect. Stabilizing Multiconverter Control is not covered at all.

In general, a Medium Voltage DC (MVDC) power system, independently from the specific application, can be seen as an example of a power electronic power system. A power electronic power system is an electrical power network in which each device, generation, or load is connected to the distribution system via a power converter. Such an idea started in the United States in the framework of the so-called All Electric Ship research program [2]. In the framework of the Electric Ship Research and Development Consortium (ESRC) a notional integrated Power System (IPS) was developed proposing DC as a way to increase flexibility in the power management. The goal was to overcome a set of challenges such as:

- Reduction of weight on board: with the use of DC it becomes possible to imagine high speed generators based on lighter electrical machines.
- Fast reconfiguration and flexible load dispatch to accommodate dynamic load management.

This study and understanding of this new type of power system created interesting new theoretical links between power system and control theory. For the first time, the dynamics of a power system were fully determined by feedback control loops instead of depending, at least partly, on the dynamic of rotating machines. Stability of a power bus becomes a control challenge related to the interaction of generator converters and load converters.

More recently, these challenges have been also introduced in terrestrial power systems with the goal to understand how DC could help increase flexibility in the distribution system. Significant effort is currently performed at RWTH Aachen in the framework of the Forschungscampus (Research Campus) Flexible Electrical Network in a tight cooperation between university and industry [3].

From an engineering point of view, this means a wider range of degrees of freedom on how the dynamics of the system will be determined. This also means that the design of each converter encompasses two different aspects: The local control of the correspondent generator or load but also the interaction with the other converters in the same bus.

The book starts with an overview of the challenges and opportunities offered by the application of DC Technology in ship power systems since this was the sector that really drove the evolution of the application in the first stage.

The second chapter presents the problem of the interaction of generation-load converters in terms of the formulation of the Constant Power Load (CPL) problem. This simple schema is a great way to introduce the dynamic interaction issue in a very simple way. A stiff control of the power demand as typically performed by power electronic driven load introduced the idea of a negative impedance reaction able to reduce the dumping capability of the network. The issue has been well known for a long time for low power DC distribution systems (such as power supply) but in the high power area it presents interesting aspects related to the bandwidth of the different converters involved in the process. Given that in MV the switching frequency of the converter is rather limited, the interaction becomes definitely more complicated than in the low power case. This analysis introduces the student to the problem of generator-load interaction in a multiconverter system.

A deeper analysis of the dynamic interaction is then presented in Chapter 4, Generation Side Control, where a full process of small-signal model synthesis is derived. In this chapter the analysis and design of a single converter is presented together with classical control solution. From the analysis of the single converter system the significant transfer functions describing the closed-loop behavior of the converter are derived and then used to look at the cascading situation. The system analysis is complemented with a review of the possible load representation. The chapter describes also practical approaches to on-line system identification techniques that can be used to extract the small-signal representation in a laboratory setting.

Chapter 5, Control Approaches for Parallel Source Converter Systems, presents a complete review of classical and modern control methods applied to the specific challenge. After reviewing basic stability criteria, the state-based approach is reviewed presenting key concepts such as controllability and observability together with pole placement approaches.

The knowledge of Chapter 5, Control Approaches for Parallel Source Converter Systems, is applied to the specific application of Generation Side Control for Ship Power Systems in the following chapter. This is one of the examples used in the book to map theory to concrete design challenges.

Chapter 7, Hardware in the Loop Implementation and Challenges, extends the analysis to a wide set of solutions able to tackle the power system control at the system level: in a nutshell the challenge can be summarized as Paralleled Source Converter Systems.

The following chapters present a large set of system simulation analysis. The different control approaches are compared and the features of the possible approaches are discussed.

The last chapter completes the simulation analysis with a review of the Hardware in the Loop technique to validate control solution.

The authors hope the readers will find this overview to be a good instrument to learn how to use control theory in power systems to continue to create a tighter interaction and greater exchange between these two fascinating fields of electrical engineering.

REFERENCES

[1] K. Ogata, Modern Control Engineering, Prentice Hall, Upper Saddle River, NJ, 2010.
[2] T. Ericsen, Advanced electrical power systems (AEPS) thrust at the office of naval research, 2003 IEEE Power Engineering Society General Meeting, 2003.
[3] A. Korompili, A. Sadu, F. Ponci, A. Monti, Flexible Electric Networks of the Future: Project on Control and Automation in MVDC grids, in: International ETG Congress 2015; Die Energiewende - Blueprints for the new energy age, Bonn, Germany, 2015, pp. 1–8.

INTRODUCTION

Shipboard power systems present a special set of challenges related to stability. In this research work, the Integrated Shipboard Power Systems (ISPSs) are the focus area. The US Navy is pushing Direct Current (DC) as the distribution technology in their future ships, which include a DC power system with multiple power elements.

The usage of DC in shipboards has certain advantages, such as the reduction of up 27% of the fossil fuel consumption in the case of the "Dina Star of Myklebusthaug Offshore" in comparison with Alternating Current (AC). Likewise, DC technology enables an adjustment of the motor speed and therefore maximum efficiency at every operating point. The efficiency gains could also be observed in the dynamic position system of a shipboard, where it maintains its position and has a high impact on its fuel consumption.

The implementation of DC also results in a reduction of noise emissions in the machine room as less vibration are present. A DC distribution system is also advantageous for the usage of batteries or other energy storage in shipboards which also contributes towards the increase in energy efficiency and the reduction of CO_2 emissions.

According to Doerry [1] there are many reasons for employing an MVDC System, including, but not limited to:

- "Power conversion equipment can operate at high frequencies, resulting in relatively smaller transformers and other electromagnetic devices."
- "Without the skin effect experienced in AC power transmission, the full cross section of a DC conductor is effective in the transmission of power. Additionally, the power factor does not apply to DC systems. Depending on the voltage selection, cable weights may decrease for a given power level."
- "Paralleling power sources only require voltage matching and do not require time-critical phase matching."

Further advantages are reduction of power system weight-to-space ratio and the reconfigurability in case of fault and enhanced power quality.

An ISPS consists typically of multiple generators, loads, power lines, power electronic devices, and filters. The fact that a small isolated distribution system with limited generation capability and low rotating mass has to accommodate fast changing load levels allows certain analogies to the terrestrial DC microgrid or other application fields as aircrafts, submarine vehicles, or space stations. In this context, the developed and presented concepts can be a vital contribution to the terrestrial DC microgrids or DC distribution systems in general, since every ship offers a greenfield deployment of new technology.

Here is a short introduction about the background and motivations of this book. The main research objectives and contributions are presented in a compact way. Finally, a structured outline is presented.

In DC ISPSs the individual loads, or groups of loads, are normally fed through power electronic converters directly connected to the distribution Medium Voltage DC (MVDC) bus. Power converters are Variable Structure System (VSSs): they are by definition nonlinear. When observing the system from the DC bus the load converters exhibit a Constant Power Load (CPL) behavior, which is also nonlinear.

In order to maintain the load power constant while being under the influence of fast current and/or voltage disturbances, the load converters are tightly controlled; as a consequence, the CPLs present towards the DC bus a negative incremental resistance behavior. Under certain conditions, this behavior may lead to instability of the MVDC bus [2]. Conditions under which the instability takes places and methods for the stability analysis of MVDC systems have been reviewed and discussed in Ref. [3].

Traditionally, the voltage stability assessment of DC systems is focused on ensuring stability with small variations around one given operating point, thus small-signal stability. The small-signal assessment has been executed in cascaded systems with the stability analysis in the frequency domain, and the Middlebrook criterion and its further developments [1]. Up to now, adequate methods for assessing stability at large-signal level are still not widespread or only available at the level of AC systems.

The MVDC ISPSs efforts have been concentrated in the design of power converters interfacing the loads towards the MVDC distribution. In order to prevent the MVDC ISPS from voltage instability, the load power converters have to compensate the small-signal negative resistance behavior of the loads, either through dedicated control algorithms [4,5] or through forcing disconnection before the MVDC system might operate out of the stable operative range [6–10].

Nevertheless, the load side stabilizing control approach on ISPSs holds an unnecessary restriction as it possesses difficulties in standardization and difficulties in using commercial off-the-shelf converters; hence, every load converter would have to include a special stabilizing control and would have to be adapted to the present system before actually inserting a load. This required adaptation would also be conflicting with the desired Plug&Play feature; in addition, the number of load converters, blue (dark gray in print version) circled in Fig. 1, is certainly larger than the number of generation-side converters, red (gray in print version) circled in Fig. 1.

Replicating a feature of terrestrial power systems on ISPSs such that the generation side is responsible for system stability, no matter which load is connected or disconnected, will thus enable the Plug&Play capability for every load, and the usage of common off-the-shelf components.

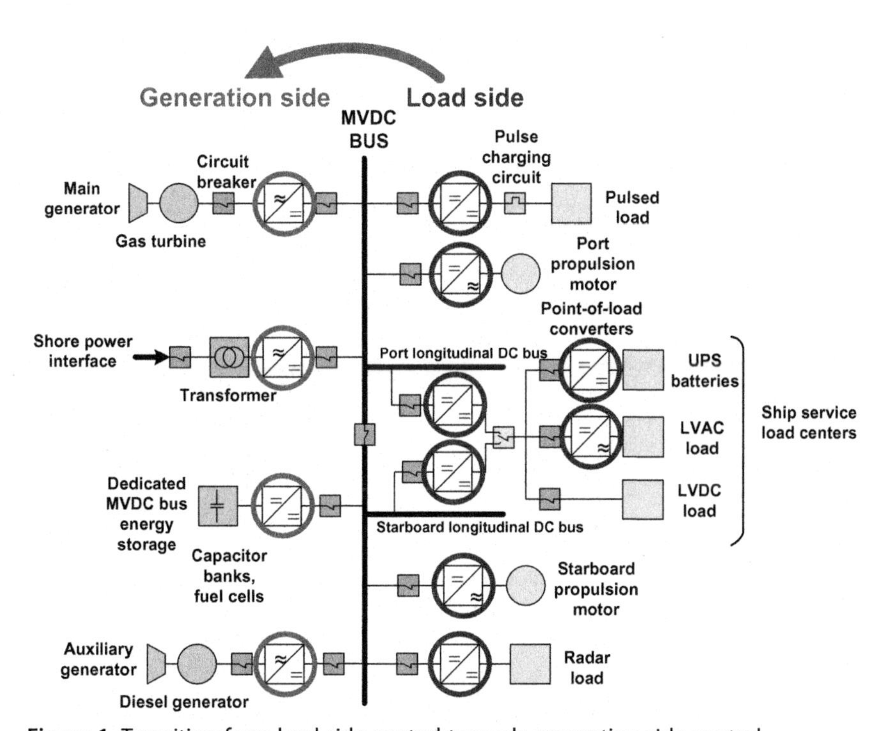

Figure 1 Transition from load side control towards generation side control.

Furthermore, the generation side stabilizing control approach allows the downsizing of output and input filters while guaranteeing the operational objectives of survivability in the presence of wide variations of load or generation capacity.

In a nutshell, this book touches on the following areas:

- A systematic overview from a system integration perspective of the interactions that might impair MVDC bus stability. The proposed approach takes into account:
    - The load type and its selected representation.
    - The converter type used to interface the load and the sources towards the MVDC bus.
    - The impact of the selected regulation strategies in each converter and the interaction between them.
    - The proof of small-signal stability.
    - The methodology for online stability assessment.
- A systematic overview of centralized and decentralized control theory and its application on MVDC bus stability. This study takes into account:
    - The usage of linear state-space control methods for stabilizing MVDC grids.
    - The usage of robust control methods for stabilizing MVDC girds.
    - The usage of nonlinear control theory and guaranteeing large-signal stability by applying Lyapunov theory.
    - The application of decentralized state estimation concepts for partitioning the network in smaller subsystems.
- Important aspects that need to be considered before implementation in a life-like environment of the selected control techniques. These include the potential challenges which may originate from a nonideal component behavior in a physical implementation, and verify that the control techniques are successful in systems which exhibit high dynamics

REFERENCES

[1] A. Korompili, A. Sadu, F. Ponci, A. Monti, Flexible Electric Networks of the Future: Project on Control and Automation in MVDC grids, in: International ETG Congress 2015; Die Energiewende - Blueprints for the new energy age, Bonn, Germany, 2015, pp. 1–8.

[2] A. Monti et al., Ship power system control: a technology assessment, in: IEEE Electric Ship Technologies Symposium, 2005., Philadelphia, PA, 2005, pp. 292–297.

[3] M. Cupelli, M. de Paz Carro, A. Monti, Hardware in the Loop implementation of a disturbance based control in MVDC grids, in: 2015 IEEE Power & Energy Society General Meeting, Denver, CO, 2015, pp. 1–5.
[4] N. Doerry, Naval Power Systems: integrated power systems for the continuity of the electrical power supply, IEEE Electrif. Mag. 3 (2) (June 2015) 12–21.
[5] IEEE Recommended Practice for 1 kV to 35 kV Medium-Voltage DC Power Systems on Ships," *IEEE Std. 1709-2010*, pp. 1, 54, Nov. 2 2010.
[6] S.D. Sudhoff, S.F. Glover, S.H. Zak, S.D. Pekarek, E.J. Zivi, D.E. Delisle, D. Clayton, Stability analysis methodologies for DC power distribution systems, in: 13th Int. Ship Control Systems Symposium (SCSS), Orlando, Florida, April 2003.
[7] A. Riccobono, E. Santi, Comprehensive review of stability criteria for DC power distribution systems, IEEE Trans. Ind. Appl. 50 (5) (Sept.-Oct. 2014) 3525–3535.
[8] J.O. Flower, C.G. Hodge, Stability and transient-behavioral assessment of power-electronics based dc-distribution systems, Part 1:the root-locus technique, in: 1st WSEAS Int. Sub-Conference on Electroscience and Technology for Naval Engineering and All Electric Ship, Athens, Greece, July 2004.
[9] S.F. Glover, S.D. Sudhoff, An experimentally validated nonlinear stabilizing control for power electronics based power systems, SAE 981255, 1998.
[10] T.L. Vandoorn, et al., Power balancing in islanded microgrids by using a dc-bus voltage reference, in: Proc. Int. Symp. Power Electron. Electrical Drives Automation, Motion (SPEEDAM), 2010, pp. 884–889.

FURTHER READING

K. Ogata, Modern Control Engineering, Prentice Hall, 2010.
T. Ericsen, Advanced electrical power systems (AEPS) thrust at the office of naval research, 2003 IEEE Power Engineering Society General Meeting, 2003.
V.N. Chuvychin, N.S. Gurov, S.S. Venkata, R.E. Brown, An adaptive approach to load shedding and spinning reserve control during underfrequency conditions, IEEE Trans. Power Systems 11 (4) (Nov. 1995) 1805–1810.
H. Seyedi, M. Sanaye-Pasand, New centralised adaptive load-shedding algorithms to mitigate power system blackouts, IET Gen. Trans. Distrib. 3 (1) (January 2009) pp. 99–114.
D. Zhiping, S.K. Srivastava, D.A. Cartes, S. Suryanarayanan, Dynamic simulation based analysis of a new load shedding scheme for a Notional Destroyer Class Shipboard Power System, in: IEEE Electric Ship Technologies Symposium, 2007. ESTS '07. pp. 95, 102, 21–23 May 2007.

CHAPTER ONE

Overview—Voltage Stabilization of Constant Power Loads

Throughout this chapter the reader will be introduced to the characteristics of the Constant Power Load (CPL) while being connected to a DC bus in Section 1.1. The Sections 1.2–1.4 present an overview on the previous research activities performed for stabilizing CPLs. Section 1.5 offers a summary of this chapter.

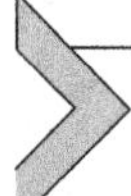

1.1 CONSTANT POWER LOAD CONNECTED TO A DC BUS

The interaction of electrical subsystems around a DC bus can lead to the instability of the latter. Often in these cases, it is associated with the occurrence of instability when CPLs are connected on the DC bus [1–14]. The phenomenon of instability is considered then as the consequence of the connection of a CPL on a DC bus.

When there is an element present which regulates the power consumption of a load (e.g., inverter/actuator, converter/battery converter, drive), it is possible to assume the aforementioned CPL behavior. The control of the load will then set the load to compensate disturbances on the bus. The load does not take into account the electrical state of the network (including DC bus). Thus, it will be seen by the DC bus as a constant power consuming element as shown in Fig. 1.1. The output of the source converter is represented by the voltage source E, which is

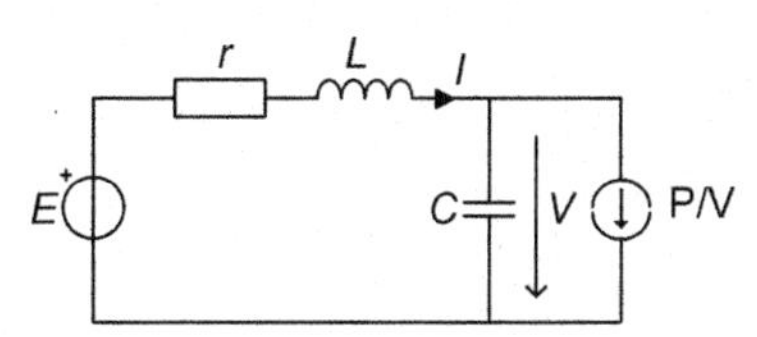

Figure 1.1 Electrical system containing a nonlinear load.

Modern Control of DC-Based Power Systems.
DOI: https://doi.org/10.1016/B978-0-12-813220-3.00001-6

interfaced through a filter circuit consisting of the equivalent series resistor of the inductor r, inductor L and capacitor C with the nonlinear load, which is represented by a current source with the characteristic P/V, V denotes the voltage drop over the capacitor.

The current absorbed by CPL on the DC bus can be modeled as Eq. (1.1) where for a given operating point (V_0, I_0), the product of the load voltage and current is kept constant. The introduction of a CPL in the network implies the appearance of a nonlinearity of the P/V type. The characteristic current–voltage curve across the CPL is represented in Fig. 1.2.

$$I = \frac{P}{V} \tag{1.1}$$

The rate of change of the current can be obtained by a linear approximation, deriving from Eq. (1.1) in the small area around the operating point. This yields the linear equivalent impedance R_{CPL} at the given operating point Eq. (1.2).

$$\frac{\partial I}{\partial V} = -\frac{P}{V_0^2} = -\frac{1}{R_{CPL}} \tag{1.2}$$

From Eq. (1.2) it can be observed that R_{CPL} is dependent on the actual voltage and current. The curve representing the current versus voltage for a CPL in Eq. (1.3) can be approximated by a straight line that is tangent to the power curve at the operating point. The equation for this line is given as follows:

$$I_{CPL} = 2I_0 - \frac{1}{R_{CPL}} V = 2\frac{P}{V_0} - \frac{P}{V_0^2} V \tag{1.3}$$

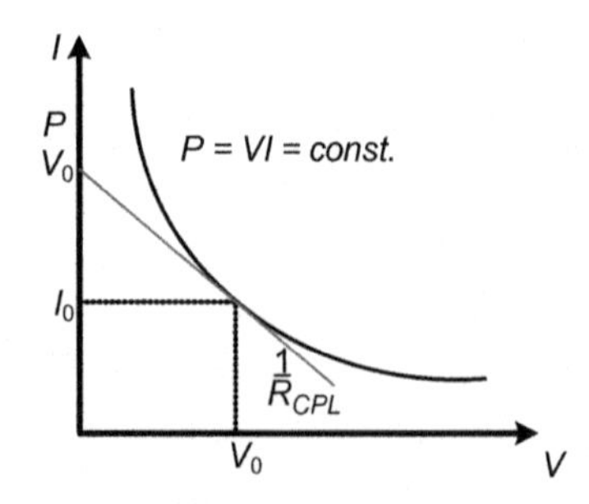

Figure 1.2 Characteristic I–V curve of a Constant Power Load.

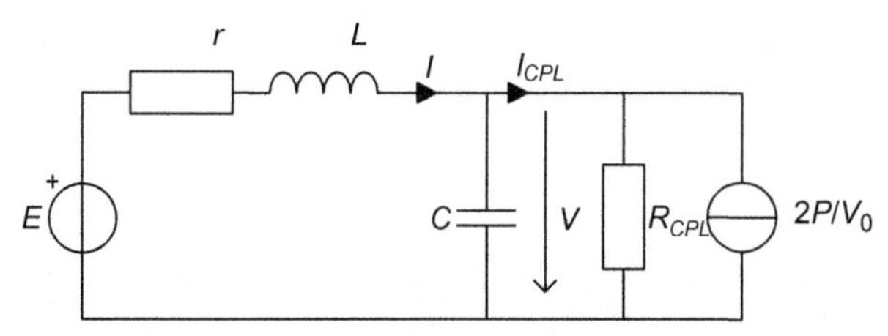

Figure 1.3 Electrical system containing a Constant Power Load, equivalent circuit around operating point.

One can conclude that a CPL around its operating point has a behavior similar to that of a negative resistance in parallel with a constant current source. The diagram corresponding to the electrical behavior of the CPL around an operating point is depicted in Fig. 1.3. The output of the source converter is represented by the voltage source E, which is interfaced through a filter circuit consisting of the equivalent series resistor of the inductor r, the inductor L, and the capacitor C; where V denotes the voltage over the capacitor.

In order to have a mathematical observation of the effect of the CPL and R_{CPL} on the network and to exclusively consider the variations of the system around the operating point lead toward omitting the DC component of Eq. (1.3), the resultant model is described in accordance to Eq. (1.4) and its characteristic polynomial is written in Eq. (1.5).

$$\begin{bmatrix} \dot{V} \\ \dot{I} \end{bmatrix} = \begin{bmatrix} \dfrac{-1}{CR_{CPL}} & \dfrac{1}{C} \\ \dfrac{-1}{L} & \dfrac{-r}{L} \end{bmatrix} \begin{bmatrix} V \\ I \end{bmatrix} \tag{1.4}$$

$$P(\lambda) = \lambda^2 + \left(\frac{r}{L} + \frac{1}{CR_{CPL}}\right)\lambda + \frac{1}{LC}\left(1 + \frac{r}{R_{CPL}}\right) \tag{1.5}$$

From the characteristic polynomial, one can deduce the relationship Eq. (1.6) which states the criteria for the eigenvalues of the system to have a negative real part. The first inequality holds because of the negative sign of R_{CPL} and usually the equivalent series resistor r is very small when comparing the absolute values, while the second will not always be respected since an increase in load power causes an increase in current which has as consequence an increase of R_{CPL} (which goes to 0).

$$\begin{cases} R_{CPL} < -r \\ R_{CPL} < -\dfrac{L}{rC} \end{cases} \tag{1.6}$$

Replacing R_{CPL} by this inequality the relationship (1.7) is obtained, which gives the local stability condition for the system.

$$P < \min\left(\frac{rC}{L}V_0^2; \frac{V_0^2}{r}\right) \tag{1.7}$$

The relation $P < V_0^2/r$ is less restrictive than the relationship $P < V_0^2/4r$ which ensures the existence of an operating point for the system while relation (1.8) gives the stability condition of the system operating points.

$$\begin{cases} R_{CPL} < -r \\ R_{CPL} < -\dfrac{L}{rC} \end{cases} \tag{1.8}$$

It is seen from Eq. (1.8) that a relationship between the stability of system, the sizing of the filter (*r*, *L*, *C*, and indirectly *E*) and the power consumed by the load can be expressed. It is seen that as the capacity of the inverter increases the system becomes more stable and vice versa for the inductance of the filter. In addition, these observations show that the negative resistance is used to excite the system because it "generates" reverse current changes which are triggered by voltage changes. So, the more power is consumed the more its conductivity increases until it compensates the overall damping of the system and becomes unstable.

The two possible assumptions regarding how to deal with a CPL—linearized around an operating point or assuming its nonlinear characteristic—will lead to two approaches for assessing the stability.

1.2 COMPENSATING CPLS BY PASSIVE COMPONENTS

In order to increase the stability in a passive way for the system depicted previously in Fig. 1.1, a filter can be added to the system. This will increase the damping of the DC bus and thus increase its stability. This setup was proposed by the authors in [15], where the authors

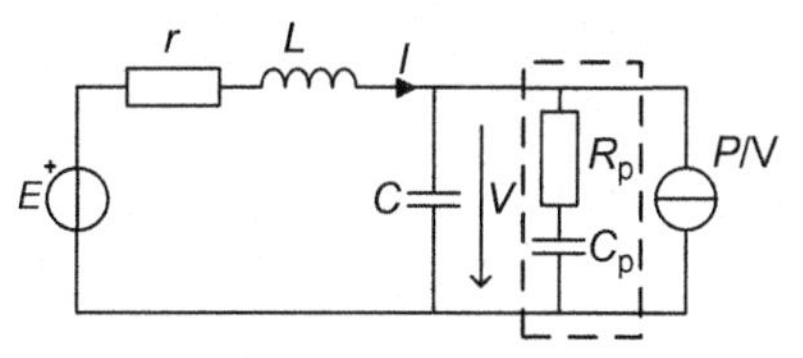

Figure 1.4 Adding a parallel RC filter.

analyzed the addition of three different filter structures: parallel RC, RL, and RL parallel series. The author show that only the structure "RC parallel" depicted in Fig. 1.4 manages to effectively stabilize the system. The addition of this filter is also studied and proposed in [16]. In both cases, the authors rightly point out that such a solution will increase the physical size of the system.

1.3 COMPENSATING CPLS WITH LOAD SIDE CONTROL

In order to increase the stability of the system without changing its structure or increasing its size and weight, it is possible to implement a stabilizing state feedback on the system load.

The destabilizing impact of the CPL can be justified by the fact that its converter control will impose a load power (e.g., mechanical power) independent of the magnitude of voltage and current provided by the network. Its control will compensate for variations induced by the load but does not take into account the changes on the DC bus. Therefore the idea is to provide the stabilizing load control the necessary information about the bus voltage. This information can be used to generate a signal ψ which when added to the reference power, will increase the stability of the system. The principle of this type of control is summarized in Fig. 1.5. This approach was demonstrated in previous publications [17–20].

The aforementioned approach needs to modify the power reference of the load. The disturbance must therefore be as small as possible in order to have the lowest impact on the load. However, it is observable that the more one wants to "stabilize" a system, the greater the stabilizing signal would be. A good compromise must therefore be found in the design of the stabilizer and this requires having easily adjustable stabilization methods.

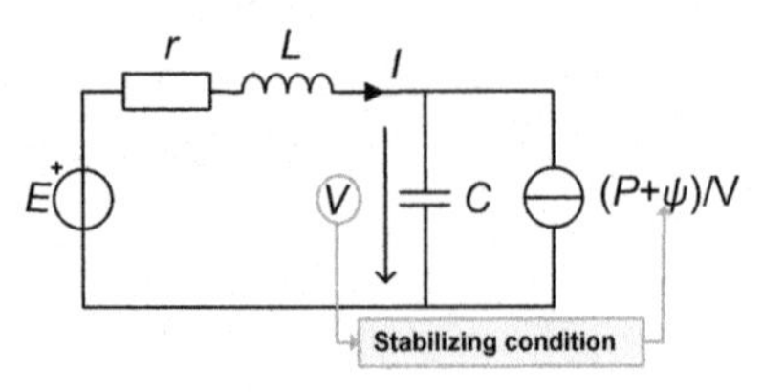

Figure 1.5 Load side stabilizing control.

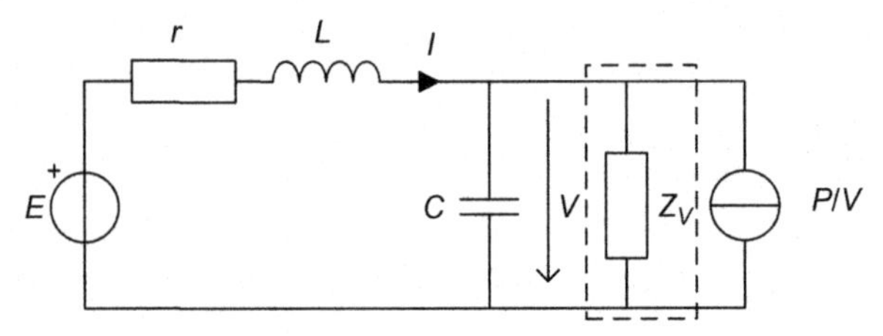

Figure 1.6 Virtual impedance.

Among the active stabilization methods acting on the load several techniques can be highlighted, which are based on the principle of emulating a virtual impedance by control [18,19,21]. Accordingly, the system equipped with its stabilizing control shall be equivalent to an electrical system with virtual impedance Z_v as shown in Fig. 1.6.

This technique presents as its main advantage a clear and instinctive sizing toward the user. That is, in order to improve the stability of the system it is sufficient to increase the size of the "virtual capacity" or "virtual resistance."

The addition of a "virtual resistance" is usually referred to as Active Damping [22], because its application increases the system damping as a real resistance. Contrary to adding a passive component, this resistance is inserted via the converter control system. Its operation principle consists of adding to the reference voltage input of a converter an additional signal. This additional signal lowers the reference voltage as the output current increases. The virtual resistance signal is usually high-pass filtered for two reasons [23]: to avoid a voltage drop in steady state and to make sure that the virtual resistance is only applied during transient conditions. This method can be simply implemented at load side, as the load converter measures its output current for control and protection purposes and the additional signal depends only on the single load supplied.

The impact of these stabilization methods on the stability of the system can be studied using either linear tools for small signal stability or nonlinear tools for large signal stability, e.g., with Linear Matrix

Inequalities [21] or the Brayton Moser Criterion [24]. The authors in [21] studied the influence of a "virtual capacity" on the large signal stability. They use the Takagi–Sugeno (TS) multimodel theory which permits to obtain a set of linear models whose fuzzy weighted combination is equivalent to the original nonlinear system [25].

Other methods available in the literature may present the implementation of a linear correction as it has been presented in Refs. [20,26]. In these cases, the impact of the stabilizing action on the stability of the new system is assessed by studying the influence of the control on the input impedance of the load.

Another approach has been proposed in Ref. [17] where an input–output linearization of the original nonlinear system was introduced. The authors claim that the advantage of this method is to offer a stabilizer whose size is independent of the operating point, thus ensuring the overall stability of the system.

Another approach relies on advanced load shedding techniques to assure stable operation of electric power systems during inadequate generation situations or voltage disturbances [27]. A real-time adaptive Load Shedding Scheme was proposed in Refs. [28,29] to consider a wide range of perturbations using real-time data. This decentralized action will be triggered by a control center which has knowledge about how vital certain loads are for operation. Most real-time load shedding techniques guarantee a fast response to the system instability [30].

1.4 COMPENSATING CPLS WITH DC BUS CONTROL

When the control of a DC–DC converter which feeds the CPL can be modified, a so-called generation side control is possible (see Fig. 1.7). In this case, the system to be stabilized possesses a new control variable—the duty cycle d. It is then possible to integrate the constraints on the stability of the system in the control of the upstream DC–DC converter. The idea of this method was introduced for the first time several years ago while considering the small signal stability of the system [31,32] but there are also applications of this idea which account for the large-signal stability [33].

More recently, this work has been complemented by new control approaches dealing with large signal stability of these systems [10,22,34–44]. These proposed methods of nonlinear control incorporate

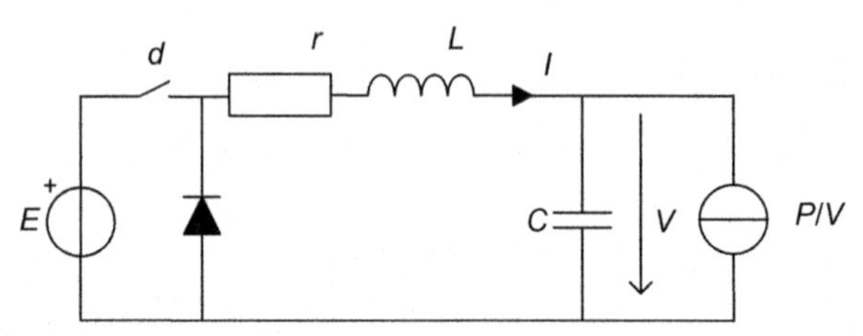

Figure 1.7 Constant power load fed by a DC–DC converter.

the stability of the system in the design of the control laws, with the exception of Ref. [10] where an adaptive state feedback was implemented which can handle disturbances up to a certain level. For example, Ref. [35] relies on the passivity properties of the system, while Refs. [36,38,39] use sliding mode control to increase the stability of the system. A boundary controller is applied in Ref. [40] to stabilize a CPL circuit. This boundary controller implements a state dependent switching. In Ref. [39] the authors use a nonlinear sliding surface control law based on state dependent switching to stabilize a microgrid.

The application of a synergetic control is presented in Refs. [41] and [42], while stabilizing CPL circuits. This nonlinear control has similarities with sliding mode, but offers an additional parameter T which defines the speed of convergence toward the manifold. Both sliding mode control and synergetic control exhibit the model order reduction characteristic while the system is on the sliding surface/manifold. One main difference between the synergetic control and sliding mode control is that the synergetic control operates with a fixed switching frequency.

In Refs. [34] and [37] the authors propose the implementation of a gain which directly compensates the nonlinear term in the CPL and therefore cancels out the source of instability. This technique is called Linearization via State Feedback, and consists of injecting through the converters a signal on their output voltage, which is capable of compensating the nonlinearity of the CPL. Other methods and techniques are also presented in Refs. [43] and [44].

In these methods it is not necessary to perform an adjustment on the CPL, either in damping, impedance change, or power change, to ensure the stability of the system. In those types of control the DC bus voltage response will be controlled to meet the constraints imposed by the CPLs for maintaining stability. Nevertheless, these methods only apply to cases of systems containing a single DC–DC converter and one CPL connected to the bus.

This book will present how some of those control concepts can be applied to systems composed of multiple loads and multiple DC–DC converters connected to the same bus.

1.5 SUMMARY

This chapter presented an introduction to the so-called negative incremental resistance and the resulting endangering of voltage stability. Afterwards, a literature overview about several methods to compensate CPLs was given. Those methods can be divided into two distinct families depicted in Fig. 1.8.

The first one is a passive compensation approach: it relies on installing a passive filter in front of the CPL. This method does not require modification of the control system and is well suited to handle cases where neither the bus feeder control nor the CPL are modifiable.

Nevertheless, the use of such methods will increase the size (weight, mass) of the system which may be very penalizing for some applications, such as aeronautics or naval systems. In these cases, one might consider the implementation of stabilizing controls on the system, which points to the second approach.

The second approach is an active approach. It relies on control techniques for compensating the CPL: depending on whether the DC bus will be controlled or not, it can be further classified as being in load side control or generation side/feeder control.

The load side control relies on adapting CPL and its impact on the stability of the network. Of course, this needs to have access to the control of the load. This method, if it is applied, must be considered during

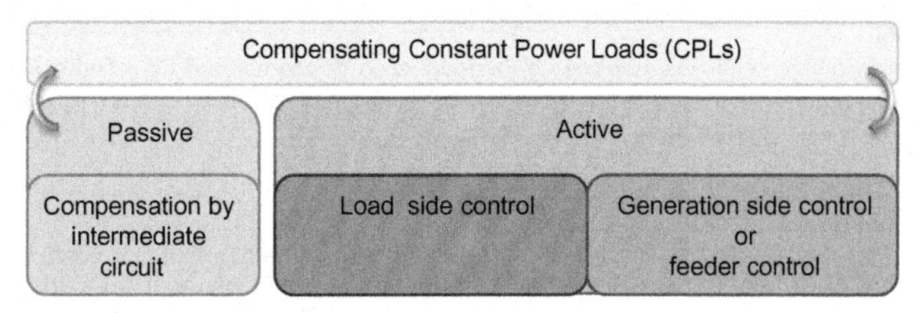

Figure 1.8 Compensating Constant Power Loads.

the implementation of load control. Nevertheless it presents the advantage of being installed on all loads of a network thus treating "locally" the impact of each CPL and could therefore be considered a decentralized control technique.

A generation side or feeder control is applicable in situations where there is access to the control of the bus supplying the converters. In this case, it is possible to implement controls laws on the bus which take into account, either through the addition of compensatory terms or specific nonlinear control laws, the destabilizing impact of the CPL on the system, and thereby increase the stability of it. Furthermore, being based on the interface converters at the generator side/feeder, it allows for using commercial off-the-shelf converters as the interface on the load side.

REFERENCES

[1] C. Rivetta, A. Emadi, G.A. Williamson, R. Jayabalan, B. Fahimi, Analysis and control of a buck DC-DC converter operating with constant power load in sea and undersea vehicles, in: Conference Record of the 2004 IEEE Industry Applications Conference, 2004. 39th IAS Annual Meeting, vol. 2, October 3–7, 2004, pp. 1146, 1153.

[2] W. Du, J. Zhang, Y. Zhang, Z. Qian, F. Peng, Large signal stability analysis based on gyrator model with constant power load, in: 2011 IEEE Power and Energy Society General Meeting, July 24–29, 2011, pp. 1, 8.

[3] K. Seyoung, S.S. Williamson, Negative impedance instability compensation in more electric aircraft DC power systems using state space pole placement control, in: 2011 IEEE Vehicle Power and Propulsion Conference (VPPC), September 6–9, 2011, pp. 1, 6.

[4] A.K. AlShanfari, J. Wang, Influence of control bandwidth on stability of permanent magnet brushless motor drive for "more electric" aircraft systems, in: 2011 International Conference on Electrical Machines and Systems (ICEMS), August 20–23, 2011, pp. 1, 7.

[5] A.P.N. Tahim, D.J. Pagano, M.L. Heldwein, E. Ponce, Control of interconnected power electronic converters in dc distribution systems, in: 2011 Brazilian Power Electronics Conference (COBEP), September 11–15, 2011, pp. 269, 274.

[6] Z. Liu, J. Liu, W. Bao, Y. Zhao, F. Liu, A novel stability criterion of AC power system with constant power load, in: 2012 Twenty-Seventh Annual IEEE Applied Power Electronics Conference and Exposition (APEC), February 5–9, 2012, pp. 1946, 1950.

[7] P. Magne, B. Nahid-Mobarakeh, S. Pierfederici, A design method for a fault-tolerant multi-agent stabilizing system for DC microgrids with Constant Power Loads, in: 2012 IEEE Transportation Electrification Conference and Expo (ITEC), June 18–20, 2012, pp. 1, 6.

[8] W. Du, J. Zhang, Y. Zhang, Z. Qian, Stability criterion for cascaded system with constant power load, IEEE Trans. Power Electron. 28 (4) (April 2013) 1843, 1851.

[9] S. Sanchez, M. Molinas, Assessment of a stability analysis tool for constant power loads in DC-grids, in: Power Electronics and Motion Control Conference (EPE/PEMC), 2012 15th International, September 4–6, 2012, pp. DS3b.2-1, DS3b.2-5.

[10] V. Arcidiacono, A. Monti, G. Sulligoi, Generation control system for improving design and stability of medium-voltage DC power systems on ships, IET Electr. Syst. Transport. 2 (3) (September 2012) 158, 167.
[11] G. Sulligoi, D. Bosich, L. Zhu, M. Cupelli, A. Monti, Linearizing control of shipboard multi-machine MVDC power systems feeding Constant Power Loads, in: 2012 IEEE Energy Conversion Congress and Exposition (ECCE), September 15–20, 2012, pp. 691, 697.
[12] S.D. Sudhoff, K.A. Corzine, S.F. Glover, H.J. Hegner, H.N. Robey Jr., DC link stabilized field oriented control of electric propulsion systems, IEEE Trans. Energy Convers. 13 (1) (March 1998).
[13] A. Emadi, B. Fahimi, Ehsani, On the concept of negative impedance instability in the more electric aircraft power systems with constant power loads. SAE J. 1999-01-2545.
[14] A. Emadi, A. Khaligh, C.H. Rivetta, G.A. Williamson, Constant power loads and negative impedance instability in automotive systems: definition, modeling, stability, and control of power electronic converters and motor drives, IEEE Trans. Vehicular Technol. 55 (4) (July 2006) 1112. 1125.
[15] M. Cespedes, L. Xing, J. Sun, Constant-Power Load System stabilization by passive damping, IEEE Trans. Power Electron. 26 (7) (July 2011) 1832, 1836.
[16] S. Girinon, H. Piquet, N. Roux, B. Sareni, Analytical input filter design in DC distributed power systems approach taking stability and quality criteria into account, in: 13th European Conference on Power Electronics and Applications, 2009. EPE '09, September 8–10, 2009, pp. 1, 10.
[17] A.-B. Awan, S. Pierfederici, B. Nahid-Mobarakeh, F. Meibody-Tabar, Active stabilization of a poorly damped input filter supplying a constant power load, in: IEEE Energy Conversion Congress and Exposition, 2009. ECCE 2009, September 20–24, 2009, pp. 2991, 2997.
[18] A.-B. Awan, B. Nahid-Mobarakeh, S. Pierfederici, F. Meibody-Tabar, Nonlinear stabilization of a DC-bus supplying a Constant Power Load, in: Industry Applications Society Annual Meeting, 2009. IAS 2009. IEEE, October 4–8, 2009, pp. 1, 8.
[19] W.-J. Lee, S.-K. Sul, DC-link voltage stabilization for reduced DC-link capacitor inverter, IEEE Trans. Ind. Appl. 50 (1) (January–February 2014) 404, 414.
[20] X. Liu, A.J. Forsyth, A.M. Cross, Negative input-resistance compensator for a Constant Power Load, IEEE Trans. Ind. Electron. 54 (6) (December 2007) 3188, 3196.
[21] P. Magne, B. Nahid-Mobarakeh, S. Pierfederici, DC-link voltage large signal stabilization and transient control using a virtual capacitor, in: 2010 IEEE Industry Applications Society Annual Meeting (IAS), October 3–7, 2010, pp. 1, 8.
[22] A.M. Rahimi, A. Emadi, Active damping in DC/DC power electronic converters: a novel method to overcome the problems of constant power loads, IEEE Trans. Ind. Electron. 56 (5) (May 2009) 1428, 1439.
[23] D. Bosich, G. Giadrossi, G. Sulligoi, Voltage control solutions to face the CPL instability in MVDC shipboard power systems, in: AEIT Annual Conference – From Research to Industry: The Need for a More Effective Technology Transfer (AEIT), 2014, September 18–19, 2014, pp. 1, 6.
[24] A. Griffo, J. Wang, D. Howe, Large signal stability analysis of DC power systems with constant power loads, in: IEEE Vehicle Power and Propulsion Conference, 2008. VPPC '08, September 3–5, 2008, pp. 1, 6.
[25] M. Takagi, M. Sugeno, Fuzzy identification of systems and its application to modeling and control, IEEE Trans. Syst. Man Cyber. 15 (15) (1988).
[26] P. Liutanakul, A.-B. Awan, S. Pierfederici, B. Nahid-Mobarakeh, F. Meibody-Tabar, Linear stabilization of a DC bus supplying a constant power load: a general design approach, Power Electron. IEEE Trans. 25 (2) (February 2010) 475, 488.

[27] T.L. Vandoorn, et al., Power balancing in islanded microgrids by using a DC-bus voltage reference, in: Proc. Int. Symp. Power Electron. Electrical Drives Automation, Motion (SPEEDAM), 2010, pp. 884–889.
[28] V.N. Chuvychin, N.S. Gurov, S.S. Venkata, R.E. Brown, An adaptive approach to load shedding and spinning reserve control during underfrequency conditions, IEEE Trans. Power Syst. 11 (4) (November 1995) 1805–1810.
[29] H. Seyedi, M. Sanaye-Pasand, New centralised adaptive load-shedding algorithms to mitigate power system blackouts, IET Gen. Trans. Distrib. 3 (1) (January 2009) 99. 114.
[30] Z. Ding, S.K. Srivastava, D.A. Cartes, S. Suryanarayanan, Dynamic simulation based analysis of a new load shedding scheme for a notional destroyer class shipboard power system, in: IEEE Electric Ship Technologies Symposium, 2007. ESTS '07, May 21–23, 2007, pp. 95, 102.
[31] D. Gadoura, V. Grigore, J. Hatonen, J. Kyyra, P. Vallittu, T. Suntio, Stabilizing a telecom power supply feeding a constant power load, in: Twentieth International Telecommunications Energy Conference, 1998. INTELEC, 1998, pp. 243, 248.
[32] J.G. Ciezki, R.W. Ashton, The application of feedback linearization techniques to the stabilization of DC-to-DC converters with constant power loads, in: Proceedings of the 1998 IEEE International Symposium on Circuits and Systems, 1998. ISCAS '98, vol. 3, May 31–June 3, 1998, pp. 526, 529.
[33] S.R. Sanders, G.C. Verghese, Lyapunov-based control for switched power converters, IEEE Trans. Power Electron. 7 (1) (January 24, 1992) 17.
[34] G. Sulligoi, D. Bosich, G. Giadrossi, Linearizing voltage control of MVDC power systems feeding constant power loads: Stability analysis under saturation, in: 2013 IEEE Power and Energy Society General Meeting (PES), July 21–25, 2013, pp. 1, 5.
[35] A. Kwasinski, P.T. Krein, Stabilization of constant power loads in Dc-Dc converters using passivity-based control, in: 29th International Telecommunications Energy Conference, 2007. INTELEC 2007, September 30–October 4, 2007, pp. 867, 874.
[36] C.H. Rivetta, A. Emadi, G.A. Williamson, R. Jayabalan, B. Fahimi, Analysis and control of a buck DC-DC converter operating with constant power load in sea and undersea vehicles, IEEE Trans. Ind. Appl. 42 (2) (March–April 2006) 559. 572.
[37] A.M. Rahimi, G.A. Williamson, A. Emadi, Loop-cancellation technique: a novel nonlinear feedback to overcome the destabilizing effect of constant-power loads, IEEE Trans. Veh. Technol. 59 (2) (February 2010) 650, 661.
[38] Y. Zhao, W. Qiao, A third-order sliding-mode controller for DC/DC converters with constant power loads, in: 2011 IEEE Industry Applications Society Annual Meeting (IAS), October 9–13, 2011, pp. 1, 8.
[39] S. Singh, D. Fulwani, Constant power loads: a solution using sliding mode control, in: IECON 2014 – 40th Annual Conference of the IEEE Industrial Electronics Society, October 29–November 1, 2014, pp. 1989, 1995.
[40] C.N. Onwuchekwa, A. Kwasinski, Analysis of boundary control for buck converters with instantaneous constant-power loads, IEEE Trans. Power Electron. 25 (8) (August 2010) 2018, 2032.
[41] I. Kondratiev, R. Dougal, Synergetic control strategies for shipboard DC power distribution systems, in: American Control Conference, 2007. ACC '07, July 9–13, 2007, pp. 4744, 4749.
[42] I. Kondratiev, E. Santi, R. Dougal, G. Veselov, Synergetic control for DC-DC buck converters with constant power load, in: 2004 IEEE 35th Annual Power Electronics Specialists Conference, 2004. PESC 04, vol. 5, June 20–25, 2004, pp. 3758, 3764.
[43] A. Khaligh, Realization of parasitics in stability of DC–DC converters loaded by constant power loads in advanced multiconverter automotive systems, IEEE Trans. Ind. Electron. 55 (6) (June 2008) 2295, 2305.

[44] A. Khaligh, A.M. Rahimi, A. Emadi, Modified pulse-adjustment technique to control DC/DC converters driving variable constant-power loads, IEEE Trans. Ind. Electron. 55 (3) (March 2008) 1133, 1146.

FURTHER READING

I. Batarseh, K. Siri, H. Lee, Investigation of the output droop characteristics of parallel-connected DC-DC converters, in: 25th Annual IEEE Power Electronics Specialists Conference, PESC '94 Record, vol. 2, June 20–25, 1994, pp. 1342, 1351.

CHAPTER TWO

Small-Signal Analysis of Cascaded Systems

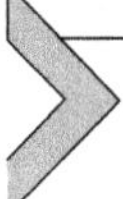

2.1 MVDC SYSTEM CONSIDERATIONS INFLUENCING VOLTAGE STABILITY

The scope of this chapter is to present to the reader how certain simplifications or assumptions impact the system behavior with respect to stability. Starting from a control oriented description of a CPL, it is then indicated to the reader how the system behavior is impacted by choosing different converter topologies and their respective control schemes. This is then later extended to different types of loads which are typically not considered in stability studies and an analysis is undertaken on the impact on a controlled system under the classic control theory (single input single output (SISO)) is altered. For the classical control theory, practical design examples for a cascaded system are given, including small-signal stability assessments at the design phase and during operation with the help of the online system identification approach.

In the medium voltage direct current (MVDC) modeling and simulation in shipboard power systems, normally the zonal or load interface converter is modeled by a step-down converter controlled as a voltage source. The step-down converter is designed and controlled for the purpose to represent the CPL characteristics as a load interface converter. However, in reality the ideal constant power load (CPL) behavior could not be valid in the MVDC system, because of the limitations mentioned as follows [1,2]:

1. Saturation of converter: When the converter saturates (i.e., duty cycle cannot exceed 1) the converter goes into open loop and therefore the forced behavior by the control is no longer valid. Therefore the normal load behavior, e.g., resistive, will be presented rather than the forced CPL behavior. This saturation exists commonly in the real system, as typically the duty cycle of a step-down converter is set for economic reasons close to the saturation limit. On one hand, this

Modern Control of DC-Based Power Systems.
DOI: https://doi.org/10.1016/B978-0-12-813220-3.00002-8

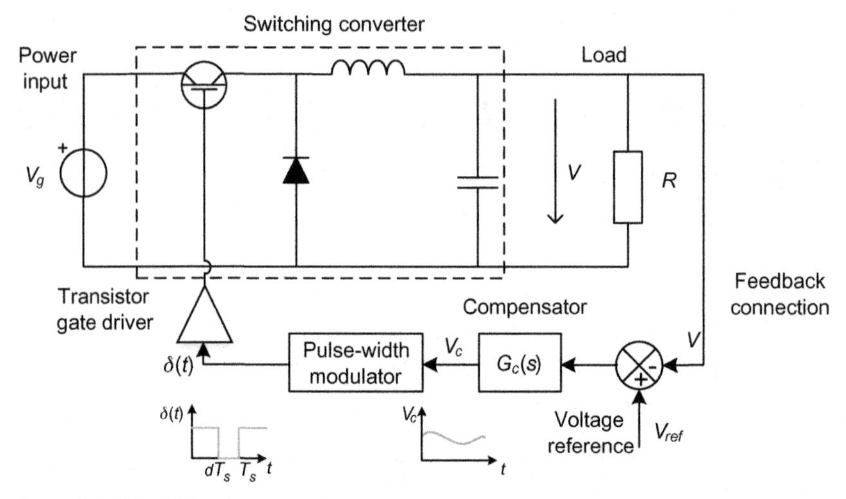

Figure 2.1 DC−DC voltage regulator system.

saturation limits the flexibility of the converters to adjust to disturbances; on the other hand, it actually could also help to stabilize the system.

2. Feedback loop gain and bandwidth: A buck converter which operates as a voltage source and supplies a resistive load is depicted in Fig. 2.1. This setup acts as a CPL and will be explained in detail in the following Section 2.2 followed by a Cascaded System Evaluation in Section 2.3 which is expanded to include the effects of a different load model in Section 2.4.

In Section 2.5 practical Proportional-Integral (PI) and Proportional-Integral-Derivative (PID) control design is presented on a stand-alone converter model. The types of controller covered are single-loop Voltage Mode Control (VMC), single-loop Current Mode Control (CMC), and double-loop VMC where the voltage control is the outer control loop and CMC (or Peak Current Mode Control (PCMC)) is the inner control loop. The procedure on how to obtain a certain targeted closed-loop performance is described in detail and an extensive simulation example is provided. Moreover, network analyzer techniques, i.e., frequency domain validation techniques, are presented to assess in the frequency domain whether the closed-loop performance of converter prototypes meets the requirements.

Section 2.6 extends the discussion about the potential stability problem of cascade systems given in Section 2.3 due to the presence of CPLs presented in Section 2.4. In particular, a comprehensive small-signal stability analysis of based on the well-known Nyquist Stability Criterion is presented. After reviewing the criterion itself and its further extensions, the focus of Section 2.6 moves to the practical aspects of stability analysis by using a cascade system example. It is shown through a simulation example how the interface impedances play a key role on the assessment of the stability of such a system. Given the fundamental importance of interface impedances in the stability analysis, the section also presents an emerging online technique to measure such impedances by using the existing converters of the cascade system. The technique is called Wideband System Identification (WSI) and allows measuring impedances in real time while the system is operating in steady state by using a short-time small-signal wideband perturbation. The WSI technique is finally contrasted in a simulation example to the well-known digital analyzer technique and practical advantages of the WSI are derived.

2.2 CONVERTER MODEL

2.2.1 Single Converter—Open Loop

In order to analyze the stability of MVDC systems, an ideal constant power assumption is often applied [3]. However, this assumption has limitations when the input characteristic of a load deviates from the CPL behavior. Furthermore, the assumption of an ideal CPL does not take into account the influence of the controller design and operating conditions on the system stability. Therefore, it is necessary to establish detailed models of the power system to fully account for the dynamic behavior and its influence on stability [4–10].

Nonetheless, until now the interdependency of the control bandwidth of the load side converter on the control goal of the generation side converter with respect to the control variable (e.g., the voltage of the DC bus) has not been sufficiently addressed; in effect the ideal CPL assumption in the form of a nonlinear controlled source inherits the time constant of the passive components of the DC link.

In the following sections it will be shown how the practical results of different converter power levels and loads, switching frequencies, and control bandwidths yield at a load behavior different from the ideal CPL behavior. The main goal is to determine under which circumstances the ideal representation is meaningful and under which it may even produce misleading results.

To be able to design the control system of the converter, it is necessary to model its dynamic behavior. Typically this includes how the variations of the input voltage, the load current, and the duty cycle affect the output voltage. As converters are nonlinear components due to the switching behavior, state-space averaging is often used to generate small-signal models. The averaging of converter circuit over the two states of the switch provides the equations of converter. This procedure neglects the switching ripple as it is considered small in well-designed converters and therefore makes the more important dynamics of the converters accessible. Because only small-signal disturbances are analyzed the averaged model is linearized at the operating point. The derivation of the linearized model via classic circuit analysis and the more common approach via state-space averaging was presented by Erickson [11]. By using this method, equivalent circuit models of DC–DC converters can be synthesized and, consequently, the canonical circuit model in Fig. 2.2 can be used to represent the physical properties of Pulse Width Modulated (PWM) DC–DC converters in Continuous Conduction Mode (CCM) [11,12]. In this model $\hat{v}_{in}$ and $\hat{v}_{out}$ corresponds to the small-signal perturbation in the input and output voltage. The small-signal perturbation in the duty cycle is represented by $\hat{d}$, while $\hat{i}_{Lo1}$ represents the load current variation. This canonical model can be used for the buck, the boost and the buck–boost converter by adapting the parameters of the model to the converter.

Canonical model parameters for the ideal buck, boost, and buck–boost converter are listed in Table 2.1. As the representation in Fig. 2.2 is a general one, it can be used for all three models by changing the values of $M(D)$, L_e, $e(s)$, $j(s)$ and therefore describes the behavior of

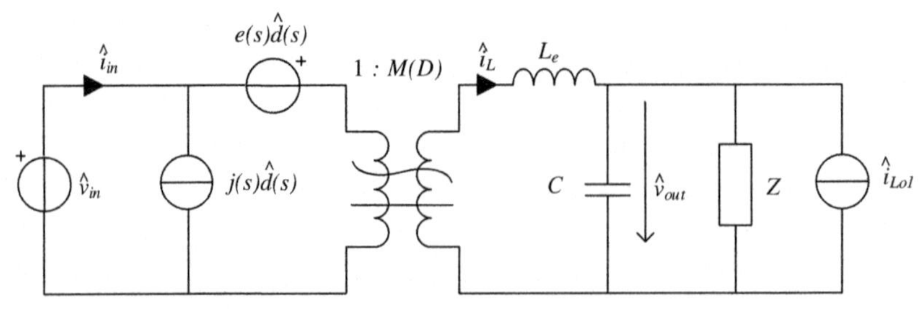

Figure 2.2 Canonical small-signal model for a terminated DC–DC converter.

Table 2.1 Canonical Model Parameters for Converters in CCM

Converter	$M(D)$	L_e	$e(s)$	$j(s)$
Buck	D	L	$\frac{v_{out}}{D^2}$	$\frac{v_{out}}{Z}$
Boost	$\frac{1}{D'}$	$\frac{L}{D'^2}$	$v_{out}\left(1-\frac{sL}{ZD'^2}\right)$	$\frac{v_{out}}{ZD'^2}$
Buck–Boost	$-\frac{D}{D'}$	$\frac{L}{D'^2}$	$-\frac{v_{out}}{D^2}\left(1-\frac{sLD}{ZD'^2}\right)$	$-\frac{v_{out}}{ZD'^2}$

the selected converter. The transformer being used in this model is an ideal transformer with the turns ratio of $1{:}\ M(D)$ which is a function of the duty cycle D. The term $e(s)\hat{d}(s)$ and $j(s)\hat{d}(s)$ represent perturbations in the duty cycle which are usually caused by a control circuit.

The derived model is called the terminated model as the load Z is already included in the model and the effects of the load are seen in the transfer functions of the converter. With this model the transfer functions of the converter can be derived by using conventional linear circuit analysis techniques. It is then possible to derive the transfer functions for a converter operating in open-loop. As it is the aim to control the output voltage of the converters to meet the specifications of the bus and the loads, the input parameters of the model are the input voltage, the duty ratio, also referred to as the control input, and the output current of the converter. This is called a voltage-output converter. Hence, the output parameters are the output voltage and the input current of the converter. The effect of the input parameters on the output parameters can be derived from Erickson's model and lead to the terminated transfer functions (2.1) to (2.7): control-to-output transfer function G_{vd-t}, line-to-output voltage transfer function G_{vg-t}, converter output impedance Z_{out-t}, control-to-inductor current transfer function G_{Ld-t}, line-to-inductor current transfer function G_{Lg-t}, and the output current-to-inductor current transfer function G_{Lo-t}. The formalism for transforming the canonical model to a mathematical model is given in [11,12]; it consists of removing all current sources which are not controlled by the input parameter from the circuit and replacing all voltage sources which are not controlled by the input parameter with a short and analyzing the resulting circuit with phasor calculus.

$$G_{vd-t}(s) = \left.\frac{\hat{v}_{out}(s)}{\hat{d}(s)}\right|_{\hat{v}_{in},\hat{i}_{Lo1}=0} = \frac{M(D)e(s)}{CL_e s^2 + \frac{L_e}{Z}s + 1} \tag{2.1}$$

$$G_{vg-t}(s) = \left.\frac{\hat{v}_{out}(s)}{\hat{v}_{in}(s)}\right|_{\hat{d},\hat{i}_{Lo1}=0} = \frac{M(D)}{CL_e s^2 + \frac{L_e}{Z}s + 1} \tag{2.2}$$

$$Z_{out-t}(s) = \frac{\hat{v}_{out}(s)}{-\hat{i}_{Lo1}(s)}\bigg|_{\hat{d},\hat{v}_{in}=0} = \frac{L_e s}{CL_e s^2 + \frac{L_e}{Z}s + 1} \tag{2.3}$$

$$G_{Ld-t}(s) = \frac{\hat{i}_L(s)}{\hat{d}(s)}\bigg|_{\hat{v}_{in},\hat{i}_{Lo1}=0} = \frac{M(D)e(s)(\frac{1}{Z} + Cs)}{CL_e s^2 + \frac{L_e}{Z}s + 1} \tag{2.4}$$

$$G_{Lg-t}(s) = \frac{\hat{i}_L(s)}{\hat{v}_{in}(s)}\bigg|_{\hat{d},\hat{i}_{Lo1}=0} = \frac{M(D)(\frac{1}{Z} + Cs)}{CL_e s^2 + \frac{L_e}{Z}s + 1} \tag{2.5}$$

$$G_{Lo-t}(s) = \frac{\hat{i}_L(s)}{\hat{i}_{Lo1}(s)}\bigg|_{\hat{d},\hat{v}_{in}=0} = \frac{1}{CL_e s^2 + \frac{L_e}{Z}s + 1} \tag{2.6}$$

$$Z_{in-t}(s) = \frac{\hat{i}_{in}(s)}{\hat{v}_{in}(s)}\bigg|_{\hat{d},\hat{i}_{Lo1}=0} = \frac{1}{M(D)G_{Lg}(s)} \tag{2.7}$$

With these equations a complete description of the buck and the boost converter is found to which the control system can be added and which then fully describes a single converter and its behavior with small-signal disturbances. A more compact notation of the Eqs. (2.1)–(2.7) is given by the following mathematical representation of the converter in (2.8):

$$\begin{pmatrix} \hat{v}_{out} \\ \hat{i}_L \end{pmatrix} = \begin{pmatrix} -Z_{out-t} & G_{vg-t} & G_{vd-t} \\ \frac{Z_{out}}{sL_e} & G_{Lg-t} & G_{Ld-t} \end{pmatrix} \begin{pmatrix} \hat{i}_{Lo1} \\ \hat{v}_{in} \\ \hat{d} \end{pmatrix} \tag{2.8}$$

This matrix of (2.8) is depicted in a control block representation of Fig. 2.2, where the open-loop small-signal model can be derived by setting $\hat{d}(s)$ equal to zero.

2.2.2 Single Converter—Closed Loop (VMC)

In order to derive a cascaded closed-loop converter system, the small-signal-closed loop transfer functions of the single converter are first required. Those transfer functions are needed for the closed-loop input and output impedance Z_{in_CL}, Z_{out_CL} and the transfer function representing the closed-loop input to output voltage perturbation G_{vg_CL} [13]. To illustrate the concept, the most common control system operation for a DC–DC converter was selected, i.e., VMC [11].

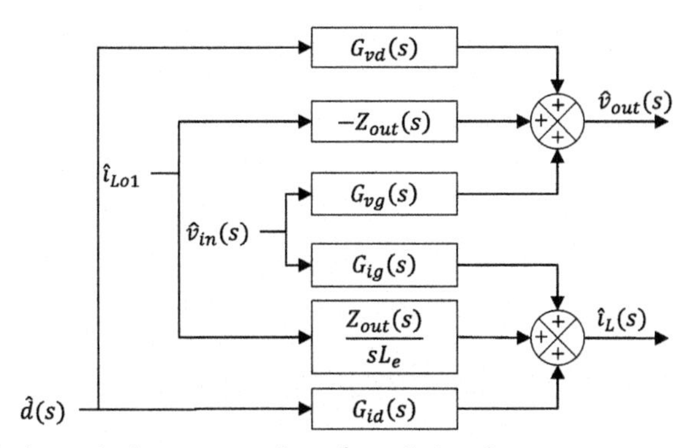

Figure 2.3 Mathematical representation of small-signal converter.

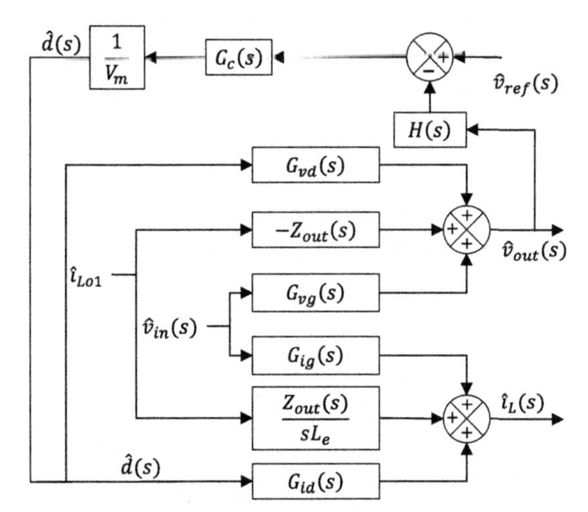

Figure 2.4 Block diagram of voltage mode control of converter.

The corresponding schematic is depicted in Fig. 2.4. The changes from Fig. 2.3 to Fig. 2.4 are: $\hat{v}_{ref}(s)$ is the perturbation in the reference voltage, the term $H(s)$ is the transfer function of the sensing network, which often can be considered as a pure gain, $G_c(s)$ is the compensator transfer function. Commonly used SISO compensators are PI, PD, or PID. The PWM gain is equal to $\frac{1}{V_m}$, where V_m corresponds to the amplitude of the DC bus.

The solution of Fig. 2.4 according to [11] yields for Eq. (2.9) the voltage output variation $\hat{v}_{out}$, where $T(s)$ is defined in general as the product of gains around the forward and feedback paths of the loop. This equation shows how the addition of a feedback loop modifies the transfer function of the system:

$$\hat{v}_{out} = \hat{v}_{ref} \frac{1}{H(s)} \frac{T(s)}{1 + T(s)} + \hat{v}_{in} \frac{G_{vg}}{1 + T(s)} - \hat{i}_{Lo1} \frac{Z_{out}}{1 + T(s)} \tag{2.9}$$

$$T(s) = H(s) G_c(s) G_{vd}(s) \frac{1}{V_m} \tag{2.10}$$

$$G_{vg\_CL}(s) = \left. \frac{\hat{v}_{out}(s)}{\hat{v}_{in}(s)} \right|_{\hat{v}_{ref}, \hat{i}_{Lo1} = 0} = \frac{G_{vg}}{1 + T(s)} \tag{2.11}$$

$$Z_{out\_CL}(s) = \left. \frac{\hat{v}_{out}(s)}{-\hat{i}_{Lo1}(s)} \right|_{\hat{v}_{ref}, \hat{i}_{Lo1} = 0} = \frac{Z_{out}}{1 + T(s)} \tag{2.12}$$

From the resulting Eq. (2.12) it can be observed that, through the feedback, the closed-loop transfer function and the output impedances are reduced by the factor of $1/(1 + T(s))$ which is called the sensitivity function [14]. It describes how the influence of disturbances in the input values is attenuated by good control design on the output voltage. An optimal controller design would ensure that the norm of $(1 + T)$ is very large over all frequencies. This would mean that the transfer functions would become insignificant and with them their effects on the output voltage.

Due to the limited bandwidth of a real controller this is not possible to realize. Consequently, the controller is designed such that the norm of T is large for frequencies smaller than a specified crossover frequency and becomes infinite for DC values. For frequencies that are above the crossover frequency, the norm of T becomes marginal compared to 1 and the closed-loop transfer functions by this means retrieve the original characteristics of the open-loop transfer functions [15]. To limit the disturbances at high frequencies the filter of the converter has to be designed to limit the bandwidth of the system so that the effects on the output voltage are limited over the whole frequency range.

The definition of the closed-loop input impedance, Z_{IN_CL}, is the ratio of perturbations in the input voltage $\hat{v}_{in}$ to the perturbations in the input current $\hat{i}_{in}(s)$, while the perturbations in the load current $\hat{i}_{Lo1}$ are set to be zero.

$$Z_{IN\_CL}(s) = \left. \frac{\hat{v}_{in}(s)}{\hat{i}_{In}(s)} \right|_{\hat{v}_{ref}, \hat{i}_{Lo1} = 0} \tag{2.13}$$

Analyzing Fig. 2.4 and setting $\hat{v}_{ref}$ and $\hat{i}_{Lo1}$ to zero the Eqs. (2.14)–(2.16) are obtained. The Eq. (2.17), which represents the relationship between input current perturbation $\hat{i}_{in}$ and inductor current perturbation $\hat{i}_L(s)$, is obtained by examining the circuit presented in Fig. 2.2.

$$\hat{d}(s) = -\hat{v}_{out1}H(s)G_c(s)\frac{1}{V_m} \tag{2.14}$$

$$\hat{v}_{out} = \hat{d}(s)G_{vd}(s) + \hat{v}_{in}(s)G_{vg}(s) \tag{2.15}$$

$$\hat{i}_L(s) = \hat{d}(s)G_{id}(s) + \hat{v}_{in}(s)G_{ig}(s) \tag{2.16}$$

$$\hat{i}_{in} = j(s)\hat{d}(s) + M(D)\hat{i}_L(s) \tag{2.17}$$

After solving this system of equations the relationship for the closed-loop input impedance (2.18) is obtained. This relationship will be further used in the cascaded converter model. It can be generalized to in respect to a closed-loop converter loaded with a generic impedance load Z which is characterized by (2.19).

$$Z_{IN\_CL}(s) = \left.\frac{\hat{v}_{in}(s)}{\hat{i}_{in}(s)}\right|_{\hat{v}_{ref},\hat{i}_{Lo1}=0} = Z_{IN}(s)\left(\frac{1+T}{1-T\frac{j(s)}{e(s)M(D)G_{ig}(s)}}\right) \tag{2.18}$$

$$Z_{IN_{CL}}(s) = \frac{(ZC_1L_es^2 + L_es + Z)V_m + ZH(s)G_c(s)M(D)e(s))}{V_m(M(D))^2(1+ZC_1s) - j(s)ZH(s)G_c(s)M(D)} \tag{2.19}$$

2.2.3 Single Converter Closed Loop (PCMC)

Another popular control mode for DC–DC converters is the PCMC [11,15]. The procedure in which the PCMC controls the converter is shown in Fig. 2.5. The inductor current I_L is compared every cycle to a control current I_{co}. As soon as $I_L > I_{co}$, the switches of the converter are actioned and the inductor current enters the downslope. This does not necessarily lead to the depicted steady state. The control current I_{co} is again determined via feedback of the output voltage.

From this technique the first advantage of the PCMC can be concluded: the inductor current is limited each cycle. A drawback of this procedure is that it introduces a stability problem in the circuit. If steady-state operation is considered and duty ratios higher than 0.5 are used then disturbances are amplified until the converter enters a subharmonic operation mode [15]. This is called the mode limit as a stable operation cannot

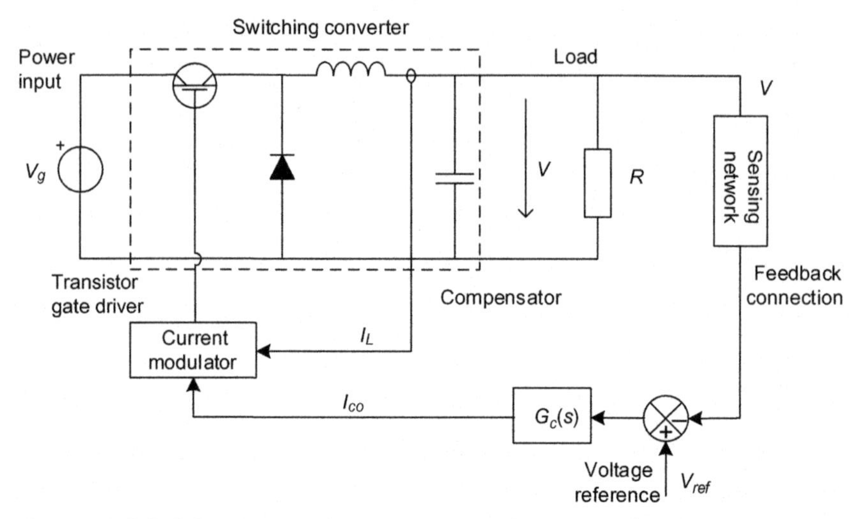

Figure 2.5 DC–DC converter PCMC structure.

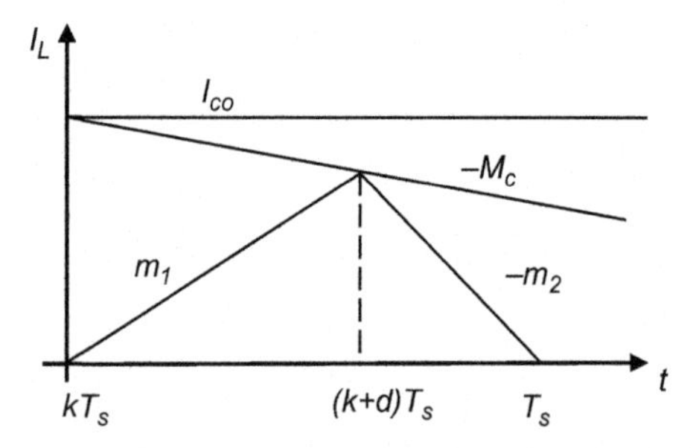

Figure 2.6 Inductor current for PCM controlled converter with compensation ramp.

be guaranteed with higher duty ratios. The mode limit can be experienced for both the buck and the boost converter, since the amplification of disturbances only depends on the ratio of the up- and downslope of the inductor current. To avoid such a limit which restricts the operation of the converter an artificial ramp is subtracted from the control current which compensates for the effect. This can be seen in Fig. 2.6.

In order to use the full duty ratio the minimum slope of the artificial ramp has to be chosen as $M_c = (m_1 + m_2)/2$ [15]. As the converter itself remains unchanged the transfer functions derived from the canonical model are still used to calculate the transfer functions in PCMC. Similar to Fig. 2.4, in Fig. 2.7 the term $H(s)$ is the transfer function of the sensing network, which often can be considered as a pure gain, $G_c(s)$ is the

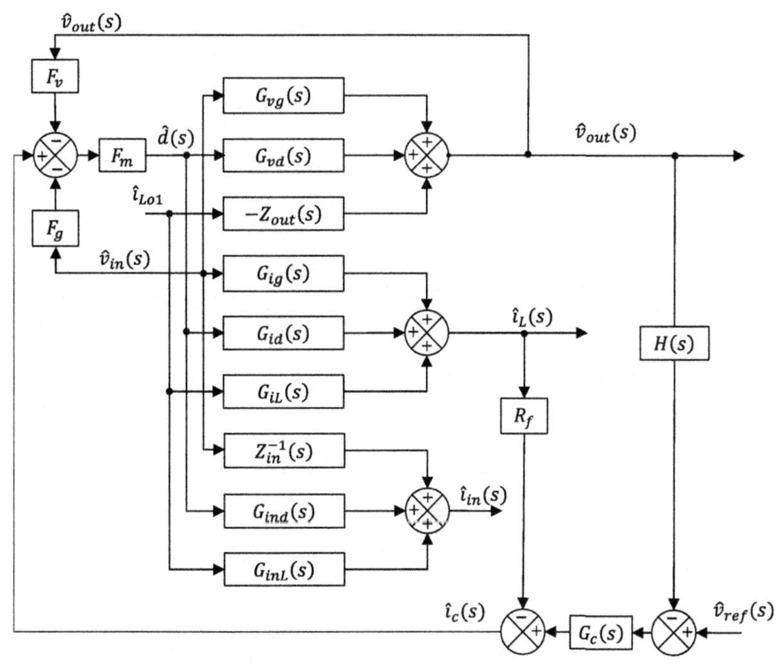

Figure 2.7 Control block diagram PCMC.

compensator transfer function. The R_f block represents the current sensing resistor which transforms the inductor current value to a voltage value.

The coefficients F_m, F_v, and F_g are gain values used to model the current modulator. The ratio F_m is related to the artificial ramp M_c and the switching frequency f_{sw} and according to [11]:

$$F_m = \frac{f_{sw}}{M_c} \tag{2.20}$$

The ratios for the factors F_v and F_g are dependent from chosen converter topology and are presented in Table 2.2.

The advantages according to [11] and [15] are that the PCMC offers a substantial damping compared on the magnitude. A drawback is that the phase constantly decreases while for the VMC a jump around the crossover frequency can be observed. A further drawback is that the PCMC has a susceptibility to noise. In [11] and [15] it is stated that the PCMC control usually increases significantly the output impedance of the converter which can be problematic in a cascaded configuration as the

Table 2.2 Gains for PCMC Coefficients [11]

	F_v	F_g
Buck	$\frac{1-2D}{2Lf_{sw}}$	$\frac{D^2}{2Lf_{sw}}$
Boost	$\frac{(1-D)^2}{2Lf_{sw}}$	$\frac{2D-1}{2Lf_{sw}}$
Buck–Boost	$\frac{-(1-D)^2}{2Lf_{sw}}$	$\frac{D^2}{2Lf_{sw}}$

Middlebrook [12] criterion states that $Z_{out} < Z_{in}$ for stable operation. Therefore often the output impedance has to be reduced via feedback of the output current to reduce the interactions with the load.

2.3 CASCADED SYSTEM (VMC)

Cascading two DC–DC converters means that the downstream converter, now referred to as converter 2 or Point of Load Converter (POL), acts as the load of the upstream converter now labeled as converter 1 or also called Line Regulating Converter (LRC). Using the canonical model presented in Fig. 2.2, and under the small-signal assumption, the system behavior is considered linear; thus, applying the two port theory leads to the topology depicted in Fig. 2.8.

Cascading implies that the output voltage of LRC is equal to the input voltage of POL and the input current of the POL equals the output current of the LRC. This correspondence can also be extended for the small-signal values of the perturbations of the voltages and currents.

Therefore, the small-signal current drawn from the LRC $\hat{i}_{Lo1}$ depends on the dynamics of the POL converter, which has to be equal to the input current $\hat{i}_{in2}$ of the POL. The same conclusion can be made for the output voltage perturbation $\hat{v}_{out1}$ of the LRC, which has to be equal to

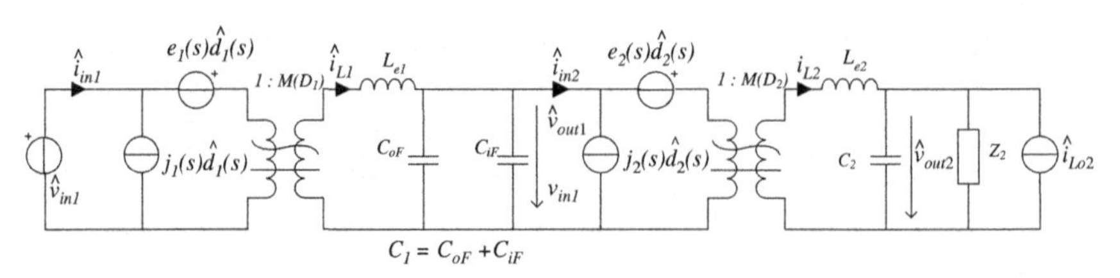

Figure 2.8 Cascaded canonical model.

the input voltage perturbation $\hat{v}_{in2}$ of the POL. It is assumed that in cascading two converters the LRC is only supplying one POL and therefore the load impedance Z, which appears in Fig. 2.2, is eliminated. And instead, the small-signal model of the converter in Fig. 2.2 is again used and leads to the representation above.

The output filter capacitor C_{oF} and input filter capacitor C_{iF} are represented by the equivalent capacitor C_1 in Fig. 2.8. The same approach as in the previous section can be used, which leads to the mathematical representation of the cascaded system. When assuming that the POL converter acts as load to the LRC converter, the small-signal current drawn from LRC $\hat{i}_{Lo1}$ can be expressed by means of the Thevenins theorem where POL converter is an impedance element which corresponds to the closed-loop input impedances Z_{IN_CL2}. This fact translates into the cascaded control block representation illustrated in Fig. 2.9 for VMC, a similar control block diagram can be drawn for the cascading of PCMC which is omitted here for clarity reasons.

$$G_{vd}(s) = \left.\frac{\hat{v}_{out}(s)}{\hat{d}(s)}\right|_{\hat{v}_{in},\hat{i}_{Lo}=0} = \frac{M(D)e(s)}{CL_e s^2 + 1} \tag{2.21}$$

Combining (2.15) and (2.21) leads to the following expression for the voltage of the DC bus capacitor in a cascaded system.

$$\hat{v}_{out1} = \hat{v}_{ref}\frac{1}{H}\frac{T}{1+T} + \hat{v}_{in}\frac{1}{H(s)}\frac{G_{vg}}{1+T} - \frac{\hat{v}_{out1}}{Z_{IN_{CL2}}}\frac{Z_{out1}}{1+T} \tag{2.22}$$

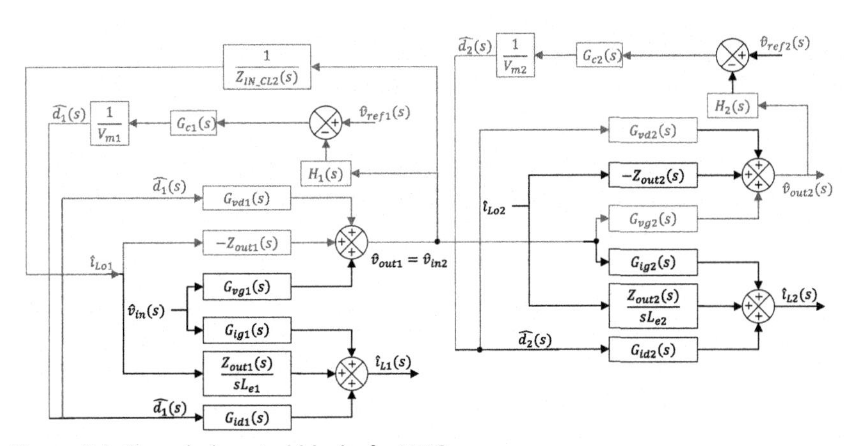

Figure 2.9 Cascaded control blocks for VMC.

Normally, a cascaded converter architecture is designed in such a way that it delivers a defined voltage $\hat{v}_{out2}$ to the load under the assumption of a predefined reference voltage $\hat{v}_{ref1}$. For the initial loop design the effect of possible perturbations in the input voltage of the cascaded system $\hat{v}_{in1}$ and the load current $\hat{i}_{Load2}$ are neglected. On further inspection of the control block diagram in Fig. 2.9, it becomes clear that for the path $\hat{v}_{ref1}$ to $\hat{v}_{out2}$ (marked in red (light gray in print version)) it is not necessary to consider the inductor currents $(\hat{i}_{L1}, \hat{i}_{L2})$. The positive feedback loop containing the block $-\frac{Z_{out1}}{Z_{INCL2}}$ can be condensed to $\frac{1}{1+\frac{Z_{out1}}{Z_{INCL2}}}$ which makes it possible to write the reference to output voltage transfer function (2.23) for the first converter:

$$\frac{\hat{v}_{out1}(s)}{\hat{v}_{ref1}(s)} = \frac{\frac{1}{V_{m1}} G_{C1} G_{vd1} \frac{1}{1+\frac{Z_{out1}}{Z_{INCL2}}}}{1+\frac{H_1}{V_{m1}} G_{C1} G_{vd1} \frac{1}{1+\frac{Z_{out1}}{Z_{INCL2}}}} \tag{2.23}$$

Inspecting the second converter loop yields the following equation:

$$\hat{v}_{out2}(s) = \left(\hat{v}_{ref2}(s) - H_2 \hat{v}_{out2}\right) \frac{1}{V_{m2}} G_{C2} G_{vd2} + G_{vg2} \hat{v}_{out1} \tag{2.24}$$

From the Eqs. (2.19), (2.22) and (2.23) it can be clearly seen that there exists interdependence on how the selection of parameters of the controllers G_{C1} and G_{C2} influence each other. It also can be seen how the dynamics of the load supplied by the second converter influences the voltage of the DC bus. This interdependence poses some interesting questions on the nontrivial design process which was further elaborated in [16].

2.4 LOAD MODELS

In this chapter the context of the canonical converters and their cascading is extended to the use case of analyzing particular load types which are connected on the POL converter. First, the CPL model is considered; afterwards, the presented methodology is used to analyze further typical load characteristics which may be connected at a POL. This work is related to the generalized load impedance model which was described in [17] under the assumption that the closed-loop buck converter is PI controlled. The destabilizing load forms, which were proposed by the authors in [17] will be extended to match a PID regulator. Those load forms represent destabilizing characteristics under certain parameter variations and dynamic conditions which exhibit a different dynamic behavior in contrast to the CPL assumption. It should also be highlighted that next to the employed Canonical Model, an alternative approach is possible which is advocated in [15], where the load Z is taken out of the canonical model and modeled through the corresponding deviation in the output current; therefore, the converter equations show the real internal dynamics of the converter without load and thus makes an interconnection of systems easier.

2.4.1 Constant Power Load Model

Due to the integration in DC grids of LRC and POL converters the efficiency of the network is increased and due to the control algorithm implemented in the converters these grids are able to handle a wide variation in either load or source [18].

As an example, a buck POL converter is used with a resistive load as shown in Fig. 2.10, where $Z_L(s) = R$. Power converters such as a buck converter are used because of their tight output voltage control capability, which enables them to respond almost immediately to system changes. On the other hand, under these conditions, the converter tends to operate as a CPL. When averaging the states and applying the canonical model with the parameters of Table 2.1 and Eq. (2.1) the linearized impedance model of the buck converter systems for a given operating point is given in (2.25). As all denominator coefficients are positive and not zero the perturbation transfer function for a single POL converter which is supplied by an ideal voltage source satisfies the Hurwitz criteria for stability.

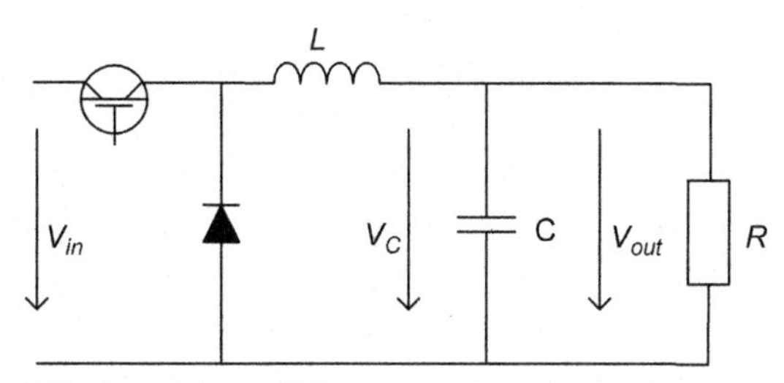

Figure 2.10 Simplified model of a CPL.

$$\frac{\hat{v}_{out}(s)}{\hat{d}(s)} = \frac{\frac{V_c}{D}}{CLs^2 + \frac{L}{R}s + 1} = \frac{\overline{v_{in}}\frac{1}{LC}}{s^2 + \frac{1}{RC}s + \frac{1}{LC}1} \tag{2.25}$$

$$Z_{IN_{CL}}(s) = \frac{(C_1 L_e s^2 Z + L_e s + Z)V_m + ZH(s)G_c(s)\frac{V_{out1}}{D}}{V_m D^2(1 + ZC_1 s) - H(s)G_c(s)V_{out1}D} \tag{2.26}$$

$$G_c(s) = \frac{K_D}{s}\left(s^2 + \frac{K_P}{K_D}s + \frac{K_I}{K_D}\right) \tag{2.27}$$

Assuming:

- A constant sensor gain;
- A PWM gain of *1*;
- Using a general PID controller (2.27) for regulating the closed-loop converters, with proportional gain K_P, integral gain K_I, and derivative gain K_D.

Those assumptions yields the following generalized expression:

$$\begin{aligned} Z_{IN_{CL}}(s) &= \frac{ZC_1 Ls^3 + (L + ZK_D\overline{v}_{in})s^2 + (Z + ZK_P\overline{v}_{in1})s + ZK_I\overline{v}_{in}}{(ZC_1 D^2 - K_D D^2\overline{v}_{in})s^2 + (D^2 - K_P D^2\overline{v}_{in1})s - K_I D^2\overline{v}_{in1}} \\ &= \frac{n_3 s^3 + n_2 s^2 + n_1 s + n_0}{d_2 s^2 + d_1 s + d_0} \end{aligned} \tag{2.28}$$

By assuming that $Z_L(s) = R$ the numerators are defined as: $n_3 = RC_1L$; $n_2 = R + ZK_D\overline{v}_{in}$; $n_1 = R + RK_P\overline{v}_{in1}$; $n_0 = RK_I\overline{v}_{in}$ while the denominators are defined as:

$$d_2 = RC_1D^2 - K_D D^2\overline{v}_{in},\ d_1 = D^2 - K_P D^2\overline{v}_{in1},\ d_0 = -K_I D^2\overline{v}_{in1} \tag{2.29}$$

The steady-state error for a step input can be calculated with the final value theorem and corresponds to the DC gain which will yield for $Z_L(s) = R$:

$$e_{ss} = \lim_{s\to 0} s\frac{1}{s} Z_{IN_{CL}}(s) = \frac{n_0}{d_0} = -\frac{R}{D^2} \tag{2.30}$$

This behaves like a negative destabilizing resistance. The authors in [19] and [20] state that the closed-loop input resistance $Z_{IN_{CL}}$ of a POL is approximately $-Z_{IN_{open}}$. The negative resistance of (2.30) pushes an open-loop pole into the right-half plane and destabilizes the transfer function of (2.25). Therefore, often in literature the approximation of a complex load is performed, which exhibits constant power behavior (2.31).

$$e_{ss} = \lim_{s \to 0} s \frac{1}{s} Z_{IN_{CL}}(s) = \frac{n_0}{d_0} = -\frac{R}{D^2} \tag{2.31}$$

What has to be noted is that by using Eq. (2.30) instead of Eq. (2.28) it is assumed that no dynamic interactions will take place between the converters, which is a very optimistic assumption, particularly when considering that the POL converters are operating at a higher switching frequency than the LRC.

In this chapter the limitations of an idealized behavior for a CPL characteristic are addressed. The CPL characteristic is dependent on the control cycle of the converter [2]. In Fig. 2.10, a buck converter supplying a resistive load is depicted. For the upcoming analysis in the frequency domain a linear behavior is assumed.

This setup can be shown in a block diagram as it is represented in Fig. 2.11, where $G_c(s)$ is the transfer function of the control loop which modulates the PWM, d is the duty cycle, $G_{vd}(s)$ is the transfer functions of the passive components in Fig. 2.10. In this analysis a voltage feedback is considered for the control system, with no voltage feed-forward nor current mode control; the resistive components in the capacitor and the inductance are neglected and thus assumed as ideal components. Delays caused by sampling and modulation are also not taken into consideration. Using the closed-loop VMC of Fig. 2.4 and introducing small perturbations Δ around an operating point (V_0, I_0) the block diagram depicted in Fig. 2.11 can be assumed. There the reader observes that on a given operating point a small disturbance can be split up into a perturbation in the duty cycle and a perturbation in the input voltage. This small-signal disturbance can be represented as two blocks which are added, as depicted in Fig. 2.11.

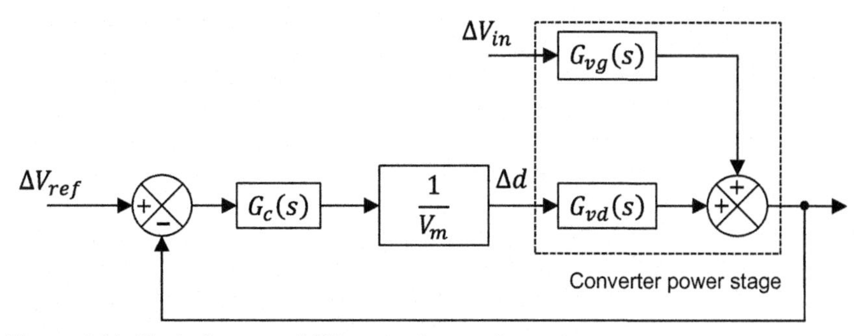

Figure 2.11 Block diagram of CPL around operating point.

The close-looped transfer function of input voltage to output voltage can be represented as:

$$\frac{\Delta V_{out}}{\Delta V_{in}} = \frac{G_{vg}(s)}{1 + G_c(s)G_{vd}(s)\frac{1}{V_M}} \tag{2.32}$$

By assuming the structure depicted above the transfer function between ΔV_{out} and ΔV_{in} is derived under the assumption that $G_{vd}(s)$ can be characterized as a second-order system and that G_c corresponds to the transfer function of a PID controller. Without restricting the generality, a constant gain of 1 for V_M is considered. The values of the gains are set to conduct a pole-zero cancellation $\left(\frac{K_p}{K_D} = \frac{1}{RC} = a; \frac{K_I}{K_D} = \frac{1}{LC} = b\right)$.

$$G_{vg}(s) = \frac{K\bar{d}}{LC(s^2 + \frac{1}{RC}s + \frac{1}{LC})}$$

$$G_{vd}(s) = \frac{V_{in}}{LC(s^2 + \frac{1}{RC}s + \frac{1}{LC})} = \frac{KV_{in}}{s^2 + as + b} \tag{2.33}$$

$$G_c(s) = \frac{K_D}{s}\left(s^2 + \frac{K_P}{K_D}s + \frac{K_I}{K_D}\right)$$

$$\frac{\Delta V_{out}}{\Delta V_{in} =} = \frac{s\,K\bar{d}}{(s + K_D K V_{in})(s^2 + as + b)}$$

In Fig. 2.12 P_{1F} and P_{2F} denote the poles of G_{vd} and $K_D K V_{in}$ is the pole resulting from the control design. The CPL characteristic is only valid for operation on the left side of $K_D K V_{in}$.

Therefore, modeling POL converters under VMC as a CPL has advantages for simplifying the system analysis. The observed system behavior in real systems or in simulations where the LRC and POL

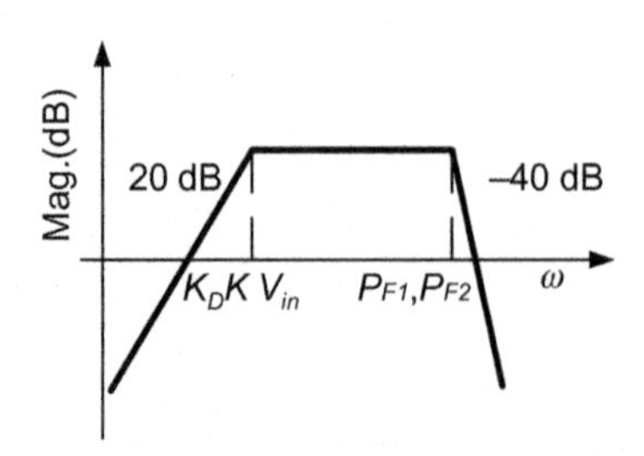

Figure 2.12 Operating limits for constant power characteristic.

converters are both represented by their switching model could lead to different stability limits. Very often in the simplified analysis stiff DC sources are considered and therefore no control loop interaction between the different converters takes place.

2.4.2 First-Order Lag Impedance

In this case the first-order lag function replaces the constant resistance where R is the DC gain of the load and τ is the time constant [17]. This type of load corresponds to a parallel circuit of a resistive and a capacitive load.

$$Z_{Load}(s) = \frac{R}{\tau s + 1}; \tau \in R^{+} \tag{2.34}$$

$$Z_{IN_{CL}}(s) = \frac{n_3 s^3 + n_2 s^2 + n_1 s + n_0}{d_3 s^3 + d_2 s^2 + d_1 s + d_0} \tag{2.35}$$

Where the nominator and denominator coefficients in (2.35):

$$n_3 = (RC_1 + \tau)L, n_2 = L + RK_D\overline{v}_{in}, n_1 = R(1 + K_P\overline{v}_{in1}), n_0 = RK_I\overline{v}_{in},$$
$$d_3 = -K_D D^2\overline{v}_{in}\tau, d_2 = D^2((RC_1 - K_D\overline{v}_{in1}) + (1 - K_P\overline{v}_{in1})\tau),$$
$$d_1 = D^2 + D^2\overline{v}_{in1}(K_P + K_I\tau), d_0 = -K_I D^2\overline{v}_{in1}$$

If $\tau = 0$, one is consequentially back at the CPL case as it is presumed that the load responds instantaneously.

2.4.3 First-Order Unstable Impedance

Replacing the load in (2.28) by an unstable impedance Z_{Load} function yields a $Z_{IN_{CL}}$ equal to (2.37).

$$Z_{Load}(s) = -\frac{b}{s - a}; \ a, b \in R^{+} \tag{2.36}$$

$$Z_{IN_{CL}}(s) = \frac{n_3 s^3 + n_2 s^2 + n_1 s + n_0}{d_3 s^3 + d_2 s^2 + d_1 s + d_0} \tag{2.37}$$

Where the nominator and denominator coefficients in (2.37):

$$n_3 = 1 - bC_1, n_2 = -(bK_D\overline{v}_{in} + La), n_1 = -(b + bK_P\overline{v}_{in1}), n_0 = -bK_I\overline{v}_{in},$$
$$d_3 = -K_D D^2\overline{v}_{in}, d_2 = D^2(1 - K_P\overline{v}_{in1} - bC_1 + aK_D\overline{v}_{in1}),$$
$$d_1 = D^2(aK_P\overline{v}_{in1} - a - K_I\overline{v}_{in1}), d_0 = K_I D^2\overline{v}_{in1}$$

As in this case the Hurwitz necessary conditions of stability are not met: instability in $Z_{IN_{CL}}$ is observed. It has to be noted that while instability can be observed its cause is different from the CPL case.

2.4.4 Nonminimum Phase Impedance

Nonminimum phase systems are quite seldom found as time continuous systems, while in DC–DC the most prominent case is the boost and buck–boost converters which exhibit right half plane zeros. They merit a consideration due to the wide implementation of digital controllers, where the minimum phase property can be lost due to the sampling. Their response to changes in the input signal are typically in opposite directions, i.e., when the input signal increases, the system output drops briefly before rising again, in contrast to systems with negative reinforcement. Due to the presence of zeros in the right complex half plane, a high feedback system will exhibit instability. Considering a first-order nonminimum phase load $Z_{Load}(s)$:

$$Z_{Load}(s) = -\frac{s-b}{s+a};\ a, b \in R^{+} \tag{2.38}$$

$$Z_{IN_{CL}}(s) = \frac{n_4 s^4 + n_3 s^3 + n_2 s^2 + n_1 s + n_0}{d_3 s^3 + d_2 s^2 + d_1 s + d_0} \tag{2.39}$$

where in (2.38):

$$n_4 = -C_1 L, n_3 = L + bC_1 L - K_D \overline{v}_{in},$$

$$n_2 = La + bK_D \overline{v}_{in} - 1 - \overline{v}_{in1} K_P, n_1 = b + b\overline{v}_{in1} K_P - K_I \overline{v}_{in},$$

$$n_0 = bK_I \overline{v}_{in}, d_3 = -C_1 D^2 - K_D D^2 \overline{v}_{in1},$$

$$d_2 = D^2(bC_1 - aK_D \overline{v}_{in1} + 1 - K_P \overline{v}_{in1})s^2,$$

$$d_1 = aD^2 + D^2 \overline{v}_{in1}(K_I - aK_P), d_0 = -aK_I D^2 \overline{v}_{in1}$$

With the previously performed analysis it has been shown that a stability analysis has to include suppositions about the dynamic behavior on top of the CPL assumption. Load impedances which have either zeros or poles in the right half plane will exhibit a drastically different behavior than the nondynamic CPL assumption.

2.4.5 Generalized Load Impedance

When using generalized impedance loads which follow the polynomial structure given in (2.40) it is possible to derive with Eq. (2.19) a generalized load input impedance Z_L of a closed-loop buck converter

$$Z_L(s) = \frac{\sum_{i=0}^{m} b_i s^i}{\sum_{j=0}^{n} a_j s^j}, \; a_j, b_i \in R \tag{2.40}$$

$$Z_{IN_{CL}}(s) = \frac{n_3 \sum_{i=0}^{m} b_i s^{i+3} + n_2 \sum_{i=0}^{m} b_i s^{i+2} + n_1 \sum_{i=0}^{m} b_i s^{i+1} + n_0 \sum_{i=0}^{m} b_i s^i + n_{d2} \sum_{j=0}^{n} a_j s^{j+2}}{d_{n2} \sum_{i=0}^{m} b_i s^{i+2} + d_2 \sum_{j=0}^{n} a_j s^{j+2} + d_1 \sum_{j=0}^{n} a_j s^{j+1} + d_0 \sum_{j=0}^{n} a_j s^j} \tag{2.41}$$

where in (2.41):

$$n_3 = C_1 L, n_2 = K_D \overline{v}_{in}, n_1 = (1 + K_P \overline{v}_{in1}), n_0 = K_I \overline{v}_{in}, n_{d2} = L, d_{n2} = C_1 D^2,$$
$$d_2 = -K_D D V_{out1}, d_1 = D^2 - K_P D^2 \overline{v}_{in1}, d_0 = -K_I D^2 \overline{v}_{in1}$$

It is observed how the distribution of the poles and zeros of $Z_{IN_{CL}}$ is influenced by the location of the poles and zeros of the generalized load, as right-handed poles and zeros of the load can lead to power system instability that are not obvious while analyzing a stable load converter.

2.5 LINEAR CONTROL DESIGN AND VALIDATION

In this section, the stand-alone converter model is taken into account one more time. It is shown how to design classical controls so that a certain dynamic closed-loop performance is obtained. Moreover, frequency domain validation techniques are presented to assess in the frequency domain whether the closed-loop performance meets the requirements.

2.5.1 Practical PI and PID Control Design

This subsection provides practical guidelines on how to design PI or PID controllers for classical linear controls, such as VMC, CMC, and outer voltage control for CMC or PCMC. For the converter under test, unit gain of the voltage and current sensors and pulse-width modulator are assumed for simplicity. The control design is performed in the *s-domain*, and no digital delays are taken into account. The idea is to provide the reader with a straightforward procedure to design basic linear controllers that is easily implementable in programming and simulation environments such as MATLAB and Simulink. The reader can refer to [21] if interested in a more formal digital control design procedure that includes all the effects that are here neglected. The reader will learn how to design PI and PID controllers to achieve the desired dynamic behavior and a null steady-state error. These specifications are translated into the desired *control bandwidth* (or *crossover frequency*) and *phase margin* of the equivalent closed-loop system. First, the CMC with its outer voltage loop design procedure is provided. Then, the VMC is given.

In CMC control, the converter has an inner PI current loop and an outer loop represented by a PI output voltage feedback control as shown in Figs. 2.13 and 2.14. First, the PI current mode (PICM) modeling is given, and then that one for the PICM voltage feedback (PICM_FB).

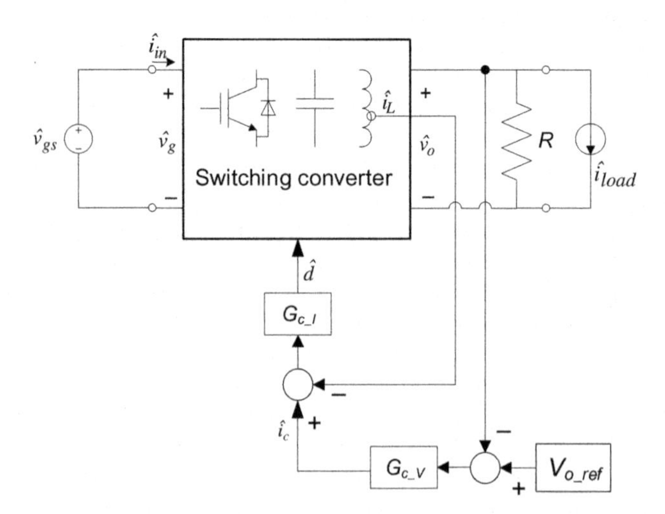

Figure 2.13 Circuit representation of a switching converter with inner PI current loop and outer PI voltage loop.

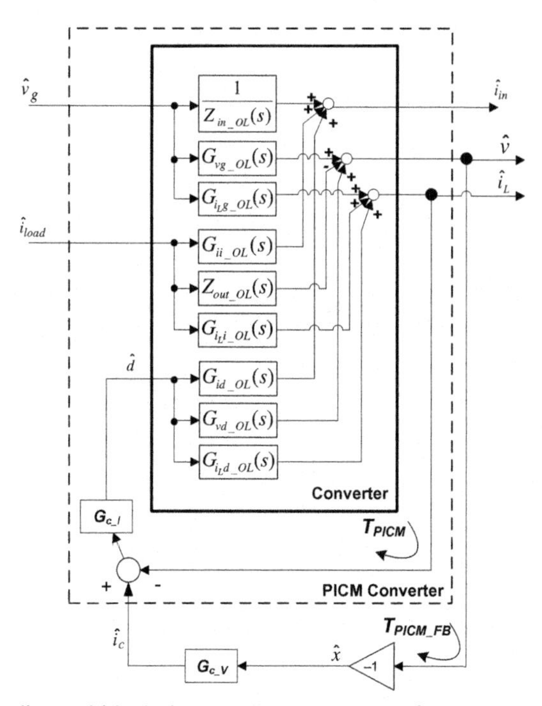

Figure 2.14 Small-signal block diagram representation of a switching converter with inner PI current loop and outer PI voltage loop.

Finally, the control design procedure for the inner current controller and outer voltage controller is presented.

2.5.1.1 Modeling and Design Procedure of Current Mode Control With PI Controllers

The small-signal linearized PICM converter model has three inputs and three outputs, as shown in Fig. 2.15. The inputs are supply voltage perturbation $\hat{v}_g$, load current perturbation $\hat{i}_{load}$, and control current $\hat{i}_c$. The first two inputs are disturbances and the third input is the control input. The three outputs are input current perturbation $\hat{i}_{in}$, output voltage perturbation $\hat{v}$, and inductor current perturbation $\hat{i}_L$. The model is

$$\begin{bmatrix} \hat{i}_{in} \\ \hat{v} \\ \hat{i}_L \end{bmatrix} = \begin{bmatrix} \frac{1}{Z_{in\_PICM}} & G_{ii\_PICM} & G_{ic\_PICM} \\ G_{vg\_PICM} & -Z_{out\_PICM} & G_{vc\_PICM} \\ G_{i_L g\_PICM} & G_{i_L i\_PICM} & G_{i_L c\_PICM} \end{bmatrix} \begin{bmatrix} \hat{v}_g \\ \hat{i}_{load} \\ \hat{i}_c \end{bmatrix} \tag{2.42}$$

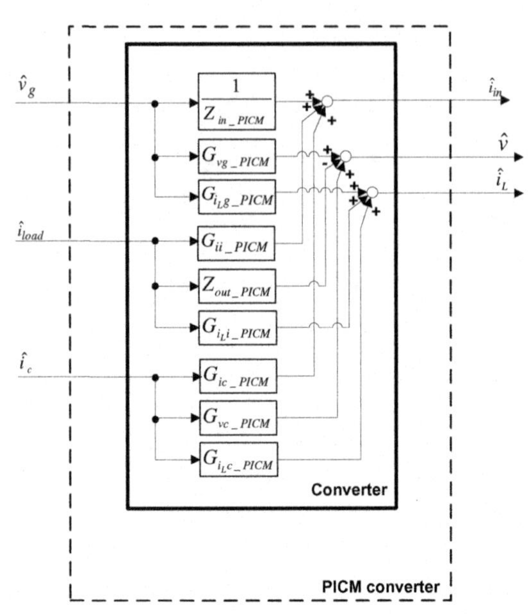

Figure 2.15 Small-signal block diagram representation of a switching converter in PICM.

The transfer functions for (2.42) are given in (2.43)−(2.51)

$$\frac{1}{Z_{in\_PICM}} = \frac{1}{Z_{in\_OL}} - \frac{G_{id\_OL}G_{i_Lg\_OL}}{G_{i_Ld\_OL}}\frac{T_I}{1+T_I} \tag{2.43}$$

$$G_{ii\_PICM} = G_{ii\_OL} - \frac{G_{id\_OL}G_{i_Li\_OL}}{G_{i_Ld\_OL}}\frac{T_I}{1+T_I} \tag{2.44}$$

$$G_{ic\_PICM} = \frac{G_{id\_OL}}{G_{i_Ld\_OL}} - \frac{G_{id\_OL}G_{i_Lg\_OL}}{G_{i_Ld\_OL}}\frac{T_I}{1+T_I} \tag{2.45}$$

$$G_{vg\_PICM} = G_{vg\_OL} - \frac{G_{vd\_OL}G_{i_Lg\_OL}}{G_{i_Ld\_OL}}\frac{T_I}{1+T_I} \tag{2.46}$$

$$Z_{out\_PICM} = \frac{1}{Z_{in\_OL}} + \frac{G_{vd\_OL}G_{i_Li\_OL}}{G_{i_Ld\_OL}}\frac{T_I}{1+T_I} \tag{2.47}$$

$$G_{vc\_PICM} = \frac{G_{vd\_OL}}{G_{i_Ld\_OL}}\frac{T_I}{1+T_I} \tag{2.48}$$

$$G_{i_Lg\_PICM} = \frac{G_{i_Lg\_OL}}{1+T_I} \tag{2.49}$$

$$G_{i_Li\_PICM} = \frac{G_{i_Li\_OL}}{1+T_I} \tag{2.50}$$

$$G_{i_{Lc}\_PICM} = \frac{T_I}{1 + T_I} \quad (2.51)$$

where $T_{PICM} = G_{c\_I} \cdot G_{i_L d\_OL}$ is the inner current control loop gain.

The PICM_FB converter model (2.52) is obtained from the PICM converter model (2.42) by imposing a control current $\hat{i}_c = - G_{c\_V} \cdot \hat{v}$.

$$\begin{bmatrix} \hat{i}_{in} \\ \hat{v} \end{bmatrix} = \begin{bmatrix} \frac{1}{Z_{in\_PICM\_FB}} & G_{ii\_PICM\_FB} \\ G_{vg\_PICM\_FB} & -Z_{out\_PICM\_FB} \end{bmatrix} \begin{bmatrix} \hat{v}_g \\ \hat{i}_{load} \end{bmatrix} \quad (2.52)$$

The transfer functions of model (2.52) are given in (2.53)−(2.56).

$$\frac{1}{Z_{in\_PICM\_FB}} = \frac{1}{Z_{in\_PICM}} \frac{1}{1 + T_{PICM\_FB}} + \frac{1}{Z_{N\_vc\_PICM}} \frac{T_{PICM\_FB}}{1 + T_{PICM\_FB}} \quad (2.53)$$

$$G_{vg\_PICM\_FB} = \left\{ \frac{G_{vg\_PICM}}{1 + T_{PICM\_FB}} \right\} \quad (2.54)$$

$$G_{ii\_PICM\_FB} = G_{ii\_PICM} + \frac{G_{ic\_PICM} Z_{out\_PICM}}{G_{vc\_PICM}} \cdot \frac{T_{PICM\_FB}}{1 + T_{PICM\_FB}} \quad (2.55)$$

$$Z_{out\_PICM\_FB} = \frac{Z_{out\_PICM}}{1 + T_{PICM\_FB}} \quad (2.56)$$

where

$$\frac{1}{Z_{N\_vc\_PICM}} = \frac{1}{Z_{in\_PICM}} - \frac{G_{ic\_PICM} G_{vg\_PICM}}{G_{vc\_PICM}} \quad (2.57)$$

And $T_{PICM\_FB} = G_{c\_V} \cdot G_{vc\_PICM}$ is the voltage FB loop gain.

After having derived the modeling, the procedure to design both the inner and outer PI controllers is presented hereafter. This procedure requires first the inner current controller to be designed, and then the outer voltage controller.

The inner current control loop gain is the following expression:

$$T_{PICM}(s) = G_{c\_I}(s) \cdot G_{i_L d\_OL}(s) \quad (2.58)$$

It is required that the loop gain of $T_{PICM}(s)$ has a certain phase margin PM_{PICM} at the control bandwidth ω_{c_PICM}. This imposes that the magnitude of the loop gain of $T_{PICM}(s)$ is unitary at the control bandwidth

ω_{c_PICM} and the phase of the loop gain of $T_{PICM}(s)$ is equal to $-180^{\circ} + PM_{PICM}$ at the control bandwidth ω_{c_PICM}. From (2.58), the following equations are derived.

$$|T_{PICM}(j\omega_{c\_PICM})| = 1 = |G_{c\_I}(j\omega_{c\_PICM})| \cdot |G_{i_L d\_OL}(j\omega_{c\_PICM})| \quad (2.59)$$

$$\begin{aligned} arg[T_{PICM}(j\omega_{c\_PICM})] &= -180^{\circ} + PM_{PICM} \\ &= arg[G_{c\_I}(j\omega_{c\_PICM})] + arg[G_{i_L d\_OL}(j\omega_{c\_PICM})] \end{aligned} \quad (2.60)$$

The latter equation can be rewritten as follows.

$$arg[G_{c\_I}(j\omega_{c\_PICM})] = -180^{\circ} + PM_{PICM} - arg[G_{i_L d\_OL}(j\omega_{c\_PICM})] \quad (2.61)$$

Notice that $arg[G_{c\_I}(j\omega_{c\_PICM})]$ is function of known quantities. Solving (2.59) and (2.60), and taking into account the current control transfer function (2.62), its control coefficients (2.63) and (2.64) are calculated.

$$G_{c\_I}(s) = K_{p\_I} + \frac{K_{i\_I}}{s} \quad (2.62)$$

$$K_{p\_I} = \cos[arg[G_{c\_I}(j\omega_{c\_PICM})]] / |G_{i_L d\_OL}(j\omega_{c\_PICM})| \quad (2.63)$$

$$K_{i\_I} = -\omega_{c\_PICM} \cdot \sin[arg[G_{c\_I}(j\omega_{c\_PICM})]] / |G_{i_L d\_OL}(j\omega_{c\_PICM})| \quad (2.64)$$

To find the coefficients of the outer voltage controller (2.65), the designer is required to use the proper plant transfer function, i.e., $G_{vc}(s)$ and substitutes to $G_{i_L d\_OL}(s)$. These coefficients are calculated as in (2.66) and in (2.67). The outer voltage control is designed so that a certain phase margin PM_{PICM_FB} at the control bandwidth $\omega_{c\_PICM\_FB}$ is obtained. Notice that, again, $arg[G_{c\_V}(j\omega_{c\_I})]$ is function of known quantities (2.68).

$$G_{c\_V}(s) = K_{p\_V} + \frac{K_{i\_V}}{s} \quad (2.65)$$

$$K_{p\_V} = \cos[arg[G_{c\_V}(j\omega_{cPICM\_FB})]] / |G_{vc}(j\omega_{c\_PICM\_FB})| \quad (2.66)$$

$$K_{i\_I} = -\omega_{c\_PICM\_FB} \cdot \sin[arg[G_{c\_V}(j\omega_{c\_I})]] / |G_{vc}(j\omega_{c\_PICM\_FB})| \quad (2.67)$$

$$arg[G_{c\_V}(j\omega_{c\_I})] = -180^{\circ} + PM_{PICM\_FB} - arg[G_{vc}(j\omega_{c\_PICM\_FB})] \quad (2.68)$$

Notice that to design the outer voltage controller $G_c(s)$ of a converter in PCMC (see Fig. 2.7), the designer can follow exactly the same procedure.

2.5.1.2 Modeling and Design Procedure of Voltage Mode Control With PID Controller

This procedure allows designing the voltage controller $G_c(s)$ of a converter in VMC (see Fig. 2.4). The procedure is very similar to that one for the outer voltage controller $G_{c\_V}(s)$ of the converter in PICM_FB. Of course, the designer is required to use the proper plant transfer function, i.e., the control-to-output transfer function $G_{vd}(s)$ defined in (2.2). However, the phase of $G_{vd}(s)$ for the basic converter topologies at high frequencies either approaches −180 degrees for the buck converter or −270 degrees for the buck–boost and boost converters [22]. Therefore, for the frequency range where the designer wants to close the voltage loop, there is no way with a PI control to "bump the phase up" to obtain a positive phase margin. The only available option is PID control. In fact, the derivative component of a PID control has the "phase lead effect" on the converter closed-loop characteristics [22]. This allows obtaining a positive phase margin. However, the derivative component slightly complicates the procedure to design the voltage controller since a third coefficient needs to be found.

The voltage control loop gain is the following expression:

$$T_{VM}(s) = G_c(s) \cdot G_{vd}(s) \tag{2.69}$$

As usual, it is required that the loop gain of $T_{VM}(s)$ has a certain phase margin PM_{VM} at the control bandwidth ω_{c_VM}. From (2.69), the following equations are derived.

$$\left|T_{VM}(j\omega_{c\_VM})\right| = 1 = \left|G_c(j\omega_{c\_VM})\right| \cdot \left|G_{vd}(j\omega_{c\_VM})\right| \tag{2.70}$$

$$\begin{aligned} arg\left[T_{VM}\left(j\omega_{c\_VM}\right)\right] &= -180^{\circ} + PM_{VM} \\ &= arg\left[G_c(j\omega_{c\_VM})\right] + arg\left[G_{vd}(j\omega_{c\_VM})\right] \end{aligned} \tag{2.71}$$

The latter equation can be rewritten as follows.

$$arg\left[G_c(j\omega_{c\_VM})\right] = -180^{\circ} + PM_{VM} - arg\left[G_{vd}(j\omega_{c\_VM})\right] \tag{2.72}$$

Notice, again, that $arg\left[G_c(j\omega_{c\_VM})\right]$ is function of known quantities. Classical PID controllers have the following transfer function

$$G_c(s) = K_p + \frac{K_i}{s} + K_d s \tag{2.73}$$

However, this PID controller has a transfer function with an increasing asymptote at high frequencies due to the derivative action [22]. This in practice has a disruptive operation of the PWM due to the propagation of the switching ripple causing failure of the closed-loop converter even though the voltage controller has been correctly designed to have the required closed-loop performance. To avoid this malfunction, high frequency poles ω_{P1} and ω_{P2} must be present in the real voltage controller to reduce the gain of the controller transfer function at high frequencies. This can be obtained with the following transfer function.

$$G_c(s) = \left(K_p + \frac{K_i}{s} + \frac{K_d s}{1 + \frac{s}{\omega_{P1}}}\right)\frac{1}{1 + \frac{s}{\omega_{P2}}} \quad (2.74)$$

The first pole ω_{P1} flattens the gain of the derivative component, while the second one ω_{P1} allows the gain to drop. As a good practical design rule, $\omega_{sw} \approx \omega_{P1} \ll \omega_{P2} \approx 10 \cdot \omega_{P1}$, where ω_{sw} is the switching frequency in *rad/s*. Solving (2.70) and (2.71) does not lead to the solution because three coefficients need to be found, i.e., K_p, K_i, and K_d, from two equations. Therefore, to solve (2.70) and (2.71), the designer needs to set one of the coefficients and solve for the others. Typically, it is convenient to set the value of the integral coefficient K_i. This is chosen relatively low, let us say $1 \ll K_i \ll 10$, to prevent antiwindup [23]. Finally, taking into account the voltage control transfer function (2.74), its control coefficients are calculated as follows.

$$K_p = \cos\left[arg\left[G_c(j\omega_{c\_VM})\right]\right] / \left|G_{vd}(j\omega_{c\_VM})\right| \quad (2.75)$$

$$K_d = \frac{\sin\left[arg\left[G_{c\_I}(j\omega_{c\_VM})\right]\right]}{\omega_{c\_VM} \cdot \left|G_{i_L d\_OL}(j\omega_{c\_PICM})\right|} + \frac{K_i}{\omega_{c\_VM}{}^2} \quad (2.76)$$

2.5.1.3 Simulation Example

A buck converter is now considered as an example. The buck converter switching frequency is 20 kHz, its input voltage is $V_g = 400\text{V}$, output voltage is $V = 200\text{V}$, and circuit parameters are $L = 3.5$ mH, $C = 50$ μF, and $R = 30\ \Omega$. A MATLAB script is provided in the Appendix that is able to perform the design of the controllers in VMC and PICM_FB.

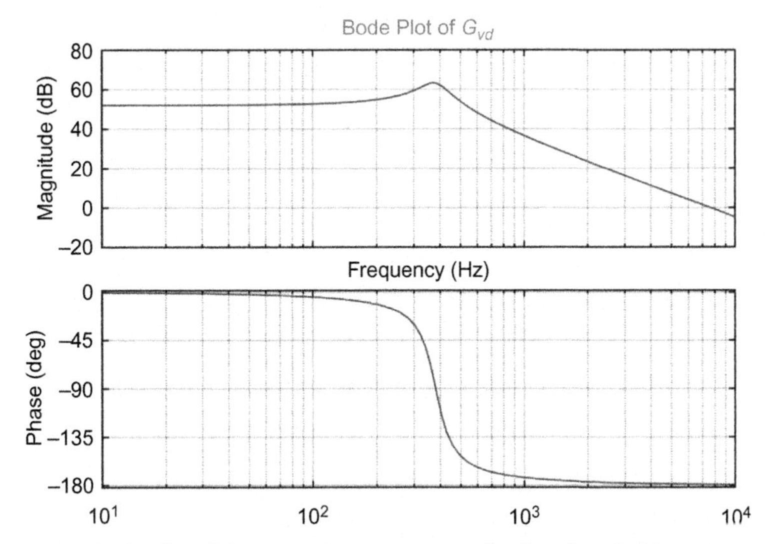

Figure 2.16 Bode plot of the control-to-output transfer function $G_{vd}(s)$.

Example 3.1: Buck converter in VMC

From (2.1), the control-to-output transfer function $G_{vd}(s)$ is plotted in Fig. 2.16. Notice that the phase of this transfer function approaches −180 degrees at high frequencies, i.e., the frequency range where it is desirable to choose the voltage control bandwidth, clearly indicating the need of a phase lead via derivative component in the voltage PID controller that has to be designed.

The PID voltage control with transfer function (2.74) is designed with crossover frequency $f_{c\_VM} = 1$ kHz and phase margin $PM\_{}_{VM} = 52$ degrees. First of all, a relatively small integral coefficient is chosen.

$$K_i = 5 \tag{2.77}$$

With the help of MATLAB, the magnitude and phase of this transfer function at the crossover frequency can be calculated.

$$\left| G_{vd}(j\omega_{c\_VM}) \right| = 67.1815 \tag{2.78}$$

$$arg\left[G_{vd}(j\omega_{c\_VM}) \right] = -1.7293e+02 \tag{2.79}$$

From (2.72), the phase of the voltage controller is

$$arg\left[G_c(j\omega_{c\_VM}) \right] = 44.9280 \tag{2.80}$$

Substituting (2.78)−(2.80) into (2.75) and (2.76), the proportional and derivative coefficients are finally calculated.

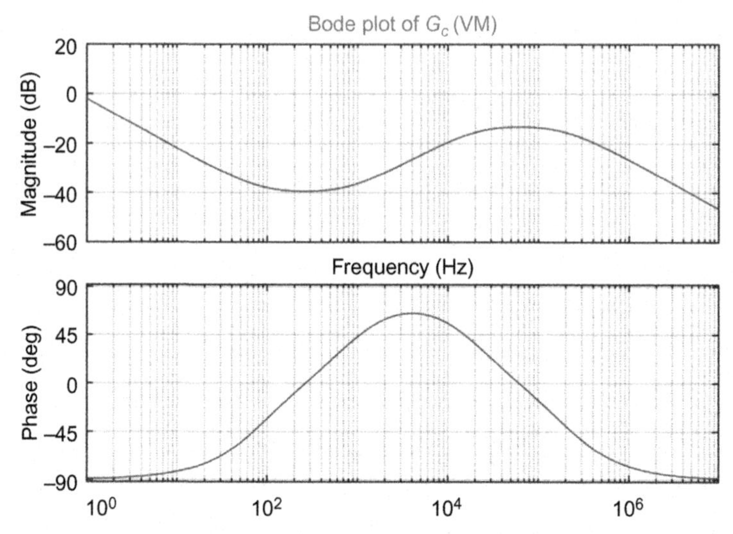

Figure 2.17 Bode plot of the voltage control transfer function $G_c(s)$.

$$K_p = 0.0105 \tag{2.81}$$

$$K_d = 1.7997e - 06 \tag{2.82}$$

Plotting (2.74) with the values of (2.77), (2.81), and (2.82) results in Fig. 2.17. Notice that the gain drops at high frequencies. The loop gain $T_{VM}(s)$ is plotted in Fig. 2.18. The phase obtained phase margin at the crossover frequency are also displayed.

Example 3.2: Buck converter in PICM_FB

The design of the inner current control and outer voltage control is a two-step approach. First, the current control is designed, then the voltage control. To design the inner current control, the control-to-inductor current transfer function $G_{i_L d}(s)$ (2.4) is the plant and it is plotted in Fig. 2.19. Notice that a simple PI controller can be deployed because the phase of this transfer function approaches −90 degrees at high frequencies, i.e., the frequency range where it is desirable to choose the current control bandwidth.

The PI current control with transfer function (2.62) is designed with crossover frequency $f_{c\_PICM} = 2$ kHz and phase margin $PM\_{PICM} = 80$ degrees.

With the help of MATLAB, the magnitude and phase of the plant transfer function at the crossover frequency can be calculated.

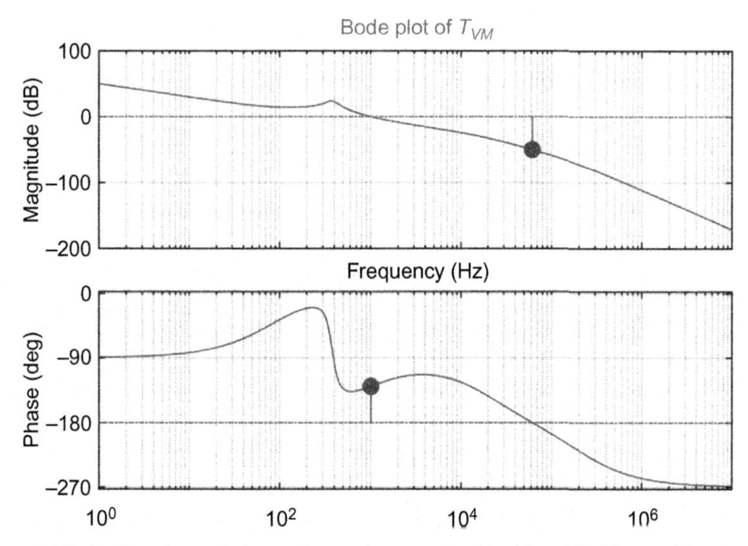

Figure 2.18 Bode plot of the voltage loop gain $T_{VM}(s)$ with Phase Margin 50.4 degrees, Delay Maring 0.000138 s, at $f = 1010$ Hz, Closed Loop stable = Yes.

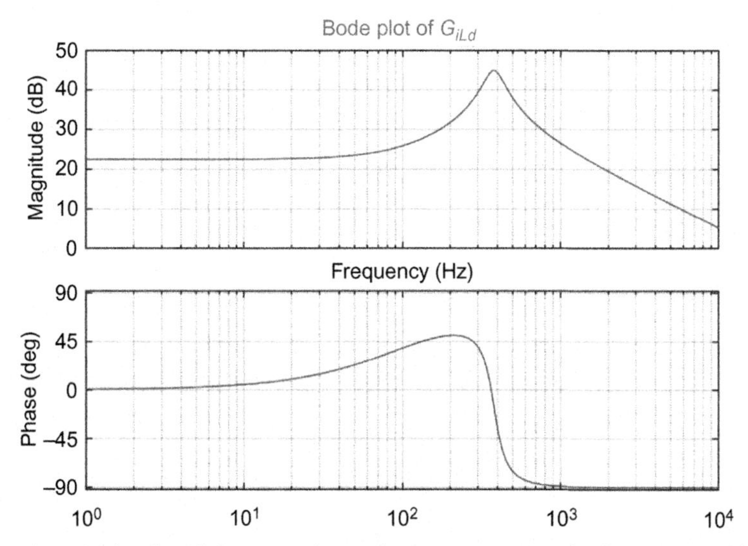

Figure 2.19 Bode plot of the control-to-inductor current transfer function $G_{i_Ld}(s)$.

$$\left|G_{i_L d}(j\omega_{c\_PICM})\right| = 9.4350 \tag{2.83}$$

$$arg\left[G_{i_L d}(j\omega_{c\_PICM})\right] = -89.8862 \tag{2.84}$$

From (2.61), the phase of the inner current controller is

$$arg\left[G_{c\_I}(j\omega_{c\_PICM})\right] = -10.1138 \quad (2.85)$$

Substituting (2.83)–(2.85) into (2.63) and (2.64), the proportional and integral coefficients can be calculated.

$$K_{p\_I} = 0.1043 \quad (2.86)$$

$$K_{i\_I} = 2.3388e+02 \quad (2.87)$$

Plotting (2.62) with the values of (2.86) and (2.87) results in Fig. 2.20. The loop gain $T_{PICM}(s)$ is plotted in Fig. 2.21. The phase obtained phase margin at the crossover frequency is also displayed.

To design the outer voltage loop, the designer needs to take into account the plant resulting from the integration of the inner current control to the converter model, schematically represented in Fig. 2.15. This transfer function is called current control input to output voltage $G_{vc}(s)$ and it is shown in Fig. 2.22.

The PI voltage control with transfer function (2.65) is designed with crossover frequency $f_{c\_PICM\_VM} = 0.1$ kHz and phase margin $PM_{\_PICM\_FB} = 80$ degrees.

Again, with the help of MATLAB, the magnitude and phase of the plant transfer function at the crossover frequency can be calculated.

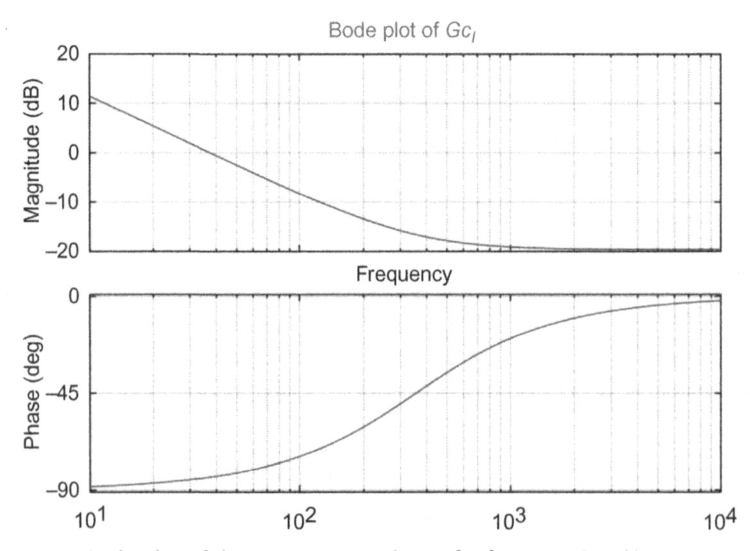

Figure 2.20 Bode plot of the current control transfer function $G_{c\_I.}(s)$.

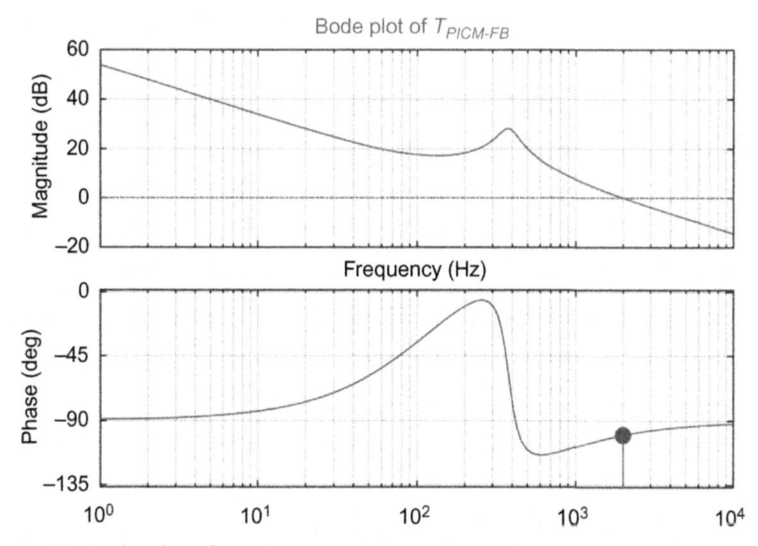

Figure 2.21 Bode plot of the inner current loop gain $T_{PICM}(s)$ with Phase Margin 80 degrees, Delay Maring 0.000111 s, at $f = 2000$Hz, Closed Loop stable = Yes.

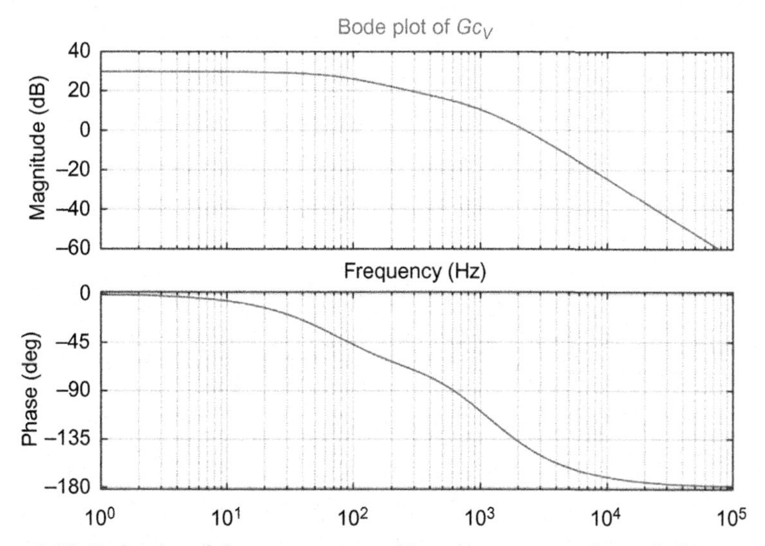

Figure 2.22 Bode plot of the current control input to output voltage $G_{vc}(s)$.

$$\left|G_{vc}(j\omega_{c\_PICM\_FB})\right| = 19.6696 \tag{2.88}$$

$$arg\left[G_{vc}(j\omega_{c\_PICM\_FB})\right] = -47.2628 \tag{2.89}$$

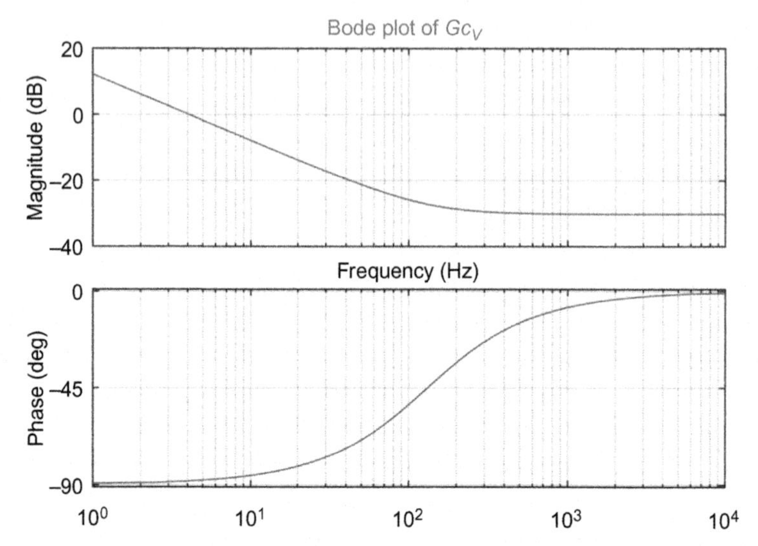

Figure 2.23 Bode plot of the voltage control transfer function $G_{c\_v}(s)$.

From (2.68), the phase of the inner current controller is

$$arg\left[G_{c\_V}\left(j\omega_{c\_PICM\_FB}\right)\right] = -52.7372 \tag{2.90}$$

Substituting (2.88)–(2.90) into (2.66) and (2.67), the proportional and integral coefficients can be calculated.

$$K_{p\_V} = 0.0308 \tag{2.91}$$

$$K_{i\_V} = 25.4228 \tag{2.92}$$

Plotting (2.65) with the values of (2.91) and (2.92) results in Fig. 2.23. The loop gain $T_{PICM\_FB}(s)$ is plotted in Fig. 2.24. The phase obtained phase margin at the crossover frequency are also displayed.

2.5.2 Network Analyzer Technique

Frequency-domain measurements to extract transfer functions of converter prototypes are a good engineering practice. This type of measurements allows one to determine if the converter system has been correctly modeled and its controller properly designed. Small-signal ac magnitude and phase measurements can be performed using an instrument known as a network analyzer, or frequency response analyzer. The key inputs and outputs of a basic network analyzer are illustrated in Fig. 2.25 along with its main features and functions.

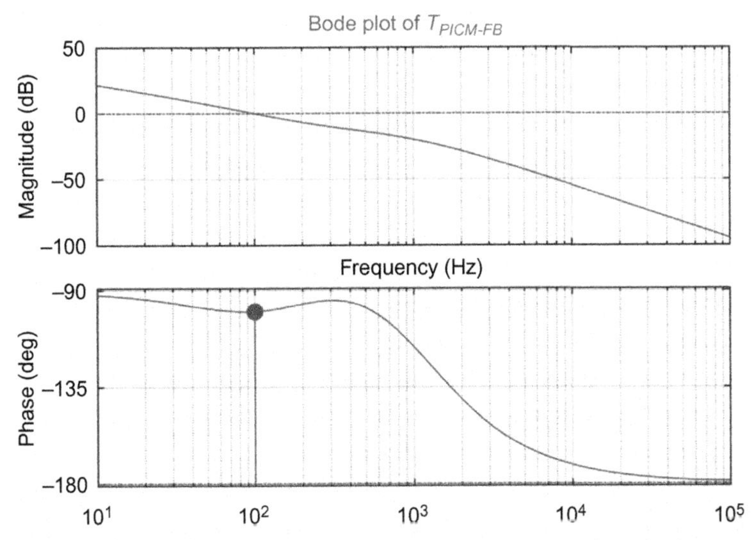

Figure 2.24 Bode plot of the inner voltage loop gain $T_{PICM\_FB}(s)$ with Phase Margin 80 degrees, Delay Maring 0.00222 s, at $f = 100$ Hz, Closed Loop stable = No.

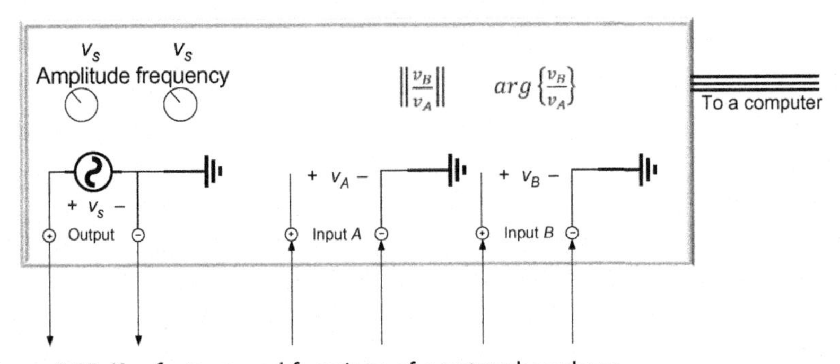

Figure 2.25 Key features and functions of a network analyzer.

The network analyzer is equipped with a sinusoidal output voltage v_s of controllable amplitude and frequency. This signal can be injected into the system under test at any desired location. The network analyzer also has two inputs, v_A and v_B (in some commercial products even more inputs are available). The negative electrodes of the output signal and input signals are connected to earth ground internally to the instrument. The network analyzer measures the components of v_A and v_B at the injection frequency, and calculates the magnitude and the phase of the quantity v_B/v_A. Many modern network analyzers automatically sweep the frequency of the injection source v_s and are able to generate several plots

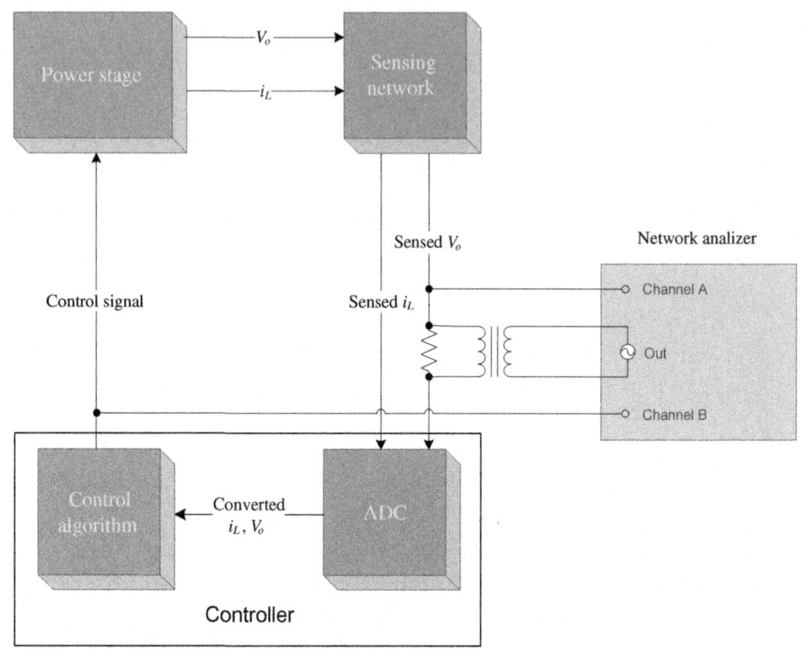

Figure 2.26 Typical test setup for measuring $G_{vd}(s)$ when the converter operates with constant duty cycle.

to display magnitude and phase of the function v_B/v_A, such as Bode plots, Nyquist plots, and other frequency-dependent plots. In power electronics applications, the network analyzer is required to perform as a narrowband voltmeter tracker while measuring v_A and v_B, otherwise switching ripples and background noise may corrupt the shapes of the sine waveforms and therefore the accurateness of frequency responses.

Let us first measure the control-to-output transfer function $G_{vd}(s)$ of a generic power converter with constant duty cycle as control signal. Fig. 2.26 illustrates the typical test setup for measuring such a transfer function. The perturbation signal is injected between the output voltage sensing network and the input of the analog to digital converter (ADC) thanks to the usage of an injection transformer and small-value resistor on the loop path. Channel B input of the network analyzer is connected to the control signal wire, while Channel A is connected to the output of the output voltage sensing network. The control-to-output transfer function $G_{vd}(s)$ is the ratio between Channel A and Channel B.

The Voltage-Mode closed-loop gain $T_{VM}(s)$ is measured according to the test setup illustrated in Fig. 2.27. In this case, the setup is basically the

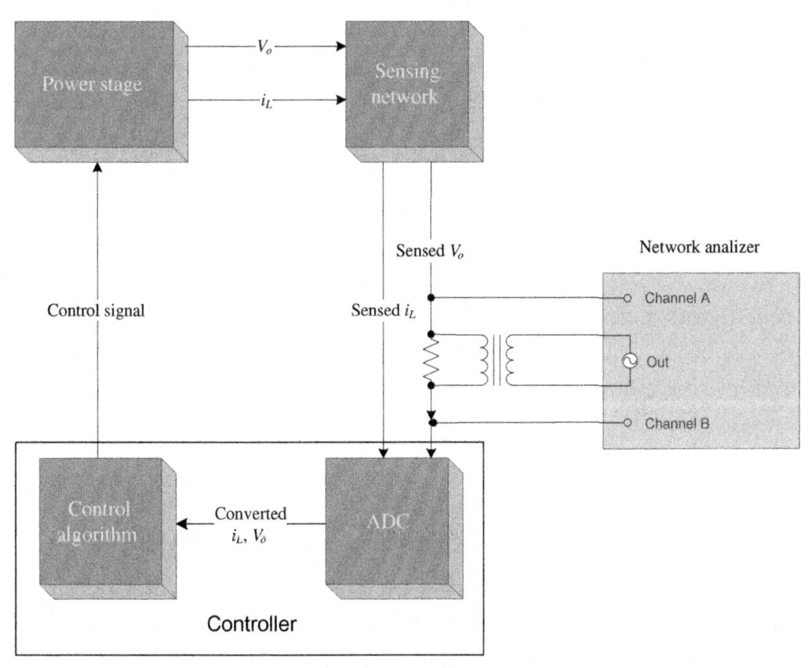

Figure 2.27 Typical test setup for measuring $T_{VM}(s)$ when the converter operates in VMC.

same as in Fig. 2.26 with the exception of Channel B input of the network analyzer connected to input of the ADC. The closed-loop gain $T_{VM}(s)$ is the ratio between Channel A and Channel B.

As this subsection has the primary goal to present a practical method to validate the modeling and control design on a converter prototype, only the setups to measure $G_{vd}(s)$ and $T_{VM}(s)$ were shown. Similar setups can be derived to measure other open-loop and closed-loop transfer functions. This can be left as an easy exercise for the reader.

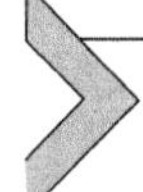

2.6 SMALL-SIGNAL STABILITY ANALYSIS OF CASCADE SYSTEMS

As a well-known challenge and as described in Chapter 1, cascade of power-electronics-based systems suffer from stability degradation caused by interactions among converters due to the CPL effect provided by the load subsystem. Typically, feedback-controlled converters, such as feedback-controlled converters and inverters, behave as CPLs at their input terminals within their control loop bandwidth [24,25]. CPLs exhibit negative incremental input impedance, which is the cause of the subsystem interaction problem and the origin of the undesired destabilizing effect [26]. Although each subsystem is independently designed to be stand-alone stable, a system consisting of the cascade of source and load subsystems may exhibit degraded stability due to subsystem interactions caused by the CPL. This is because the subsystem interaction affects the bandwidth, the phase, and the gain margin of each individual converter subsystem [27]. In the past, the subsystem interaction problem was not significant because an individual subsystem such as a tightly regulated converter operated under quasiideal conditions: low source impedance at its input and mainly passive loads at its output [28]. In modern cascade systems, the subsystem interaction is a serious issue, as is shown in this section.

2.6.1 The Nyquist Stability Criterion and Its Practical Usage

To address stability issues of a cascade system, several authors have studied the linearized system under steady-state conditions by defining the source subsystem and the load subsystem at an arbitrary interface within the overall system. Fig. 2.28 shows the equivalent system broken down into two subsystems assumed to be individually stable. The total input-to-output transfer function is (Laplace variable "*s*" is omitted hereafter for simplicity)

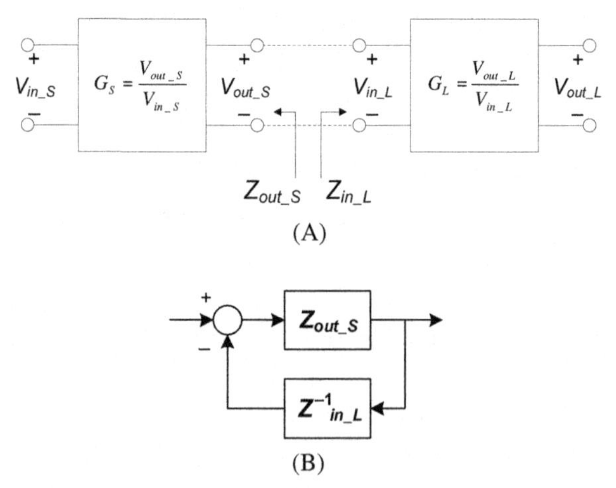

Figure 2.28 Equivalent source subsystem interaction with the equivalent load subsystem (A) and equivalent MIMO feedback block diagram (B).

$$G_{SL} = \frac{V_{out\_L}}{V_{in\_S}} = G_S G_L \cdot \frac{Z_{in\_L}}{Z_{in\_L} + Z_{out\_S}} = G_S G_L \cdot \frac{1}{1 + T_{MLG}} \quad (2.93)$$

where the minor loop gain T_{MLG} is defined as

$$T_{MLG} = \frac{Z_{out\_S}}{Z_{in\_L}} \quad (2.94)$$

It is called minor loop gain because it can be shown that the output impedance of the source subsystem Z_{out_S} and input impedance of the load subsystem Z_{in_L} form a type of negative feedback system as depicted in Fig. 2.28.

Since G_S and G_L are stable transfer functions, the minor loop gain term is the one responsible for stability. Therefore, a necessary and sufficient condition for stability of the cascade system can be obtained by applying the Nyquist Criterion to T_{MLG}, i.e., the interconnected system is stable if and only if the Nyquist contour of T_{MLG} does not encircle the $(-1, j0)$ point. From the control theory of linear systems, two quantities called Gain Margin (GM) and Phase Margin (PM) can be defined. These quantities quantify "how far" the Nyquist contour of T_{MLG} is from the critical $(-1, j0)$ point, and, therefore, "how far" the system is from being unstable. The quantities GM and PM are also related to the dynamic response of the system: the higher the damping, the larger GM and PM are (and vice versa the lighter the damping, the smaller GM and PM are) [22]. Specifying GM and PM provides the engineers a way to design for system stability with certain stability margins linked to the desired dynamic time domain performance. Based on this concept, what is practically needed is to guarantee that the Nyquist diagram of T_{MLG} does not encircle the $(-1, j0)$ point with sufficient stability margins. For this reason, many practical stability criteria for the cascade system of Fig. 2.28 were proposed. These stability criteria define various boundaries between forbidden and allowable regions for the polar plot of T_{MLG}. The boundaries are defined by a certain GM and PM and are shown in Fig. 2.29. The forbidden regions are the ones that include the $(-1, j0)$ point. System stability can be ensured by keeping the contour of T_{MLG} outside the forbidden regions. Based on the definition of the forbidden regions, design formulations can be specified which relate the desired GM and PM to the system parameters. Note that these criteria give only sufficient, but not necessary stability conditions.

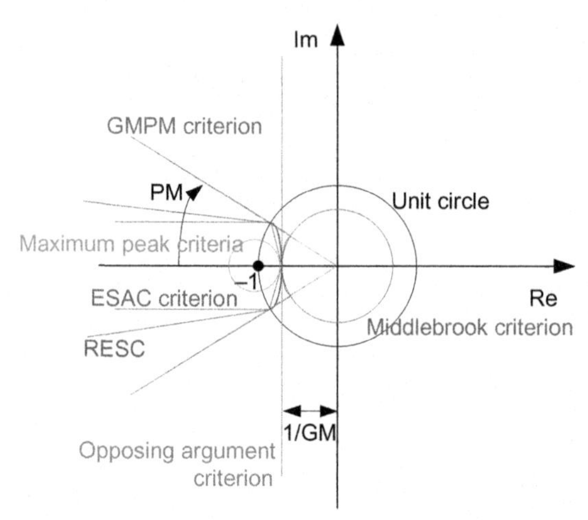

Figure 2.29 Stability criteria: boundaries between forbidden regions and allowable regions.

To assess overall system stability, several stability criteria for DC systems based on forbidden regions for the minor loop gain have been proposed in the literature, such as the Middlebrook Criterion [29], and its various extensions, such as the Gain and Phase Margin (GMPM) Criterion [30], the Opposing Argument Criterion [31–33], the Energy Source Analysis Consortium (ESAC) Criterion [34,35], and its extension the Root Exponential Stability Criterion (RESC) [36]. All these criteria have been reviewed in [37], which presents a discussion, for each criterion, of the artificial conservativeness of the criterion in the design of DC systems, and the design specifications that ensure system stability. Shortcomings of all these criteria (also discussed in [37]) are that they lead to artificially conservative designs, encounter difficulties when applied to multiconverter systems (more than two interconnected subsystems) especially in the case when power flow direction changes, and are sensitive to component grouping. Moreover, all these criteria, with the exception of the Middlebrook Criterion, are not conducive to an easy design formulation. Another significant practical difficulty present with all the prior stability criteria is the minor loop gain online measurement [38]. It requires two separate measurements, source subsystem output impedance and load subsystem input impedance, and then some postprocessing. Due to the complexity in the calculation, this approach is not suitable for online stability monitoring. (Only in the work [39], a practical approach to measure the stability margins of the minor loop gain was proposed. However, such an approach, based on the Opposing Argument Criterion, fails when used with other less conservative criteria.)

More recent stability criteria include the Three-Step Impedance Criterion (T-SIC) [40], the Unified Impedance Criterion (UIC) [41], and the Maximum Peak Criteria (MPC) [42,43]. The T-SIC [40] relaxes the conservativeness of previous criteria because it does not assume that G_S in (2.93) is necessarily a stable transfer function, which is typical of regulated source subsystem. For this reason, the impedance criterion should not be applied on the minor loop gain defined in (2.94), but rather on an extended minor loop gain defined in [40]. All previous stability criteria were developed for source subsystem interaction alone or load subsystem interaction alone by using the Middlebrook's Extra Element Theorem (EET) [44]. Derived by using the 2EET [45], the UIC [41], particularly suitable for cascade connected subsystems, constructs the minor loop gain considering the simultaneous interaction of both source and load subsystems. The last proposed stability criterion in order of time is the MPC [42,43] which defines the minimum forbidden region for the minor loop gain among all prior stability criteria (Fig. 2.29). Such a forbidden region is determined by the maximum allowable peak of the sensitivity function, providing a direct measure of the stability robustness. However, as also demonstrated in [42,43], the state of the stability robustness strongly suffers from the interface where the minor loop gain is measured.

2.6.1.1 Simulation Example

As an illustrative example, an averaged model simulation of a cascade of a buck converter with a Voltage Source Inverter (VSI) in Fig. 2.30 is considered. The values of voltages and components for both buck converter and VSI are also reported in Fig. 2.30. The controllers of both the buck converter and VSI are designed according to the procedure presented in Section 2.5.

The load VSI, modeled in synchronous dq coordinates [46], is controlled by an inner PI current mode (PICM subscript) loop with crossover frequency $f_{c\_PICM} = 1$ kHz and phase margin $PM_{\_PICM} = 80$ degrees, and an outer PI voltage (PICM_FB subscript) loop with crossover frequency $f_{c\_PICM\_FB} = 0.1$ kHz and phase margin $PM_{\_PICM\_FB} = 80$ degrees. Due to its importance in the small-signal stability analysis of the cascade system, Fig. 2.31 depicts how the input impedance of the VSI is modified by effect of the control action with respect to the open-loop (OL subscript) case. Notice that by the addition of the PI current loop and then of the outer PI

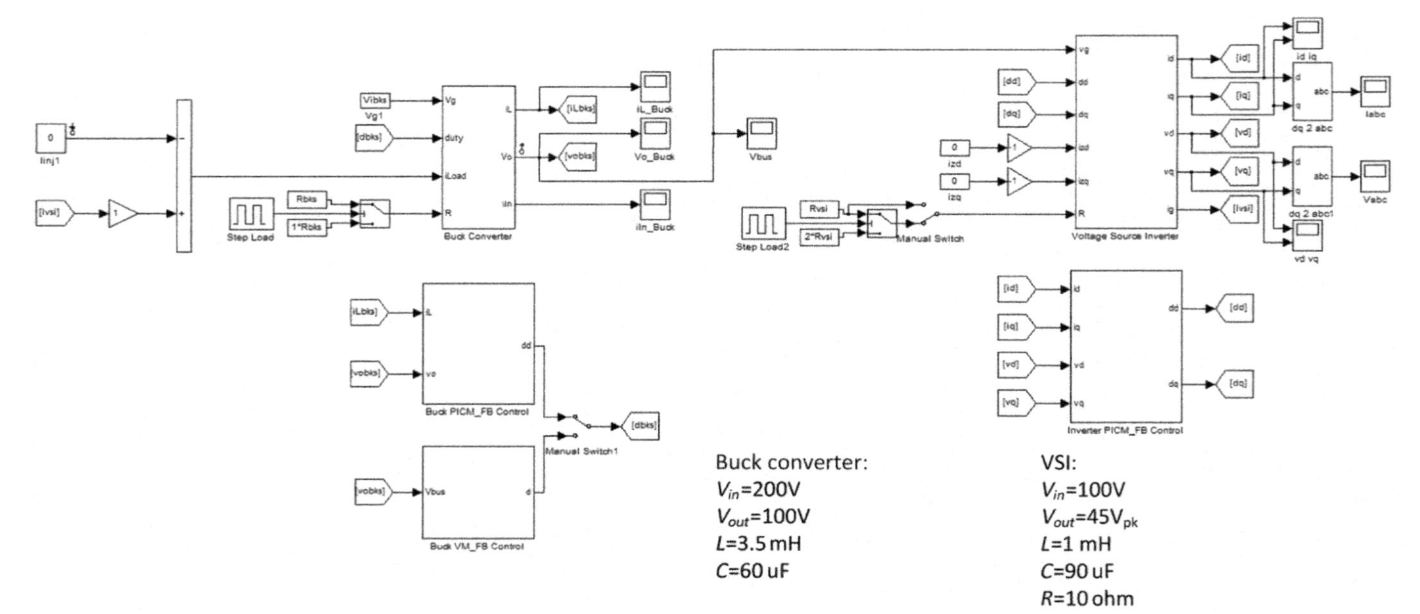

Figure 2.30 Averaged model simulation in Simulink of the cascade of a buck converter and a VSI.

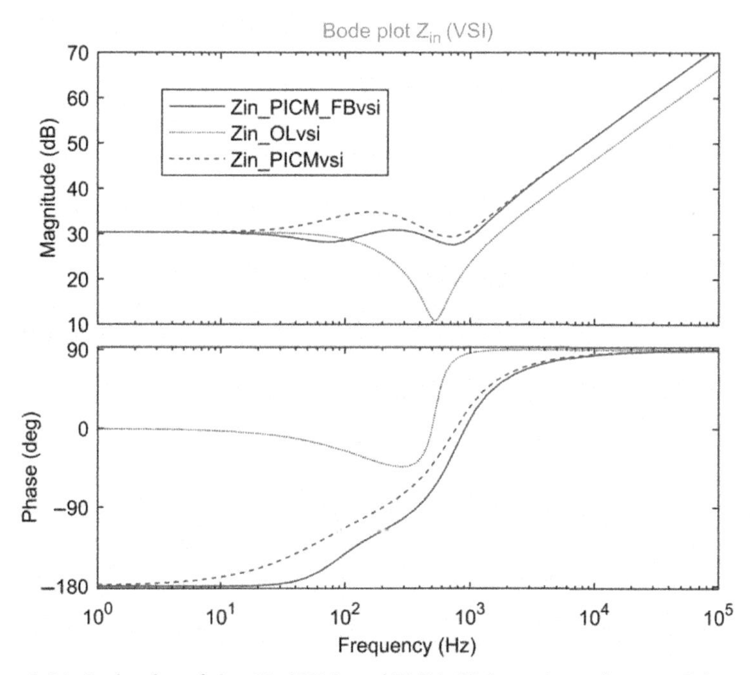

Figure 2.31 Bode plot of the OL, PICM, and PICM_FB input impedances of the VSI.

voltage loop, the PICM-controlled VSI, and the PICM_FB-controlled VSI behave as a CPL within the control bandwidth.

Two different types of control are implemented on the source buck converter to show their effect on its output impedance. Notice that the buck converter control was designed to take into account a load resistance that sinks the same amount of power that the VSI would do in its place. After the controller design is complete, the buck converter resistive load is removed and the VSI is connected. The first type of control that will be analyzed is a current mode, i.e., an inner PI current loop with crossover frequency $f_{c\_PICM} = 1$ kHz and phase margin $PM_{\_PICM} = 80$ degrees, and an outer PI voltage loop with crossover frequency $f_{c\_PICM\_FB} = 0.1$ kHz and phase margin $PM_{\_PICM\_FB} = 70$ degrees. The second type of control is a PID voltage mode (VM_FB subscript), i.e., a single voltage loop with crossover frequency $f_{c\_VM} = 0.5$ kHz and phase margin $PM_{\_VM} = 52$ degrees.

The current mode control case is analyzed first. Figs. 2.32 and 2.33 depict how the output impedance of the buck converter with and without resistive load, respectively, is modified by effect of the control action with respect to the open-loop (OL subscript) case. Notice how the

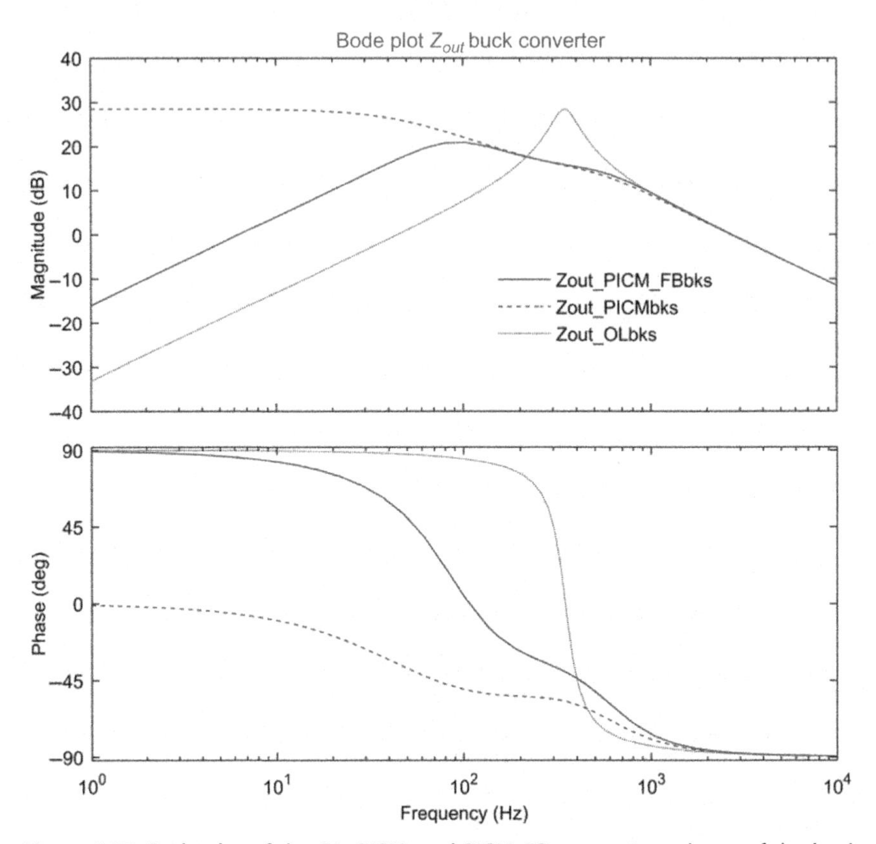

Figure 2.32 Bode plot of the OL, PICM, and PICM_FB output impedance of the buck converter with resistive load.

addition of the PI current loop and then of the outer PI voltage loop modify the phase of the output impedance at low frequencies.

The case of VM_FB-controlled buck converter produces a different result. Figs. 2.34 and 2.35 depict how the output impedance of the buck converter with and without resistive load, respectively, is modified by effect of the control action with respect to the open-loop (OL subscript) case. Notice that the addition of the PID voltage loop has the effect of making a negative incremental impedance within the control bandwidth.

In the next two subsections, the stability of the system is assessed by using the Nyquist Criterion applied to the minor loop gain in frequency-domain simulations as well as time-domain simulations. Two cases are analyzed: a stable one and an unstable one, depending on the types of the controllers implemented for the source buck converter. Table 2.3 summarizes the results that are presented in the next two subsections.

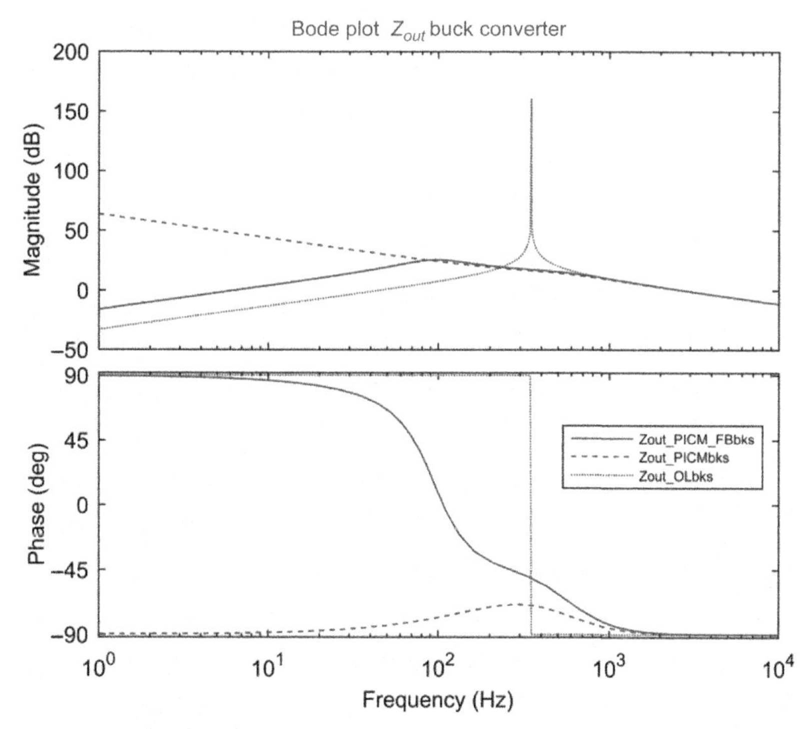

Figure 2.33 Bode plot of the OL, PICM, and PICM_FB output impedance of the buck converter with resistive load removed.

2.6.1.1.1 Stable Cases

The case of the cascade of a PICM_FB-controlled buck converter and a PICM_FB-controlled VSI is given first. The Nyquist Stability Criterion on the minor loop gain $T_{MLG} = Z_{out\_PICM\_FB}/Z_{in\_PICM\_FB}$ confirms system stability since the contour does not encircle the $(-1, j0)$ point, as shown in Fig. 2.36. The time-domain simulation is reported in Fig. 2.37 which shows the transient of the bus voltage and VSI three-phase output voltage in correspondence of a VSI symmetric three-phase load step from 20Ω to 10Ω. A stable performance is evident.

The case of the cascade of a VM_FB-controlled buck converter and a PICM_FB-controlled VSI is now presented. The Nyquist Stability Criterion on the minor loop gain $T_{MLG} = Z_{out\_VM\_FB}/Z_{in\_PICM\_FB}$ predicts system stability since the contour stays does not encircle the $(-1, j0)$ point, as shown in Fig. 2.38. The time-domain simulation is reported in Fig. 2.39 which shows the transient of the bus voltage and

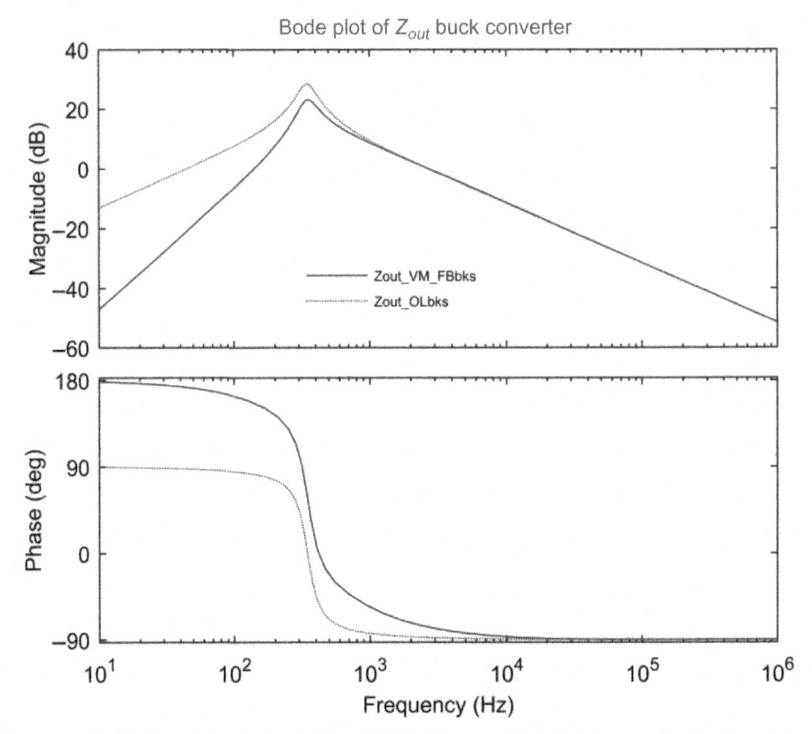

Figure 2.34 Bode plot of the OL and VM_FB output impedance of the buck converter with resistive load.

VSI three-phase output voltage in correspondence of a VSI symmetric three-phase load step from 20Ω to 10Ω. A stable performance is evident.

2.6.1.1.2 Unstable Cases

To make the system consisting of the cascade of a PICM_FB-controlled buck converter and a PICM_FB-controlled VSI unstable, the buck converter outer voltage loop phase margin is reduced to $PM\_{PICM\_FB} = 50$ degrees. The resulting system is unstable, as confirmed by using the Nyquist Stability Criterion on the minor loop gain $T_{MLG} = Z_{out\_PICM\_FB}/Z_{in\_PICM\_FB}$ since the contour encircles the $(-1, j0)$ point, as shown in Fig. 2.40. The time-domain simulation is reported in Fig. 2.41 which show the transient of the bus voltage and VSI three-phase output voltage in correspondence of a VSI symmetric three-phase load step from 20Ω to 10Ω. An unstable performance is evident.

To make the system consisting of the cascade of a VM_FB-controlled buck converter and a PICM_FB-controlled VSI unstable, the source buck

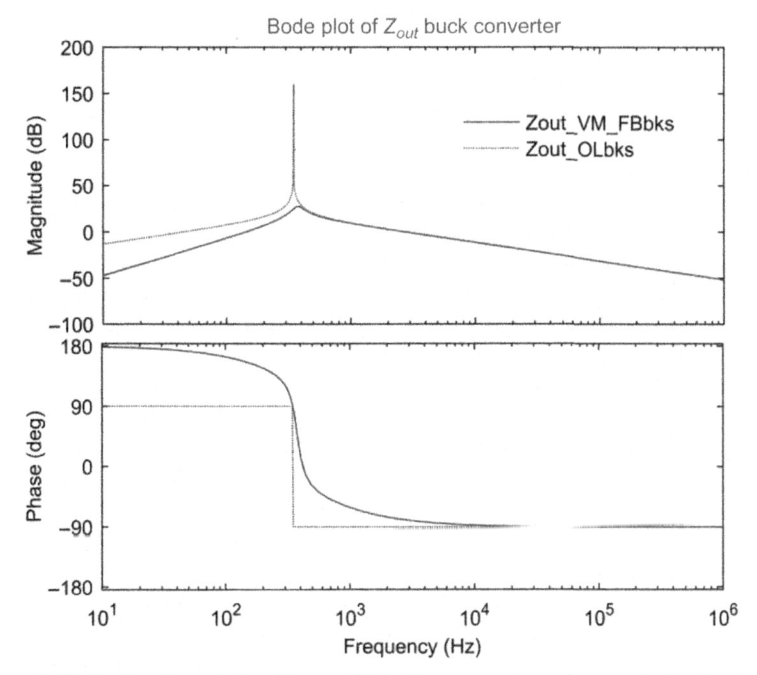

Figure 2.35 Bode plot of the OL and VM_FB output impedance of the buck converter with resistive load removed.

Table 2.3 Summary of Results for Stable and Unstable Cases

Case	Type of Control	Stable?	Figures
Stable	PICM_FB Buck PICM_FB VSI	Yes	2.36, 2.37
	VM_FB Buck PICM_FB VSI	Yes	2.38, 2.39
Unstable	PICM_FB Buck PICM_FB VSI	No	2.40, 2.41
	VM_FB Buck PICM_FB VSI	No	2.42, 2.43

converter outer voltage loop cross-over frequency is reduced to $f_{c\_VM\_FB} = 0.2$ kHz. The system is unstable, as confirmed by the Nyquist Stability Criterion on the minor loop gain $T_{MLG} = Z_{out\_VM\_FB}/Z_{in\_PICM\_FB}$ since the contour encircles the $(-1, j0)$ point, as shown in Fig. 2.42. The time-domain simulation is given Fig. 2.43 which show the transient of the bus voltage and VSI three-phase output voltage in

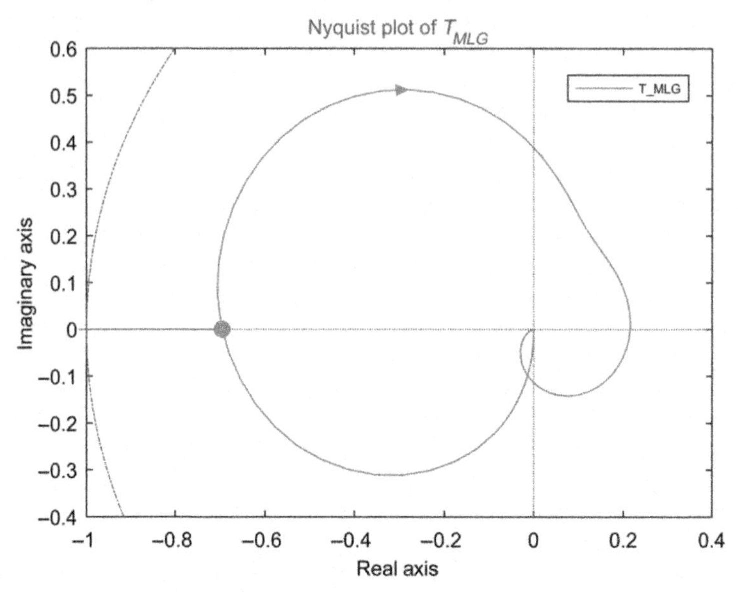

Figure 2.36 Nyquist plot of the minor loop gain for the cascade of PICM_FB-controlled buck converter and PICM_FB-controlled VSI (stable case).

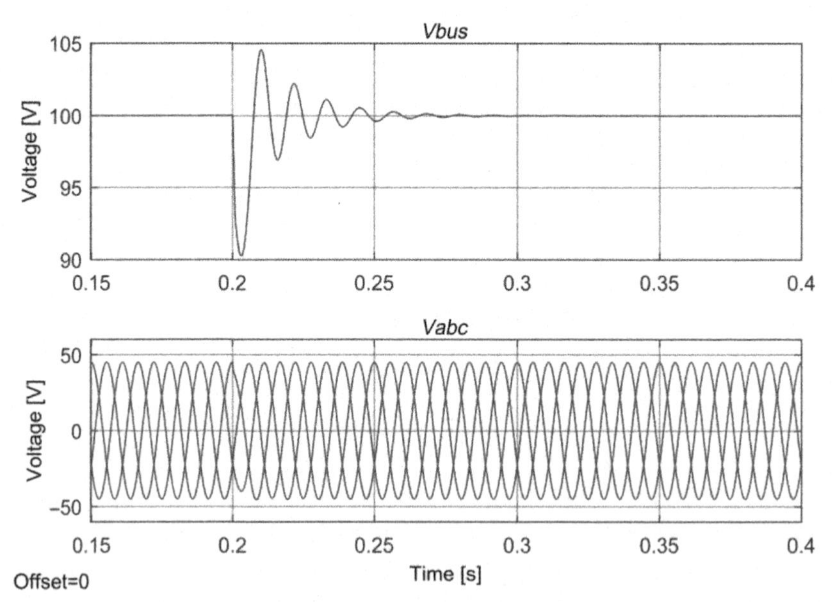

Figure 2.37 Bus voltage and VSI three-phase output voltage transient in correspondence of a symmetric VSI load step from 20Ω to 10Ω for the cascade of PICM_FB-controlled buck converter and PICM_FB-controlled VSI (stable case).

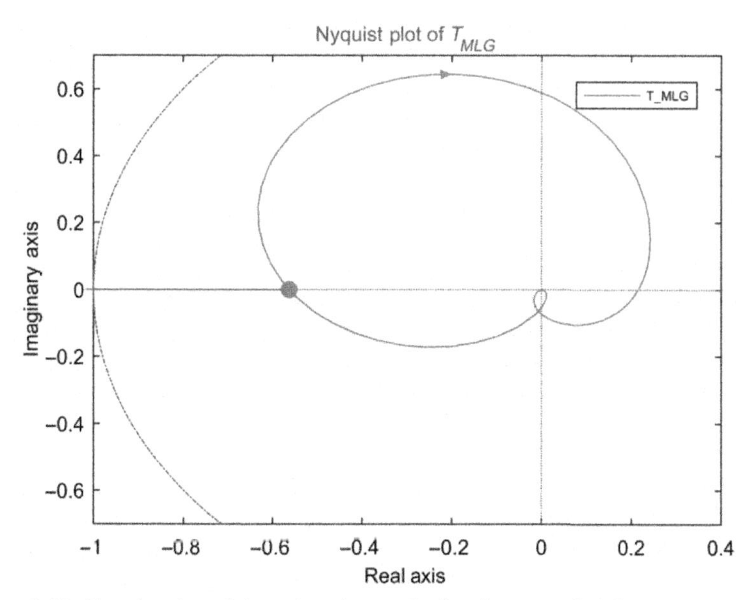

Figure 2.38 Nyquist plot of the minor loop gain for the cascade of VM_FB-controlled buck converter and PICM_FB-controlled VSI (stable case).

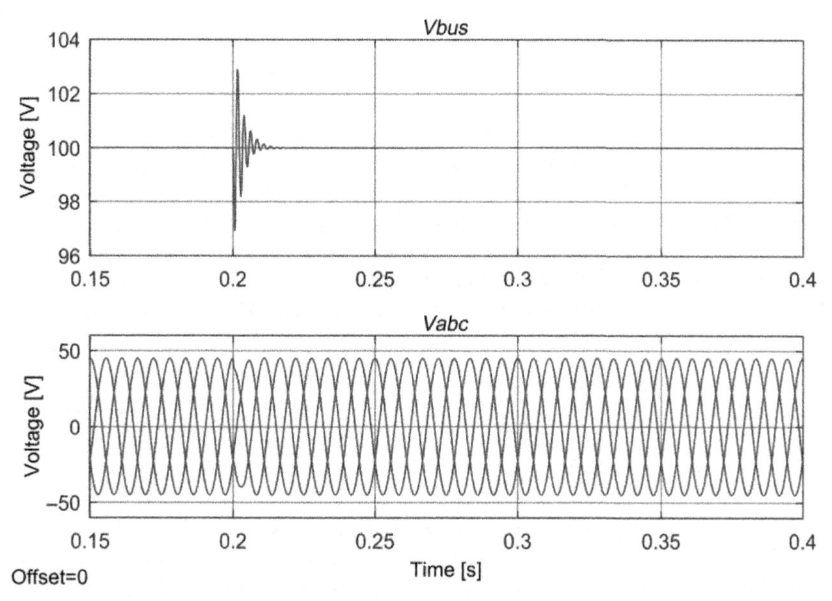

Figure 2.39 Bus voltage and VSI three-phase output voltage transient in correspondence of a symmetric VSI load step from 20Ω to 10Ω for the cascade of VM_FB-controlled buck converter and PICM_FB-controlled VSI (stable case).

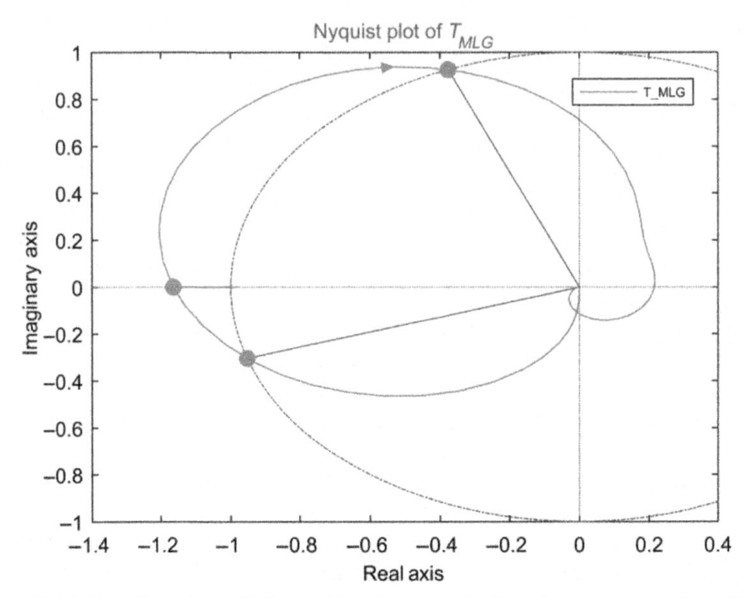

Figure 2.40 Nyquist plot of the minor loop gain for the cascade of PICM_FB-controlled buck converter and PICM_FB-controlled VSI (unstable case).

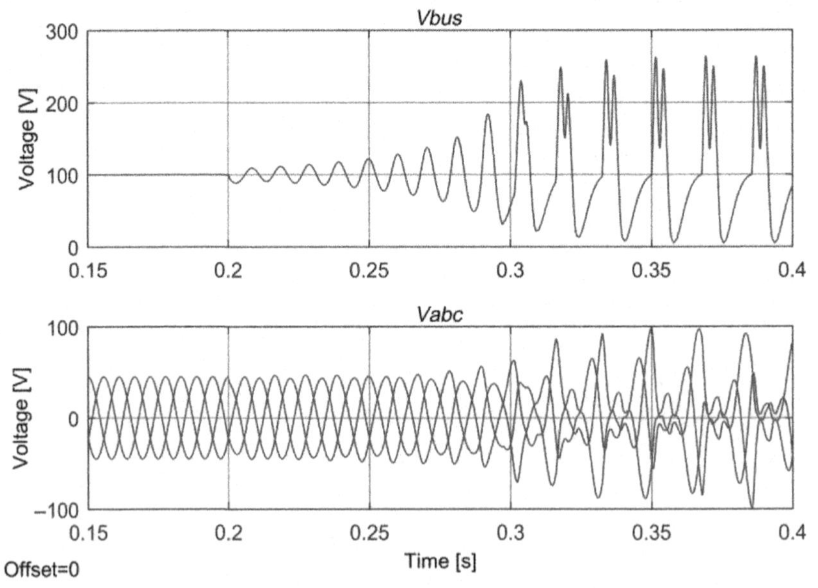

Figure 2.41 Bus voltage and VSI three-phase output voltage transient in correspondence of a symmetric VSI load step from 20Ω to 10Ω for the cascade of PICM_FB-controlled buck converter and PICM_FB-controlled VSI (unstable case).

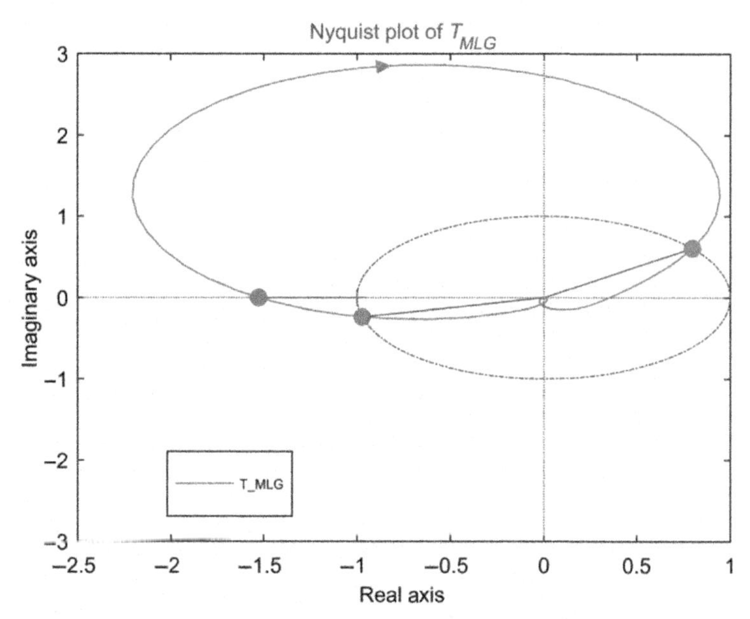

Figure 2.42 Nyquist plot of the minor loop gain for the cascade of VM_FB-controlled buck converter and PICM_FB-controlled VSI (unstable case).

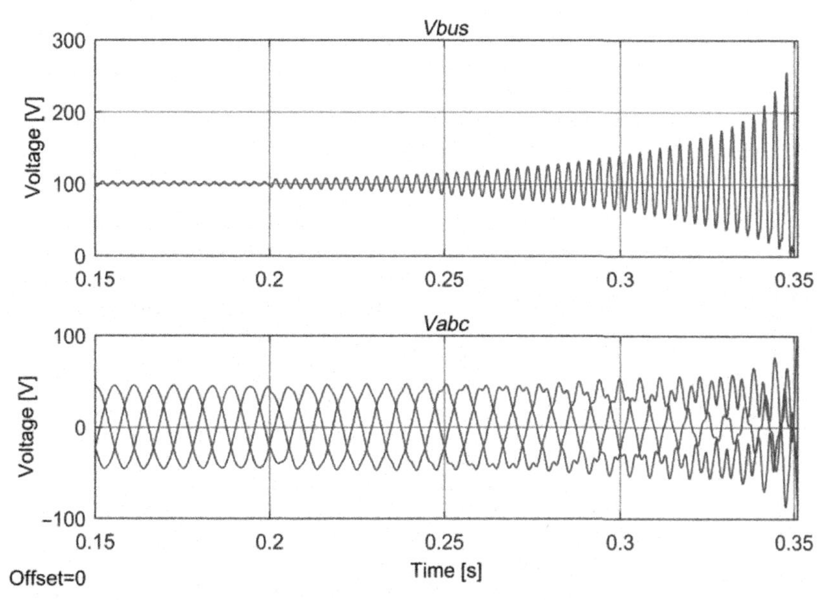

Figure 2.43 Bus voltage and VSI three-phase output voltage transient in correspondence of a symmetric VSI load step from 20Ω to 10Ω for the cascade of VM_FB-controlled buck converter and PICM_FB-controlled VSI (unstable case).

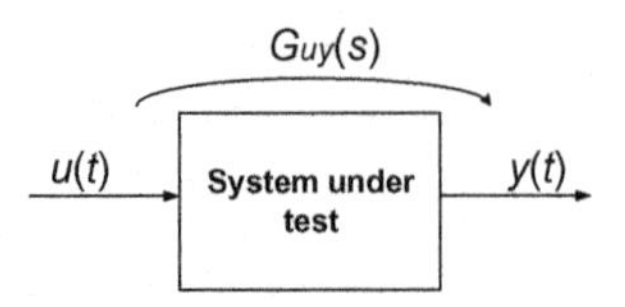

Figure 2.44 A dynamic system under test to find the input-to-output transfer function $G_{uy}(s)$.

correspondence of a VSI symmetric three-phase load step from 20Ω to 10Ω. An unstable performance is evident.

2.6.2 Online Wideband System Identification Technique

In order to apply the Nyquist Stability Criterion to the minor loop gain defined in (2.94), it is necessary to know the interface impedances of the cascade system. Offline analytical methods to study the stability provide some limitations because in real systems the impedances continuously vary over time and in relation to too many parameters. Therefore, online methods to measure impedances are required in order to monitor system stability margins in real time and take stabilizing actions if needed. The online technique to measure impedances[1] presented in this section has the following characteristics. First, it is able to complete the impedance measurement in a short time, allowing fast response to system variations. Additionally, it utilizes the existing power converters of the cascade system and their sensors to perform the impedance measurement (it is not needed to add specialized equipment with the associated extra cost, size and weight).

Before focusing on how to measure an impedance, let us briefly review the fundamental principle for measuring a transfer function of a dynamic system. In order to measure the input-to-output transfer function $G_{uy}(s)$ of the dynamic system under test shown in Fig. 2.44, an excitation signal with a desired frequency content is applied to the system input $u(t)$ and the system response $y(t)$ is measured. The time-domain measurements of the signals $u(t)$ and $y(t)$ are processed to calculate the $G_{uy}(s) = y(s)/u(s)$ in the frequency domain. The excitation signal can be either narrowband or broadband as shown in Fig. 2.45. An example of a narrowband signal is a sine wave, which ideally has power only at a single frequency. An example of a wideband signal is white noise, which has power over a wide frequency range. The pseudo-random binary signal (PRBS) shown in Fig. 2.45 is a digital approximation of white noise and

[1] Other small-signal transfer functions can also be measured.

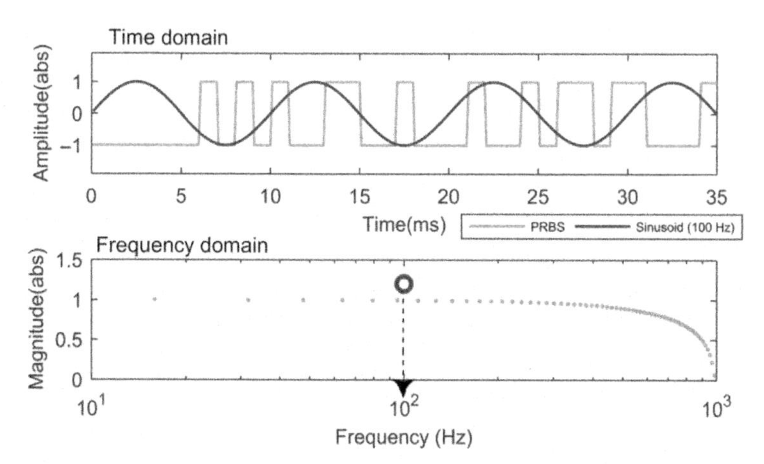

Figure 2.45 A wideband PRBS signal (*red* (light gray in print version)) and a narrowband sinusoidal signal (*blue* (dark gray in print version)) in the time domain (top) and in the frequency domain (bottom)

has an approximately flat signal content in the frequency domain (the signal energy drops to zero at signal generation frequency which is 1000 Hz in this example). Specialized equipment used to measure transfer functions of dynamic systems, such as network analyzers, frequently uses narrowband excitation signals. This gives measurement with accuracy and reduced noise, but the measurement typically requires long time, because the measurement is performed one frequency at a time and the excitation frequency must be swept across the frequency range of interest. Conversely, wideband signals, such as the PRBS, excite all frequencies of interest simultaneously, allowing a much faster measurement using Fast Fourier Transform (FFT) techniques. This latter solution is described hereafter in this section, since a fast measurement technique is desired.

The online method to measure system impedances by using the wideband excitation signal PRBS is called the *online WSI technique* [47,48]. The online WSI technique can generate the wideband excitation which can be controlled in both magnitude and duration for both voltage and current at the impedance measurement point. Then, FFT techniques can be applied on the measured voltage and current to identify the impedance of the system under test. This technique to measure impedances is used in conjunction with existing power electronics converters which serve as power amplifiers for the excitation signal PRBS. Moreover, since the converters have digital control and sensing, they may also be used to monitor the evolution of impedances on the top of their power conversion function. In fact, it is possible to integrate these digital network analyzer techniques [48] into the converter controllers allowing them to be used as online monitors without any extra power hardware.

Fig. 2.46 shows the cascade of two converters. Each of the converters has its own digital controller. As an example, PI current-voltage nested-loop controllers are deployed for both the source and load converters. The excitation signal PRBS is injected to the duty cycle and to both the current and voltage reference signals of the converter controller. Both voltage and current responses during the duration of the injection are measured. Processing those measurements from the time domain to the frequency domain returns the source impedance. Notice that both the PRBS generator and processing are implemented in an embedded controller that can be integrated to control platform of the load converter. Moreover, the measure is performed during the steady-state operation of the cascade system. It is, in fact, required that the excitation does not perturb the steady state too much. Therefore, the injected PRBS requires proper scaling to obtain the desired controlled perturbation amplitude at the impedance measurement point. This can be accomplished by properly choosing the scaling factors *a*, *b*, and *c* before being added to the duty cycle, current reference, and voltage reference, respectively. A good choice for the scaling factors is to obtain a perturbation amplitude between 5% and 10% of their steady-state values. It is possible to show that, to measure the load impedance, the injection must occur onto the

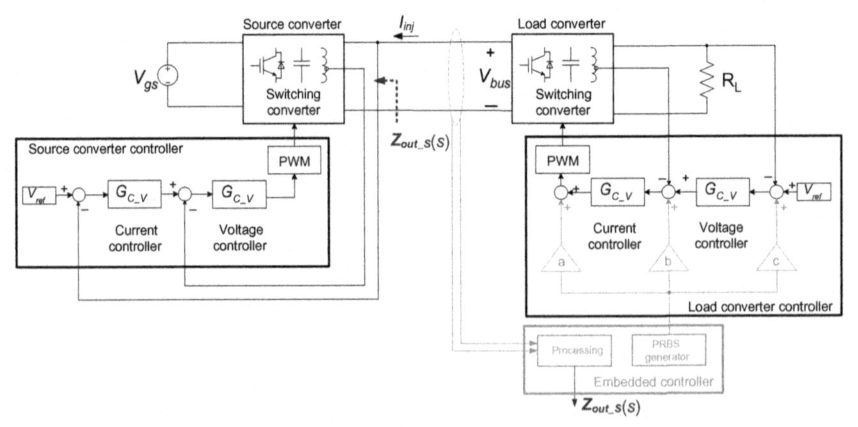

Figure 2.46 Conceptual cascade of converters showing the injection of PRBS over the load converter control and embedded control platform for online identification of source impedance.

source converter control. The reader can easily draw the schematic to measure the load impedance and this is left as an exercise.

The reason why the excitation signal PRBS is added to the duty cycle and to both the current and voltage reference signals of the converter controller is linked to the frequency response of the closed-loop converter that serves as power amplifier for the PRBS. By using the analysis in [22], it can be shown that the voltage at the impedance measurement point is proportional to the PRBS injected at the various stages of the control loop according to the following relationship:

$$v \propto \begin{cases} \dfrac{1}{1+T_I} PRBS_1 & \text{if PRBS is injected to duty cycle} \\ \dfrac{T_I}{1+T_I} PRBS_2 & \text{if PRBS is injected to current reference signal} \\ \dfrac{T_V}{1+T_V} PRBS_3 & \text{if PRBS is injected to voltage reference signal} \end{cases} \tag{2.95}$$

where $PRBS_1$, $PRBS_2$, and $PRBS_3$ represent the excitation signal at the duty cycle, current reference, and voltage reference, respectively. The quantity T_I is the current control loop gain and the quantity T_I is the voltage control loop gain.

The controllers under consideration in this example are PI controllers. If correctly designed, the (current or voltage) loop gain T is very large in magnitude within the control bandwidth, while it is very small beyond the control bandwidth [22]. Therefore, the following quantities can be approximated as

$$\frac{T}{1+T} \approx \begin{cases} 1 \; for \| T \| \gg 1 \\ T \; for \| T \| \ll 1 \end{cases} \tag{2.96}$$

$$\frac{1}{1+T} \approx \begin{cases} \dfrac{1}{T} \; for \| T \| \gg 1 \\ 1 \; for \| T \| \ll 1 \end{cases} \tag{2.97}$$

By using the results of (2.96) and (2.97), the following statements are true:

- If only $PRBS_1$ is applied, v has smaller amplitude than $PRBS_1$ within the frequency range where T_I is very large, i.e., within the control bandwidth, while it has the same amplitude at larger frequencies.

- If $PRBS_2$ (or $PRBS_3$) is applied, v has the same amplitude of the injected disturbance within the frequency range where T_I (or T_V) is very large, i.e., within the control bandwidth, while is reduced in amplitude at larger frequencies.

In other words, if PRBS is injected to the duty cycle only, PRBS, seen as a disturbance, is rejected within the bandwidth of the selected control loop (either current or voltage), while it is not rejected beyond the bandwidth of the selected control loop. On the other hand, if PRBS is injected to the current or voltage reference signals, it is not rejected within the bandwidth of the selected current or voltage control loop, respectively. Therefore, injecting PRBS to the duty cycle and to either the current or the voltage reference signals ensures that PRBS is not rejected by the selected control action over a wide frequency range.

2.6.2.1 The Implementation of the WSI Technique

The implementation of the WSI algorithm in an embedded controller is divided into multiple stages as depicted in Fig. 2.47. The stages are detailed below.

2.6.2.1.1 PRBS Generation

To generate the PRBS, a x-bit linear feedback shift-register (LFSR) is implemented at a rate that can be chosen. Fig. 2.48 shows an example of a 15-bit LFSR. The XOR-ed value of bit 14 and bit 15 are fed back to the beginning of the register. The last value of the register is shifted to achieve a white-noise approximation with zero mean [47]. The PRBS signal is then properly scaled and added to the duty cycles and the control reference signals of the inverter, as shown in Fig. 2.46. The amplitude of

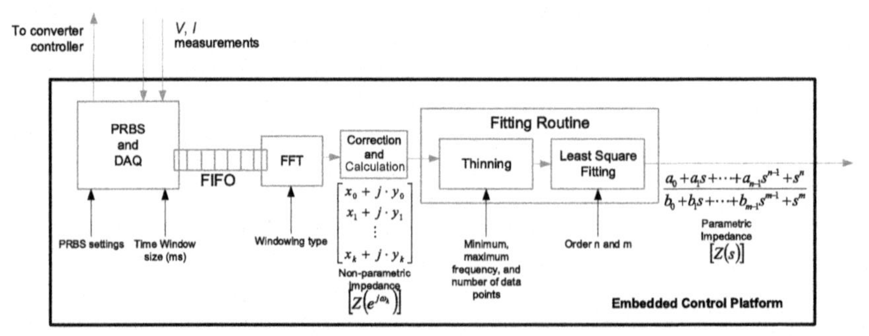

Figure 2.47 Implementation of the routines of the WSI technique in an embedded controller.

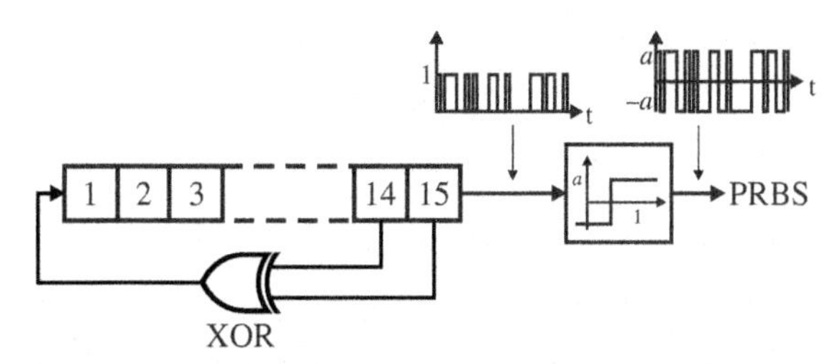

Figure 2.48 Block diagram of LFSR for PRBS generation.

PRBS can be selected; it is recommended to produce a perturbation between 5% and 10% of the steady-state operating point as explained earlier.

2.6.2.1.2 Data Acquisition

While PRBS is being generated and injected, the current and voltage signals at the impedance measurement point are sampled with a sampling rate that can be chosen. The acquired time window during which PRBS is injected and voltages and currents are measured can also be chosen. The voltage and current samples require to be directly written in the memory according to a "First-In First-Out" (FIFO) architecture which allows buffering the sampled values and processing them as a batch after the data acquisition (DAQ) is completed. The PRBS signal is only injected during the DAQ phase and turned off immediately after the complete time window is acquired.

2.6.2.1.3 Fast Fourier Transform

The proper voltage and current serve to build the impedance to be identified. Voltage and current spectra are calculated by performing an FFT algorithm. The type of windowing can be selected.

2.6.2.1.4 Calculation of Nonparametric Impedance

Dividing voltage and current spectra yields the nonparametric grid impedance as follows

$$Z\left[e^{j\omega_n}\right] = \frac{FFT\{v[n]\}}{FFT\{i[n]\}} \tag{2.98}$$

where $v[n]$ and $i[n]$ are the voltage and current samples. The nonparametric impedance consists of N complex data points:

$$N = t_{window} \cdot f_{sampling} \tag{2.99}$$

where t_{window} is the observation window during which PRBS is injected and voltage and current are measured and $f_{sampling}$ is the sampling frequency of $v[n]$ and $i[n]$.

2.6.2.1.5 Fitting Routine

The fitting routing consists of a real-time algorithm able to return the parametric impedance from the nonparametric data set. The fitting consists of two steps: data thinning and least squares fitting. The thinning enforces equal weighting over the full frequency window of the nonparametric data. The thinning technique is used to obtain a logarithmically spaced subset of the data points. In order to do so, the lower and upper frequency boundaries for the data thinning routine have to be specified. All data points outside this window are disregarded for further processing. The least squares fitting routine matches the thinned data to a polynomial function and it is an algorithm based on Levy [49]. The order of numerator n and denominator m must be selected. If they are not known a priori, the order can be tentatively estimated by increasing n and m until the fitting result matches the nonparametric impedance. The result of the fitting routine is the parametric impedance, given by the coefficients of a polynomial function below.

$$Z(s) = \frac{a_0 + a_1 s + \cdots + a_{n-1} s^{n-1} + a_n s^n}{b_0 + b_1 s + \cdots + b_{m-1} s^{m-1} + b_m s^m} \tag{2.100}$$

2.6.2.2 Performance of the WSI Technique and Overcoming Practical Challenges

The real-time performance of the WSI technique is given by the execution time of all the routines described in the previous subsection starting from when the PRBS injection is activated and ending when the parametric impedance result is available. The execution time of the implemented WSI technique consists of the chosen time window and the impedance calculation time. The impedance calculation time $t_{calculation}$ comprises the execution time of the FFT algorithm, which is of complexity $O(N \cdot log(N))$, and the fitting routine, which depends on exit conditions of the least squares fitting algorithm [49]. Therefore, to measure the impedance of the system under test, the online WSI technique takes $t_{window} + t_{calculation}$.

A major challenge in impedance identification consists of maintaining the small-signal condition and, at the same time, guaranteeing a measurable perturbation within the frequency range of interest. In theory, the maximum identifiable frequency is defined by the Nyquist frequency, i.e., half the switching frequency of the inverter [47,48]. The lower frequency boundary, instead, is equal to the inverse of the measured time window t_{window} during which the PRBS is injected. In practice, the upper boundary depends on the amplitude of the injected PRBS signal. In fact, the classic output filters of power electronic converters attenuate the high frequency excitation, eventually leading to a signal level that is comparable with the noise floor ($< -60dB$) and therefore too small to be identified. As a consequence, the attenuation at high frequencies introduced by the output filters dictates the minimum amplitude of the injected PRBS signal. Therefore, to obtain a good identification as close as possible up to the Nyquist frequency, despite the attenuation of the output filters, and maintaining at the same time the small-signal condition, a good choice is to perturb voltage and current with magnitude between 5% and 10% of their steady-state value. As an alternative solution, a different perturbation signal may be used; an alternative to white noise is blue noise which increases the high-frequency content of the PRBS without affecting the low-frequency content. The implementation of a blue noise filter for identification purposes in described in [48].

The choice of the time window is another practical challenge worthy of discussion. As the time window is directly linked to the number N of complex data points of the nonparametric impedance according to (2.99), for a fixed sampling rate, a long enough time window should be selected in order to have enough data points to capture eventual sharp features of the identified impedance, such as lightly damped resonances. On the other hand, the time window should be chosen short enough to avoid increasing impedance calculation times. Moreover, how short the time window chosen should be is in relation to the characteristics of the system under test. For the cascade of power electronic converters, it is important to set the time window short enough in order to catch with fast changing impedances, e.g., due to load steps or system reconfiguration.

The finite duration of the time window poses also some limits on the FFT algorithm. The FFT assumes infinite periodicity of the signal to be transformed. However, such an assumption is not practical because all the signal acquisitions are limited in time. During the acquisition, the FFT returns additional spurious frequency components around the existing harmonic content because of the discontinuities at the edges of the time window. This well-known problem is called spectral leakage [50]. In order to minimize the effect of spectral leakage, it is recommendable to use windowing (e.g., Hanning) of the same length as the acquired signal.

Another practical aspect is the resolution of the digital to analog (D/A) and analog to digital (A/D) converters to output the PRBS signal and acquire the voltage and current measurements, respectively. Low resolutions increase the minimum viable amplitude of the excitation. To get satisfactory identification results, D/A and A/D converters are required to have enough number of bits. For example, with 16-bit D/A and A/D converters which give a resolution of 6.10mV on a voltage measurement of $\pm$ 200V and a resolution of 0.31mV on the injected PRBS of $\pm$ 10V.

2.6.2.3 Simulation Example

This simulation example shows the performance in the time domain and frequency domain of two identification techniques. The classic technique with digital network analyzer is presented first. Then, the WSI technique is presented. The output impedance of the source converter of the cascade system shown in Fig. 2.49 is taken into consideration. The switch selects the excitation signal to be added to the duty cycle of the load converter, which operates in open-loop for simplicity. This comes either from the narrowband sine sweep generator or from the wideband PRBS generator. The two techniques take the same voltage and current measurements from the impedance measurement point. A similar processing is implemented in the two techniques which consists of implementing (2.98). Therefore, both techniques calculate the nonparametric impedance.

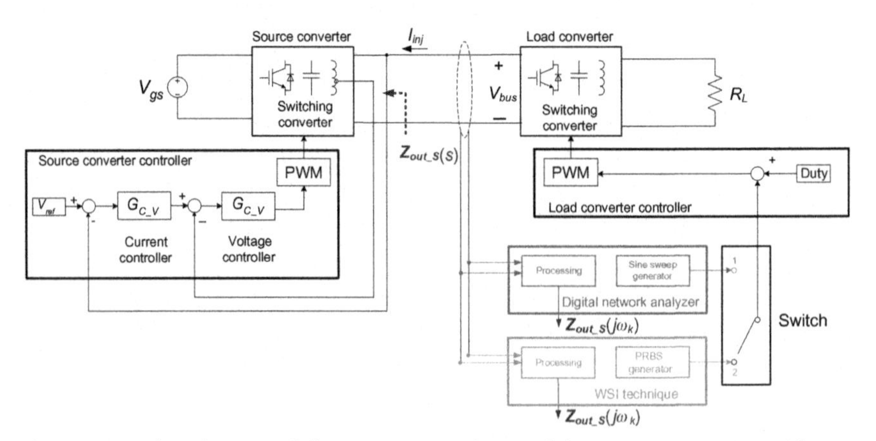

Figure 2.49 Identification of the output impedance of the source converter with two techniques: the digital network analyzer and the Wideband System Identification technique (WSI).

The digital network analyzer techniques of this example generates 38 small-signal sine waves with increasing frequency at steps of 40 Hz starting from 50 Hz. Therefore, the maximum frequency is 1570Hz. Every frequency is kept constant for 0.1s to make sure that system transient due to the frequency step is extinguished and therefore the system response is correctly measured. Fig. 2.50 shows the voltage and current measurements during the sine sweep. Notice that to sweep from 50Hz to 1570Hz with the step of 0.1s takes 3.8s during which the cascade system must be maintained in steady state. Fig. 2.51 depicts the Bode plot of the identified output impedance of the source converter compared to the analytic transfer function. Very good matching is evident.

The WSI techniques of this example generates 15-bit small-signal PRBS at a rate of 200kHz. Fig. 2.52 shows the voltage and current measurements during PRBS injection. Notice that the perturbation at the impedance measurement point takes 1.6s during which the cascade system must be maintained in steady state. Fig. 2.53 depicts the Bode plot of the identified output impedance of the source converter compared to the analytic transfer function. Very good matching is again evident. The spurious identification at high frequency is beyond the Nyquist frequency

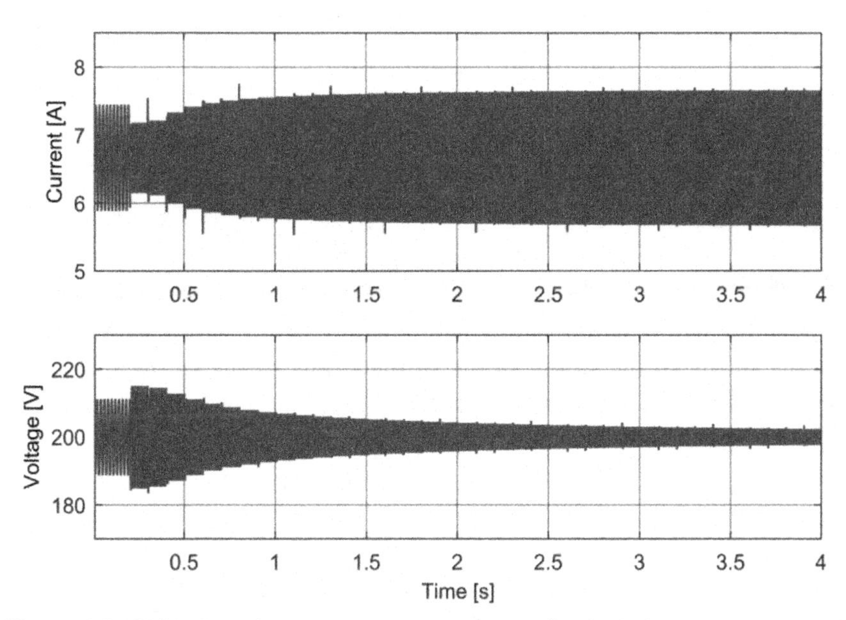

Figure 2.50 Voltage and current measurements at the impedance measurement point with the Digital Network Analyzer technique.

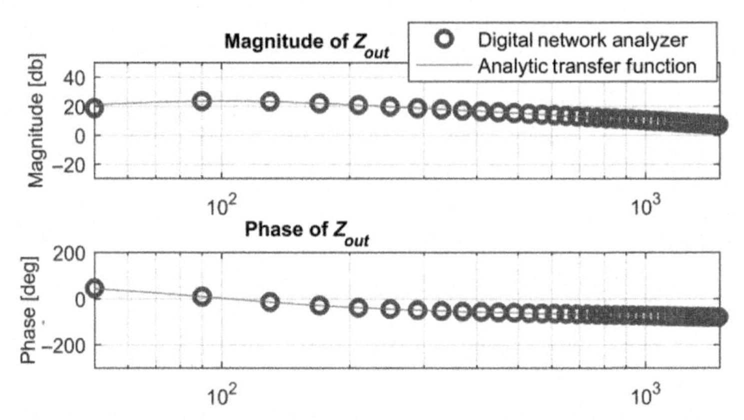

Figure 2.51 Bode plot of the identified output impedance of the source converter compared to the analytic transfer function with the Digital Network Analyzer technique.

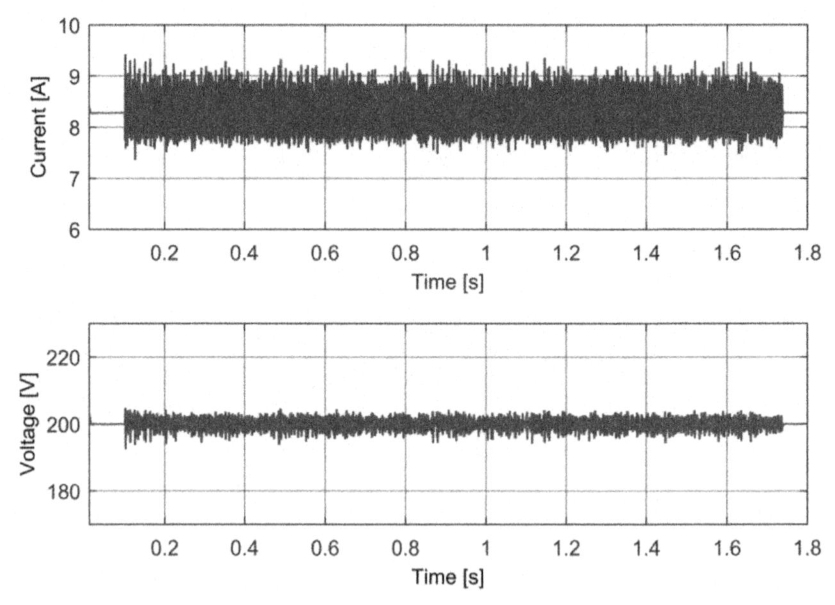

Figure 2.52 Voltage and current measurements at the impedance measurement point with the Wideband System Identification technique.

which is $f_{sw}/2 = 10\text{kHz}$ in this example. As a concluding remark, the WSI is able to excite many more frequencies in the system than the digital network analyzer technique in half of the injection time. This is a practical advantage in stability analysis applications due to the strict requirement of fast impedance measurements over a wide frequency range.

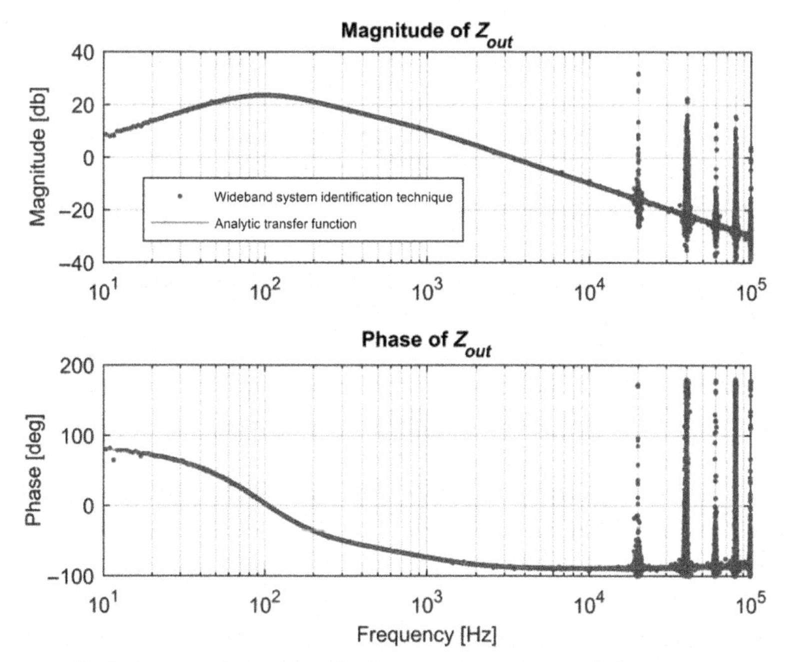

Figure 2.53 Bode plot of the identified output impedance of the source converter compared to the analytic transfer function with the Wideband System Identification technique.

2.7 SUMMARY

In this chapter a dynamic modeling of power converters in a cascaded arrangement based on small-signal properties was presented; this helps drawing conclusions for DC distribution systems. The terminal characteristics of tightly controlled power converters are an important factor for stability analysis and controller design. They were modeled taking into account the conventional averaging approach for DC–DC converters, which requires negligible current ripple and takes basically the DC term of the Fourier series [11].

Furthermore, the authors in [15] investigated the terminal characteristics such as the impedance for different converters looking at two aspects. The first aspect is that the control loop (VMC or PCMC) changes the input and output impedances of the converters, therefore the closed-loop impedances should be used. The second aspect is that the loading effect of the load is overlooked as it alters the dynamics of the upstream converter. The influences of different load models were presented.

A special section was devoted to the so-called CPL. There, its behavior was derived and its boundary conditions explained. An important fact that could be drawn from this analysis is that when using dynamic models an ideal CPL representation does not represent the worst case condition. Additionally, the CPL does not exhibit infinite bandwidth but takes over the bandwidth characteristics of the LRC.

An extension of the procedure presented in this chapter to other converter topologies is possible. For example, if unidirectional power flow is sufficient according to the system requirements, Single Active Bridge (SAB) converters could be used; their modeling in small-signal was presented by Demetriades [51]. When the system requires bidirectional power flow, Dual Active Bridge (DAB) converters could be considered. There exist different modeling approaches for DAB in literature, the primary difference among them lies in the set of state variables selected. In [52], De Doncker introduced the DAB model as a first-order system; after, Demetriades [53] proposed a second-order model; in [54] Alonso presented a third-order model which is further expanded by Krismer up to a fifth-order model [55]. Recently, Qin [56,57] presented an open-loop third-order model based on the DAB generalized state-space modeling and a sixth-order model which takes Equivalent Series Resistance into account. The generalized state-space modeling approach, which was introduced in [58,59], uses more terms of the Fourier series than the conventional averaging approach in the state variables; thus enabling a full order continuous time state-space average model which could include the transformer current [56,57].

The aforementioned models can be used to derive the different transfer functions of SAB and DAB converters in analogy to the Eqs. (2.1)–(2.7), and following a similar procedure to the presented in this chapter, although using the phase shift angle instead of the duty cycle as the control variable.

In this context, it can be said that even when changing the converter topology, the methodology presented throughout this chapter would remain unaltered.

Furthermore, we gave an introduction to the linear control design and validation procedure for voltage and current mode control, followed by a simulation example. We highlighted the different stability criteria and their application when combining different sources and load converters. With the online wideband system identification technique, we displayed a powerful tool for the engineers, as this can be incorporated into a real-world converter setup and generate the stability analysis during operation.

REFERENCES

[1] A.M. Rahimi, A. Emadi, An analytical investigation of DC/DC power electronic converters with constant power loads in vehicular power systems, IEEE Trans. Vehicul. Technol. 58 (6) (July 2009) 2689–2702.

[2] L. Zhu, M. Cupelli, H. Xin, A. Monti, A discussion on the possible overestimation of the adverse effect of constant power loads on system stability, in Proceedings of "The 5th International Conference on Grand Challenges in Modeling and Simulation (GCMS 2012)." Sponsored by the Society for Modeling and Simulation International (SCS) in cooperation with ACM SIGSIM. Genoa, Italy. ISBN 978-1-61839-983-0, Simulation Series, vol. 44, no 11, pp. 37–43. July 8–11, 2012.

[3] C.H. Rivetta, A. Emadi, G.A. Williamson, R. Jayabalan, B. Fahimi, Analysis and control of a buck DC-DC converter operating with constant power load in sea and undersea vehicles, IEEE Trans. Ind. Applicat. 42 (2) (March–April 2006) 559–572.

[4] W. Du, J. Zhang, Y. Zhang, Z. Qian, F. Peng, Large signal stability analysis based on gyrator model with constant power load. 2011 IEEE Power and Energy Society General Meeting, pp. 1–8, July 24–29, 2011.

[5] K. Seyoung, S.S. Williamson, Negative impedance instability compensation in more electric aircraft DC power systems using state space pole placement control, 2011 IEEE Vehicle Power and Propulsion Conference (VPPC), pp. 1–6, September 6–9, 2011.

[6] A.K. AlShanfari, J. Wang, Influence of control bandwidth on stability of permanent magnet brushless motor drive for "more electric" aircraft systems, 2011 International Conference on Electrical Machines and Systems (ICEMS), pp. 1–7, 20–23 Aug. 2011.

[7] A.P.N. Tahim, D.J. Pagano, M.L. Heldwein, E. Ponce, Control of interconnected power electronic converters in dc distribution systems, 2011 Brazilian Power Electronics Conference (COBEP), pp. 269–274, September 11–15, 2011.

[8] Z. Liu, J. Liu, W. Bao, Y. Zhao, F. Liu, A novel stability criterion of AC power system with constant power load, 2012 Twenty-Seventh Annual IEEE Applied Power Electronics Conference and Exposition (APEC), pp. 1946–1950, February 5–9, 2012.

[9] P. Magne, B. Nahid-Mobarakeh, S. Pierfederici, A design method for a fault-tolerant multi-agent stabilizing system for DC microgrids with Constant Power Loads, 2012 IEEE Transportation Electrification Conference and Expo (ITEC), pp. 1–6, June 18–20, 2012.

[10] W. Du, J. Zhang, Y. Zhang, Z. Qian, Stability criterion for cascaded system with constant power load, IEEE Trans. Power Electr. 28 (4) (April 2013) 1843–1851.

[11] R.W. Erickson, D. Maksimović, Fundamentals of Power Electronics, 2nd ed., Kluwer Academic, Norwell, MA, 2001.

[12] R.D. Middlebrook, S. Cuk, A general unified approach to modelling switching-converter power stages, Int. J. Electr. 42 (6) (Jun. 1977) 521–550.

[13] A. Ioinovici, Power Electronics and Energy Conversion Systems, John Wiley & Sons, Chichester, West Sussex ; Hoboken, 2012.

[14] S. Skogestad, I. Postlethwaite, Multivariable Feedback Control: Analysis and Design, 2nd ed., Wiley, Chichester, 2005.

[15] T. Suntio, Dynamic Profile of Switched-Mode Converter, Wiley, Chichester, 2009.

[16] R. Ahmadi, M. Ferdowsi, Controller design method for a cascaded converter system comprised of two DC-DC converters considering the effects of mutual interactions, 2012 Twenty-Seventh Annual IEEE Applied Power Electronics Conference and Exposition (APEC), pp. 1838–1844, February 5–9, 2012.

[17] J.R. LeSage, R.G. Longoria, W. Shutt, Power system stability analysis of synthesized complex impedance loads on an electric ship, 2011 IEEE Electric Ship Technologies Symposium (ESTS), pp. 34–37, April 10–13, 2011.
[18] J.G. Ciezki, R.W. Ashton, Selection and stability issues associated with a navy shipboard DC zonal electric distribution system, IEEE Trans. Power Deliv. 15 (2) (Apr 2000) 665–669.
[19] C. Rivetta, A. Emadi, G.A. Williamson, R. Jayabalan, B. Fahimi, Analysis and control of a buck DC-DC converter operating with constant power load in sea and undersea vehicles, Conference Record of the 2004 IEEE Industry Applications Conference, 2004. 39th IAS Annual Meeting., vol. 2, pp. 1146–1153, Oct. 3–7, 2004.
[20] M.K. Zamierczuk, R.C. Cravens, A. Reatti, Closed-loop input impedance of PWM buck-derived DC-DC converters, in 1994 IEEE International Symposium on Circuits and Systems, 1994. ISCAS'94, 1994, vol. 6, pp. 61–64.
[21] S. Buso, P. Mattavelli, Digital Control in Power Electronics, Mogan & Claypool Publishers, Princeton, NJ, 2006.
[22] R.W. Erickson, D. Maksimovic, Fundamentals of Power Electronics, 2nd ed., Kluwer Academic Publisher, 2001.
[23] A. Baetica, Integrator windup and PID controller design, Caltech, accessed 20 February 2016, available at https://www.cds.caltech.edu/~murray/wiki/images/6/6f/Recitation_110_nov_17.pdf.
[24] A. Emadi, A. Khaligh, C.H. Rivetta, G.A. Williamson, Constant power loads and negative impedance instability in automotive systems: definition, modeling, stability, and control of power electronic converters and motor drives, IEEE Trans. Vehicular Technology 55 (Jul. 2006) 1112–1125.
[25] A. Emadi, A. Ehsani, Dynamics and control of multi-converter DC power electronic systems, Power Electronics Specialists Conference, 2001. PESC. 2001 IEEE 32nd Annual, vol. 1, pp. 248–253, 2001.
[26] Y. Jang, R.W. Erickson, Physical origins of input filter oscillations in current programmed converters, IEEE Trans. Power Electronics 7 (4) (Jul. 1992) 725–733.
[27] A. Emadi, B. Fahimi, M. Ehsani, On the concept of negative impedance instability in advanced aircraft power systems with constant power loads, Society of Automotive Engineers (SAE) Journal, Paper No. 1999-01-2545, 1999.
[28] A.B. Jusoh, The instability effect of constant power loads, in Proc. IEEE National Power and Energy Conference, 2004, pp. 175–179.
[29] R.D. Middlebrook, Input Filter Considerations in Design and Application of Switching Regulators, IEEE IAS Annual Meeting, 1976.
[30] C.M. Wildrick, F.C. Lee, B.H. Cho, B. Choi, A method of defining the load impedance specification for a stable distributed power system, Power Electr. IEEE Trans. 10 (3) (May 1995) 280–285.
[31] X. Feng, Z. Ye, K. Xing, F.C. Lee, D. Borojevic, Impedance specification and impedance improvement for DC distributed power system, Power Electronics Specialists Conference, 1999. PESC 99. 30th Annual IEEE, vol. 2, pp. 889–894, 1999.
[32] X. Feng, Z. Ye, K. Xing, F.C. Lee, D. Borojevic, Individual load impedance specification for a stable DC distributed power system, Applied Power Electronics Conference and Exposition, 1999. APEC '99. Fourteenth Annual, vol. 2, pp. 923–929, March 14–18, 1999.
[33] X. Feng, J. Liu, F.C. Lee, Impedance specifications for stable DC distributed power systems, Power Electr. IEEE Trans. 17 (2) (March 2002) 157–162.
[34] S.D. Sudhoff, S.F. Glover, P.T. Lamm, D.H. Schmucker, D.E. Delisle, Admittance space stability analysis of power electronic systems, Aerospace . Electr. Syst. IEEE Trans. 36 (3) (Jul 2000) 965–973.

[35] S.D. Sudhoff, S.F. Glover, Three-dimensional stability analysis of DC power electronics based systems, Power Electronics Specialists Conference, 2000. PESC 00. 2000 IEEE 31st Annual, vol. 1, pp. 101–106, 2000.
[36] S.D. Sudhoff, J.M. Crider, Advancements in generalized immittance based stability analysis of DC power electronics based distribution systems, Electric Ship Technologies Symposium (ESTS), 2011 IEEE, pp. 207–212, April 10–13, 2011.
[37] A. Riccobono, E. Santi, Comprehensive review of stability criteria for DC distribution systems, Energy Conversion Congress and Exposition (ECCE), 2012 IEEE, pp. 3917–3925, September 15–20, 2012.
[38] Y. Panov, M. Jovanovic, Practical issues of input/output impedance measurements in switching power supplies and application of measured data to stability analysis, Applied Power Electronics Conference and Exposition, 2005. APEC 2005. Twentieth Annual IEEE, vol. 2, pp. 1339–1345, March 6–10, 2005.
[39] X. Feng, Lee, F.C., On-line measurement on stability margin of DC distributed power system, Applied Power Electronics Conference and Exposition, 2000. APEC 2000. Fifteenth Annual IEEE, vol. 2, pp. 1190–1196, 2000.
[40] X. Wang, R. Yao, F. Rao, Three-step impedance criterion for small-signal stability analysis in two-stage DC distributed power systems, IEEE Power Electr. Lett. 1 (Sept. 2003) 83–87.
[41] H.Y. Cho, E. Santi, Modeling and stability analysis in multi-converter systems including positive feedforward control, in Proc. IEEE 34th Annual Conference IECON08, Nov. 2008, pp. 839–844.
[42] S. Vesti, J.A. Oliver, R. Prieto, J.A. Cobos, T. Suntio, Simplified small-signal stability analysis for optimized power system architecture, Applied Power Electronics Conference and Exposition (APEC), 2013 Twenty-Eighth Annual IEEE, vol., no., pp. 1702–1708, March 17–21, 2013.
[43] S. Vesti, T. Suntio, J.A. Oliver, R. Prieto, J.A. Cobos, Impedance-based stability and transient-performance assessment applying maximum peak criteria, Power Electr. IEEE Trans. 28 (5) (May 2013) 2099–2104.
[44] R.D. Middlebrook, Null double injection and the extra element theorem, IEEE Trans. Educat. 32 (Aug. 1989) 167–180.
[45] R.D. Middlebrook, The two extra element theorem, in Proc. IEEE 21st Annu. Frontiers in Education Conference, pp. 702–708, 1991.
[46] A. Riccobono, E. Santi, Positive Feed-Forward control of three-phase voltage source inverter for DC input bus stabilization with experimental validation, in press for publication in the IEEE Transactions on Industry Applications on April, 16, 2012.
[47] B. Miao, R. Zane, D. Maksimovic, System identification of power converters with digital control through cross-correlation methods, IEEE Trans. Power Electr. 20 (5) (2005) 1093–1099.
[48] A. Barkley, E. Santi, Online monitoring of network impedances using digital network analyzer techniques, Applied Power Electronics Conference and Exposition, 2009. APEC 2009. Twenty-Fourth Annual IEEE, Washington, DC, pp. 440–446, 2009.
[49] E.C. Levy, Complex-curve fitting, IRE Trans. Autom. Control AC-4 (1) (May 1959) 37–43.
[50] National Instruments 2017, Understanding FFTs and Windowing, White Paper, accessed 17 October 2017 available at http://www.ni.com/white-paper/4844/en/.
[51] G.D. Demetriades, On Small-Signal Analysis and Control of the Single- and Dual-Active-Bridge Topologies, PhD Dissertation, KTH Stockholm, 2005.
[52] R.W.A.A. De Doncker, D.M. Divan, M.H. Kheraluwala, A three-phase soft-switched high-power-density DC/DC converter for high-power applications, IEEE Trans. Ind. Applicat. 27 (1) (Jan/Feb 1991) 63–73.

[53] G.D. Demetriades, H.-P. Nee, Dynamic modeling of the Dual-Active Bridge topology for high-power applications, IEEE Power Electronics Specialists Conference, 2008. PESC 2008, pp. 457–464, June 15–19, 2008.
[54] A.R. Alonso, J. Sebastian, D.G. Lamar, M.M. Hernando, A. Vazquez, An overall study of a Dual Active Bridge for bidirectional DC/DC conversion, 2010 IEEE Energy Conversion Congress and Exposition (ECCE), pp. 1129–1135, September 12–16, 2010.
[55] F. Krismer, J.W. Kolar, Accurate small-signal model for the digital control of an automotive bidirectional dual active bridge, IEEE Trans. Power Electr. 24 (12) (Dec. 2009) 2756–2768.
[56] H. Qin, J.W. Kimball, Generalized average modeling of dual active bridge DC–DC converter, IEEE Trans. Power Electr. 27 (4) (April 2012) 2078–2084.
[57] H. Qin, Dual active bridge converters in solid state transformers, PhD Dissertation, Missouri University of Science and Technology, 2012.
[58] S.R. Sanders, J.M. Noworolski, X.Z. Liu, G.C. Verghese, Generalized averaging method for power conversion circuits, IEEE Trans. Power Electr. 6 (Feb. 1991) 251–259.
[59] V.A. Caliskan, G.C. Verghese, A.M. Stankovic, Multifrequency averaging of DC/DC converters, IEEE Trans. Power Electr. 14 (1) (Jan 1999) 124–133.

CHAPTER THREE

Background

This chapter will provide a short recall of control theory fundamentals, which will be used in the designed controller throughout Chapter 6, Simulation. Sections 3.1–3.3 summarize the most important concepts of classic control theory. Sections 3.4 and 3.5 introduce the basic methods of modern control theory which focuses on systems represented in the state-space. In Section 3.6, we introduce the droop control concept which is necessary for operating parallel converter systems as is shown in Chapter 6, Simulation.

3.1 FREQUENCY RESPONSE APPROACHES

The frequency response approaches, which are based on the concept of frequency response, provide powerful tools for analysis and design of linear, time-invariant (LTI) systems. These approaches were developed by Bode, Nyquist, and Nichols, among other contributors, in the 1930s and 1940s [1].

Frequency response of a system can be defined as its steady-state response to a sinusoidal input signal as the frequency of the signal varies from zero to infinity. For a stable, LTI system, the frequency response can be described by

$$G(j\omega) = Me^{j\phi} \tag{3.1}$$

where M is the ratio of the amplitudes of the output and input sinusoids and ϕ is the phase shift of the output sinusoid with respect to the input sinusoid. If $G(s)$ is the transfer function of the system, it can be shown that the frequency response can be obtained by replacing s with $j\omega$. When the transfer function of the system is not known, the frequency response can be usually obtained using a signal generator and precise measurement equipment.

Modern Control of DC-Based Power Systems.
DOI: https://doi.org/10.1016/B978-0-12-813220-3.00003-X

The most widely-used methods to plot the frequency response of a system are:

- Nyquist or polar plots; and
- Bode or logarithmic diagrams.

Each of these representations can be used to analyze the stability and design of the control system to obtain the desired system behavior.

3.2 NYQUIST STABILITY CRITERION

Detailed information about how to plot the Nyquist or polar plots can be found in Ref. [1]. The stability analysis of control systems using the Nyquist plots is based on the Nyquist stability criterion. This criterion provides a means to determine the stability of the closed-loop transfer function based on the open-loop poles and the open-loop frequency response of the system.

According to the Nyquist stability criterion, for an LTI system with the forward transfer functions $G(s)$ and feedback transfer function $H(s)$, the number of the zeros (Z) of the $1 + G(s)H(s)$ in the right-half s-plane (which will be equal to the poles of the closed-loop system) can be given by

$$\boldsymbol{Z} = \boldsymbol{N} + \boldsymbol{P} \tag{3.2}$$

where

- $\boldsymbol{N}$ is the number of the clockwise encirclements of the point $-1 + j0$ by the locus of $G(s)H(s)$, as s varies along a closed contour in the s plane, which encloses the entire right-half s-plane in a clockwise direction (i.e., Nyquist path).
- $\boldsymbol{P}$ is the number of the poles of $G(s)H(s)$.

Using Nyquist plots, it is not only possible to determine whether a system is stable or unstable, but also to infer the degree of stability of a stable system and information on how to improve the stability, if necessary.

3.3 BODE DIAGRAMS

Bode diagrams or logarithmic plots are composed of two plots, i.e., the plot of the magnitude of the open-loop transfer function ($20\log\left|G(j\omega)H(j\omega)\right|$), and the phase angle plot, both versus the frequency in a logarithmic scale. Using a log-magnitude plot, it is possible to convert the multiplication of the factors of the $G(j\omega)H(j\omega)$ into addition and use a simple method for sketching the log-magnitude diagrams. Thus, knowing the contribution of basic factors that can appear in an arbitrary open-loop transfer function, one can readily sketch the overall log-magnitude diagram of the system as the addition of the individual curves corresponding to each of the factors.

The main factors that appear in an arbitrary open-loop transfer function are as follows:

- Constant gain K
- Integral and derivate factors
- First-order factors
- Second-order or quadratic factors

The following table provides a summary of the contribution of each of the above factors on the Bode diagrams. It should be noted that for the first-order and quadratic factors, the provided summary lists an approximate rather than the exact contribution of these factors around the corner frequency. The corner frequency is the frequency at which, the asymptotes of the magnitude diagram meet, which equals $\omega = \frac{1}{T}$ and $\omega = \omega_n$ for the first- and second-order factors, respectively.

Factor	Impact on Magnitude Plot	Impact on the Phase Plot
Constant gain K	Constant contribution of $20\log\|K\|$	Nothing
Derivative factors $(j\omega)^{+1}$	Straight line with slope of +20 dB/dec and magnitude of 0 dB at $\omega = 1$	A constant contribution of $+90°$ at all frequencies
Integral factors $(j\omega)^{-1}$	Straight line with slope of −20 dB/dec, and magnitude of 0 dB at $\omega = 1$	A constant contribution of $-90°$ at all frequencies
First order factors in the numerator $(1+j\omega T)^{+1}$	For $\omega \leq \frac{1}{T}$: 0 dB For $\omega > \frac{1}{T}$: Straight line with slope of +20 dB/dec	For $\omega \leq \frac{1}{10T}$: 0° For $\frac{1}{10T} < \omega \leq \frac{10}{T}$: gradual decrease from 0° to +90° For $\omega > \frac{10}{T}$: 0°
	For $\omega \leq \frac{1}{T}$: 0 dB	For $\omega \leq \frac{1}{10T}$: 0°

(*Continued*)

(Continued)

Factor	Impact on Magnitude Plot	Impact on the Phase Plot
First order factors in the denominator $(1+j\omega T)^{-1}$	For $\omega > \frac{1}{T}$: Straight line with slope of -20 dB/dec	For $\frac{1}{10T} < \omega \leq \frac{10}{T}$: gradual decrease from $0°$ to $-90°$ For $\omega > \frac{10}{T}$: $0°$
Quadratic factors in the numerator	For $\omega \ll \omega_n$: 0 dB	For $\omega \ll \omega_n$: $0°$
$[1+2\xi\left(\frac{j\omega}{\omega_n}\right)+\left(\frac{j\omega}{\omega_n}\right)^2]^{+1}$	Near ω_n: Resonant peak, whose magnitude is determined by the value of ξ	Near ω_n: gradual increase from $0°$ to $+180°$; the increased profile heavily dependent on the value of ξ
	For $\omega \gg \omega_n$: $+40$ dB/dec	For $\omega \gg \omega_n$: $+180°$
Quadratic factors in the denominator	For $\omega \ll \omega_n$: 0 dB	For $\omega \ll \omega_n$: $0°$
$[1+2\xi\left(\frac{j\omega}{\omega_n}\right)+\left(\frac{j\omega}{\omega_n}\right)^2]^{-1}$	Near ω_n: A resonant peak, whose magnitude is determined by the value of ξ	Near ω_n: gradual decrease from $0°$ to $-180°$; the decreased profile heavily dependent on the value of ξ
	For $\omega \gg \omega_n$: -40 dB/dec	For $\omega \gg \omega_n$: $-180°$

For systems with nonminimum-phase transfer functions, i.e., transfer functions with neither zeros nor poles in the right-half s-plane, the concepts of phase margin and gain margin can be used to determine the stability.

The phase margin can be defined as follows:

$$\boldsymbol{PM} = 180° + \varnothing \tag{3.3}$$

where $\varnothing$ is the phase angle of the open-loop transfer function at the frequency in which the magnitude of the open-loop transfer function is equal to zero.

Gain margin can be also defined as the inverse of the gain of the open-loop transfer function at ω_c, which is the frequency at which the phase angle of the open-loop transfer function is equal to -180 degrees:

$$\boldsymbol{GM} = \frac{1}{|G(j\omega_c)H(j\omega_c)|} \tag{3.4}$$

In decibels, GM can be expressed as:

$$\boldsymbol{GM_{dB}} = 20\log\frac{1}{|G(j\omega_c)H(j\omega_c)|} = -20\log|G(j\omega_c)H(j\omega_c)| \tag{3.5}$$

For a nonminimum phase system to be stable, it is necessary that both *PM* and *GM* are positive. Thus, knowledge of positive *PM* or positive *GM* alone cannot be used to confirm the stability of a nonminimum-phase system.

As a rule of thumb, for a *PM* between 30 and 60 degrees and a *GM* larger than 6 dB, a nonminimum-phase system is not only stable but has also satisfactory behavior even in spite of reasonable variations in system parameters [1].

To illustrate the above-discussed concepts, let us consider a system with the following transfer function:

$$\boldsymbol{G}(s)\boldsymbol{H}(s) = \frac{1}{s(s+1)(s+6)}$$

The Bode plots for this system are shown in Fig. 3.1. It can be observed that in the phase diagram, the pole at the origin is the dominant factor for small frequencies, which has resulted in a phase of -90 degrees. Then the other two poles start to gradually contribute to the phase diagram as the frequency increases until the phase approaches -270 degrees for large frequencies. In the magnitude diagram, the constant gain of 1/6 and the pole at the original are the effective factors for lower frequencies and then each of the other two poles gradually start contributing -20 dB/dec to the magnitude diagram.

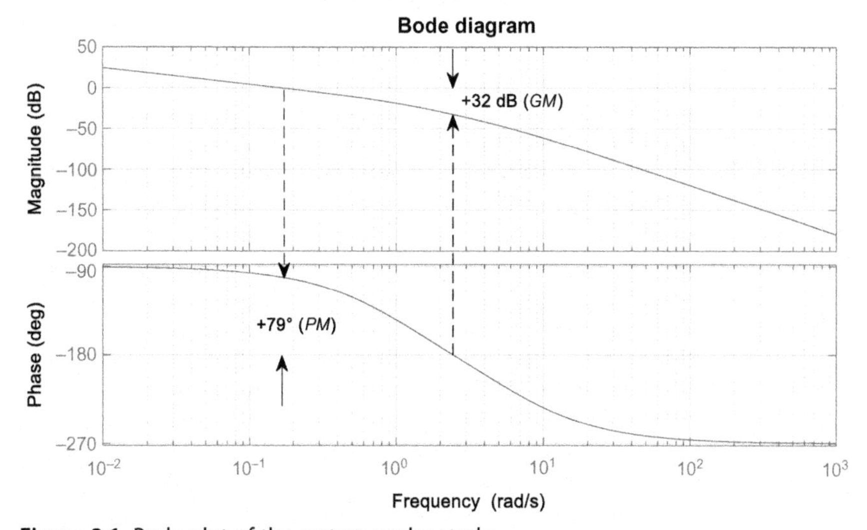

Figure 3.1 Bode plot of the system under study.

The phase margin and the gain margins both have positive values, which indicate that the closed-loop system is stable. If a very large gain (more precisely, a gain larger than 32 dB) is added to the open-loop transfer function, then the magnitude diagram will be shifted upwards while the phase diagram remains untouched. In this case, depending on the value of the gain, it can be that the phase margin becomes negative, thus leading to an unstable closed-loop behavior.

The very interesting aspect of using Bode (and Nyquist) diagrams is that they provide a very intuitive insight into the impact of each factor in the transfer function of the system on the overall system behavior. This is particularly useful when designing a control system as the designer would have a qualitative understanding of the pros and cons of adding each type of controller to the system by considering their contribution to the frequency response of the system.

Another important point is that when using frequency response approaches, the user will have to translate the desired behavior usually in terms of damping ratio, phase margin, and gain margin and then select the controller to satisfy these requirements. With this effort, the user hopes to move the location of the closed-loop poles in such a way that the desired overall system is fulfilled. However, there is no direct control on the location of the closed-loop poles during this procedure. In the state-space approach, the situation is different and it is possible to place the closed-loop poles in arbitrary locations if certain preconditions are met. These points will be discussed in more detail in the next sections.

3.4 LINEAR STATE-SPACE

In traditional control theory, the assumption is that the output has enough information to control the system. This may, however, not be the case since the output could also contain redundant or too little information. In contrast, modern control theory considers the states of the controlled system. The idea is that if the state of the system can be fed back, the amount of information that is needed to control the system is minimized. This becomes clear with the definition of the term state:

> *The state of a dynamic system is the smallest set of variables (called state variables) such that knowledge of these variables at $t = t_0$, together with knowledge of the input for $t \geq t_0$, completely determines the behavior of the system for any time $t \geq t_0$ [1].*

Any state can be represented as a point in the state-space which is defined by the state variables. The state variables are the coordinate axes of the state-space and their values indicate the position of a state in the state-space. Although the state-space representation is not unique and any linear system can be described by an infinite number of state-space representations, all these representations consist of the same number of state variables. This number is equal to the number of integrators in the system because their outputs are the state variables of the system. Apart from state variables, state-space analysis considers input and output variables. In general, a linear time varying system can be described as follows in the state-space:

$$\begin{aligned} \dot{\boldsymbol{x}}(t) &= \boldsymbol{A}(t)\boldsymbol{x}(t) + \boldsymbol{B}(t)\boldsymbol{u}(t) \\ \boldsymbol{y}(t) &= \boldsymbol{C}(t)\boldsymbol{x}(t) + \boldsymbol{D}(t)\boldsymbol{u}(t) \end{aligned} \tag{3.6}$$

where $\boldsymbol{A}(t)$ is called the state matrix, $\boldsymbol{B}(t)$ the input matrix, $\boldsymbol{C}(t)$ the output matrix, and $\boldsymbol{D}(t)$ the direct transmission matrix. $\boldsymbol{x}(t)$ is the vector of state variables, $\boldsymbol{u}(t)$ the vector of inputs, and $\boldsymbol{y}(t)$ the vector of outputs. In case of a time invariant system, the matrices describing the system are constant. Since state-space models are described in vector and matrix form, the analysis of multiinput multioutput is inherently supported.

From Eqs. (3.6), one major advantage of the state-space representation can be directly concluded: in this equation, regardless of the order of the system, the first derivate of each state variable is expressed in terms of other state variables and the system inputs. Therefore, although the size of the matrix A increases with an increase in the order of the system, the state-space equations will still have the same form. This means that with the increase in the system order, only the size of the equations increase, but the same matrix approaches can be applied to tackle the system in the state-space approach.

The existence of a solution to a control system described by a given state-space representation is tied to its controllability and observability. The concepts of controllability and observability are briefly described in the following subsections.

3.4.1 Controllability

A system is controllable at time t_0 if there is an unconstrained input $\boldsymbol{u}(t)$ which can transfer the system from any initial state $\boldsymbol{x}(t_0)$ to any other state $\boldsymbol{x}(t_1)$ within a finite time as visualized in Fig. 3.2 for a second order

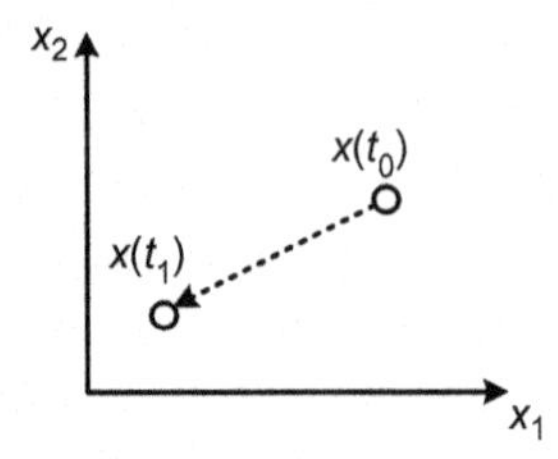

Figure 3.2 Transition from state $\boldsymbol{x}(t_0)$ to state $\boldsymbol{x}(t_1)$.

system. A system which is controllable in every state is, by definition, completely state controllable. $\boldsymbol{x_2}$

It can be shown that a linear system described by $\boldsymbol{A}_{n\times n}$ and with the input matrix $\boldsymbol{B}_{n\times m}$ is completely state controllable if and only if the controllability matrix $\boldsymbol{W}_c$ as defined below is full rank:

$$\boldsymbol{W}_c = [\boldsymbol{B}|\boldsymbol{AB}|\boldsymbol{A}^2\boldsymbol{B}|\ldots|\boldsymbol{A}^{n-1}\boldsymbol{B}] \tag{3.7}$$

If the rank of $\boldsymbol{W}_c = n$, i.e., if $\boldsymbol{W}_c$ is full rank, it means that there are n independent directions which are controllable in the state-space using the input vector $\boldsymbol{u}(t)$. Therefore, it is possible to go from any $\boldsymbol{x}(t_0)$ to any $\boldsymbol{x}(t_1)$ in any time by simply adjusting the speed for each direction properly.

For rank $\boldsymbol{W}_c < n$, at least one direction cannot be controlled. This can be acceptable if the desired direction is controllable. If the natural behavior of the system is not sufficient, there is the possibility to change the input channel to the system and describe the system with a different state-space model. It is important to note that the intrinsic nature of the system is expressed by $\boldsymbol{A}$ but $\boldsymbol{B}$ is related to our input channel selection.

It should be also noted in that practice, there are maximum available or allowable input values, for instance to avoid saturation of actuators. Therefore, it is not possible to apply unconstrained inputs as mentioned in the definition of controllability. To take into consideration such limitations when dealing with real systems, optimal control approaches, which allow embedding limitations on inputs in the mathematical formulation of the problem, can be applied.

3.4.2 Observability

A system is observable at time t_0 if it is possible to determine the state $\boldsymbol{x}(t_0)$ of the system from the observation of its output over a finite time interval $t_0 \leq t \leq t_1$. Complete observability is given if the system is observable in every state. This means that each state transition affects all output variables after some time. This is important, especially in state feedback control, since not all state variables might be accessible from outside and they have to be estimated from the available output variables.

It can be shown that the complete state observability of a linear system described by $\boldsymbol{A}_{n \times n}$ and the output matrix $\boldsymbol{C}_{m \times n}$ can be determined by its observability matrix which is defined as:

$$\boldsymbol{W_o} = \begin{bmatrix} \boldsymbol{C} \\ \boldsymbol{CA} \\ \vdots \\ \boldsymbol{CA}^{n-1} \end{bmatrix} \tag{3.8}$$

The system is completely observable if and only if the rank of $\boldsymbol{W_o} = n$.

If rank $\boldsymbol{W_o} < n$, the system is not completely observable, which is not acceptable for control purposes, whereas it could be fine if complete controllability was not given. To solve this issue, it should be checked if there are alternatives for placing the measurement sensors in the system. In this way, the output matrix $\boldsymbol{C}$ can be changed.

3.4.3 Pole Placement

For a completely state-controllable system, it is possible to place the closed-loop poles in arbitrary locations using the state feedback technique. In this way, by means of pole placement, it can be defined exactly how the system reacts to a specific input or disturbance. In contrast to traditional control design, where the damping ratio and undamped natural frequency of the dominant closed-loop poles is shaped, pole placement directly adjusts all closed-loop poles.

The main idea in the pole placement technique is to feed the states of the system back to the input with some properly selected coefficients k. Therefore, this technique is also referred to as the state feedback method. The state feedback matrix $\boldsymbol{K}$ composed of state feedback gains k is calculated in such a way that the resulting system has the desired eigenvalues.

In the presence of state feedback, the control signal is given by:

$$\boldsymbol{u} = -\boldsymbol{Kx} \tag{3.9}$$

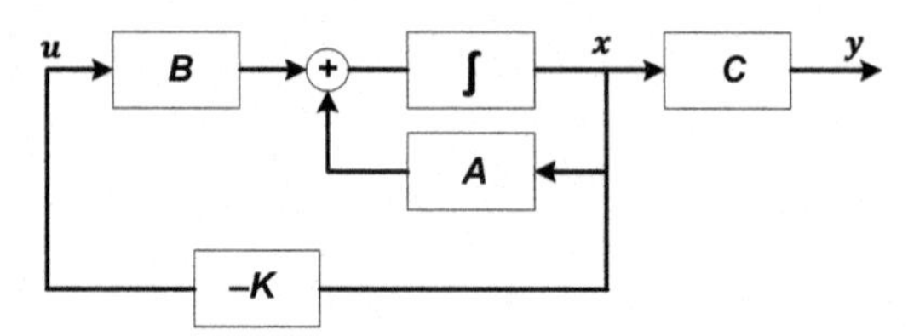

Figure 3.3 State feedback to place the poles.

where K is the state feedback gain matrix. The size of K depends on the number of desired eigenvalues. Thus, given a system as defined in (3.6), the state transition equation can be written as:

$$\dot{x}(t) = Ax(t) - BKx(t) = (A - BK)x(t) \tag{3.10}$$

Eq. (3.10) can be visualized as shown in Fig. 3.3.

However, there are prerequisites to pole placement. Apart from the condition of complete state controllability, the preassumption in pole placement technique is that all the system's states are available for being fed back to the input. Consequently, the system states must be either directly measured or estimated using a so-called state estimator or observer. Besides, the system must be completely state-controllable. The latter can be verified using the method described in Section 3.4.

The most intuitive way to determine the values of K is the *direct substitution method.* This method works well for small systems. Given the desired poles on one hand and Eq. (3.10) on the other hand, the characteristic polynomial of the desired system can be written as:

$$P(\lambda) = \prod_{i=1}^{n} \left(s - \mu_i\right) = |sI - A + BK| \tag{3.11}$$

where μ_i are the desired eigenvalues of the system. This equation can be solved for $K = [k_1 k_2 \ldots k_n]$.

For large systems, different methods may offer an easier way to calculate the state feedback gains. Two such methods, namely the *transformation matrix method* and *Ackermann's Formula* are briefly described in the following.

The transformation matrix method requires the coefficients $a_1 \ldots a_n$ of the original characteristic polynomial:

$$|sI - A| = s^n + a_1 s^{n-1} + \ldots + a_{n-1}s + a_n \tag{3.12}$$

Then, the transformation matrix can be calculated using the following equation:

$$T = W_c H \tag{3.13}$$

where W_c is the controllability matrix and H is calculated from the coefficients of the original characteristic polynomial:

$$H = \begin{bmatrix} a_{n-1} & a_{n-2} & \cdots & a_1 & 1 \\ a_{n-2} & a_{n-3} & \cdots & 1 & 0 \\ \vdots & \vdots & 1 & \vdots & \vdots \\ a_1 & 1 & \cdots & 0 & 0 \\ 1 & 0 & \cdots & 0 & 0 \end{bmatrix} \tag{3.14}$$

knowing the desired closed-loop poles $\mu_1 \ldots \mu_n$, the desired characteristic polynomial is calculated to determine the coefficients $\alpha_1 \ldots \alpha_n$:

$$(s - \mu_1) \ldots (s - \mu_n) = s^n + \alpha_1 s^{n-1} + \ \ldots \ + \alpha_{n-1} s + \alpha_n \tag{3.15}$$

The required state feedback gain matrix K can be calculated as follows:

$$K = [\alpha_n - a_n | \ldots | \alpha_1 - a_1] T^{-1} \tag{3.16}$$

Another option to determine the state feedback gain matrix is Ackermann's formula:

$$K = \begin{bmatrix} 0 & 0 & \ldots & 1 \end{bmatrix} W_c^{-1} \Phi(A) \tag{3.17}$$

where W_c is the controllability matrix. $\Phi(A)$ is the characteristic equation of A and can be obtained from the following equation:

$$\Phi(A) = A^n + \alpha_1 A^{n-1} + \ldots + \alpha_{n-1} A + \alpha_n I \tag{3.18}$$

As before, α_i are the coefficients of the desired characteristic polynomial after applying the state feedback K and moving the poles to the desired places.

3.5 OBSERVER

As mentioned in the previous section, the underlying preassumption for using the pole placement technique is the access to all state

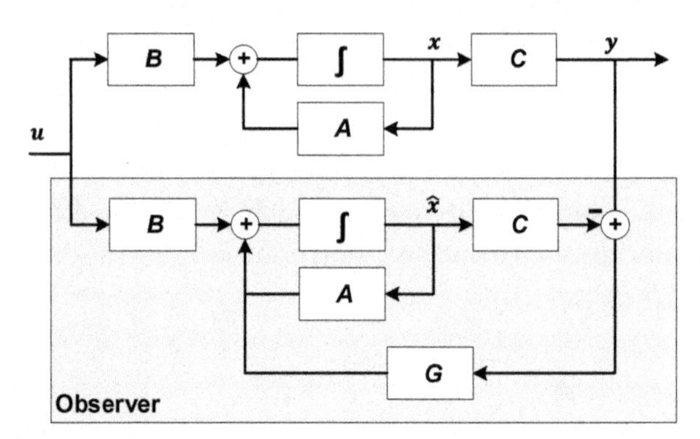

Figure 3.4 Observer scheme.

variables. In practice, one or more of the state variables may not be measured for any of the following reasons:

- The State might not be measurable, e.g., magnetic flux;
- The measurement reliability may be low, e.g., encoders for speed in electric machines;
- The sensor is not installed to reduce cost; or
- The state variable has no physical meaning.

The estimation of unmeasured state variables is referred to as observation or state estimation. A component which performs the estimation or observation of the state variable, is called an observer. The necessary and sufficient condition for the design of a state observer is that the observability condition, as described in Section 3.4.2, is satisfied.

As shown in Fig. 3.4, the state observer reconstructs the state variables using the measurements of the control variables and the system outputs.

The model of the observer is basically identical to the plant model, except for the additional term which includes the estimation error to compensate for model inaccuracies and the lack of the initial conditions:

$$\dot{\hat{x}} = A\hat{x} + Bu + G(y - C\hat{x}) = (A - GC)\hat{x} + Bu + Gy \tag{3.19}$$

where $\hat{x}$ is the observed or estimated state vector and G is the observer gain matrix.

$A\hat{x} + Bu$ is the term calculated based on our knowledge of the system model and the previously estimated state. However, we know that this term does not offer a perfect estimation of the system state. Therefore, a compensation term $G(y - C\hat{x})$ is added to account for the inaccuracies associated with the system model, i.e., the matrices A and B as well as the lack of knowledge about the initial system state. The larger the estimation error, the larger will be also the compensation term. If $\boldsymbol{y} = \hat{\boldsymbol{y}}$, then the state estimation is perfect and there will be no compensation term.

In order to determine how quickly the estimated state converges to the real state, let us define the error vector as:

$$\boldsymbol{e} = \boldsymbol{x} - \hat{\boldsymbol{x}} \tag{3.20}$$

By substituting the variables $\boldsymbol{x}$ and $\hat{\boldsymbol{x}}$ from (3.6) and (3.19) and derivation, the error vector dynamics can be obtained:

$$\dot{\boldsymbol{e}} = (\boldsymbol{A} - \boldsymbol{GC})\boldsymbol{e} \tag{3.21}$$

Therefore, the dynamic of the observer are determined by the eigenvalues of the term $(\boldsymbol{A} - \boldsymbol{GC})$.

Impact of observer in a closed-loop system with state-feedback: Using the estimated state of the system ($\hat{\boldsymbol{x}}$) rather than the actual system state ($\boldsymbol{x}$) in the pole placement technique affects the behavior of the overall closed-loop system. Using Eqs. (3.4) and (3.19), it can be shown that the dynamics of the closed-loop estimated-state feedback system is governed by the following equation:

$$\begin{bmatrix} \dot{\boldsymbol{x}} \\ \dot{\boldsymbol{e}} \end{bmatrix} = \begin{bmatrix} \boldsymbol{A} - \boldsymbol{BK} & \boldsymbol{BK} \\ 0 & \boldsymbol{A} - \boldsymbol{GC} \end{bmatrix} \begin{bmatrix} \boldsymbol{x} \\ \boldsymbol{e} \end{bmatrix} \tag{3.22}$$

Consequently, the characteristics equation can be written as:

$$|s\boldsymbol{I} - \boldsymbol{A} + \boldsymbol{BK}||s\boldsymbol{I} - \boldsymbol{A} + \boldsymbol{GC}| \tag{3.23}$$

This means that the closed-loop poles of a system with state feedback and a state observer are composed of the poles from the state feedback when no observer is present plus the poles from the state observer design. In other words, the dynamics of the closed-loop system is determined by both the poles of the observer and the new poles from pole placement.

From a practical point of view, the above conclusion means that the observer and the state-feedback can be designed separately. Furthermore, in order to minimize the impact of the presence of the state observer on the overall dynamics of the systems, its poles are usually placed farther from the imaginary axis on the left-hand side plane compared to the desired closed-loop poles, which are placed using the state feedback.

3.6 DROOP

Paralleling a number of converters typically offers advantages over a single high power converters, such as low component stresses or increased reliability or ease of maintenance and repair, etc. [2,3]. One basic objective of parallel-connected converters is to share the load current between the converters. To achieve this goal some form of control has to be used to equalize the currents among the converters [4].

An ideal voltage source can always maintain its output voltage at nominal voltage, regardless of the current supplied. If several ideal sources are connected in parallel, it is unclear which voltage source contributes how much to the total generation, since theoretically each source can take over the complete generation (up to the maximum power) regardless of the voltage level. Theoretically, the source with the highest initial voltage would take over the complete load. From this theoretical thought it follows that ideal voltage sources connected in parallel cannot provide load sharing. Furthermore, the Kirchhoff voltage and current laws have to be obeyed when connecting sources together. The Kirchhoff voltage law states that the sum of voltages of all branches forming a loop must be equal to zero. This again means that two ideal voltages sources cannot be connected in parallel, even if the voltage sources would be theoretically of the same magnitude, the current values would be undefined [5]. Kirchhoff current law dictates that the sum of currents in a node must equal zero. While in practice ideal voltage and current sources do not exist, power electronics controller power sources are a very close approximation of this behavior.

A real voltage source, on the other hand, has an internal resistance that drops the voltage as the current increases. Normally this is undesired as a voltage level independent of the current is requested: on the other hand, the resistive effect creates automatically a power sharing effect. In effect, it can be shown that the power sharing is dependent on the value of the internal resistance. Acting on the slope of the characteristic allows the power sharing to be refined.

While adding a resistance in series to the source is not a practical approach because of the induced losses, simulating the resistive behavior with the control brings the sharing capability and no physical losses. This type of regulation is called the Voltage Droop control. In summary, this means that the output voltages of the voltage sources connected in parallel have a linear correlation with the power or current generated.

3.6.1 Voltage Droop

In general, a voltage/current dependence can be created by adding in series a virtual output impedance, which corresponds to an increase in

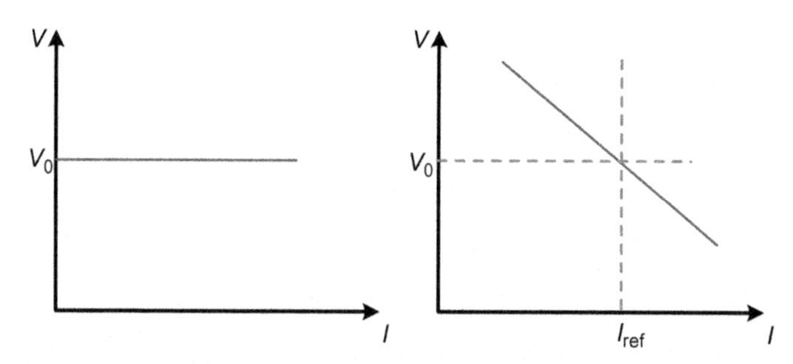

Figure 3.5 V–I characteristic of an ideal voltage source (left) and real voltage source droop curve (right).

the natural internal resistance of the source [6]. The resulting curve is also called V–I-characteristic or Droop-characteristic and is referred to as such in this work. In a nonideal source, the droop characteristic is a sloping straight line passing through a reference point, often referred to as a set point. This point specifies how much current or power is to be generated at the nominal voltage. The slope of these straight lines corresponds to the serial output resistance, as shown in Fig. 3.5.

To explain more precisely the natural load distribution of several parallel converters with steep droop characteristic, a simple example network as shown in Fig. 3.6 is used. It consists of two converters Conv1 and Conv2, which are connected in parallel to each other on a busbar. The voltage on the busbar is V_0, and a load is connected to the same busbar that requires the current $I_0 = \frac{P_{load}}{V_0}$. In this example cables are treated as resistances.

If both voltage sources have exactly the same droop characteristic, both will deliver the same current $\frac{I_0}{2}$ under load. However, if the droop characteristic is different, e.g., at different operating points of the sources, the load distribution is as shown in Fig. 3.7.

Due to the unequal reference power, i.e., the generated power at nominal voltage, the droop characteristics of the sources are shifted relative to each other. In this example, this leads to an unequal distribution of the supplied currents with the voltage V_0. Fig. 3.7 also shows that the slope of the droop characteristic determines the load distribution. The steeper the curve is, the narrower the currents I_1 and I_2 are. However, a steep droop characteristic also has disadvantages: due to the steepness, the voltage regulation of the sources becomes worse [7]. From this consideration it follows that in the case of the voltage droop control, a compromise between voltage regulation and load sharing has to be made.

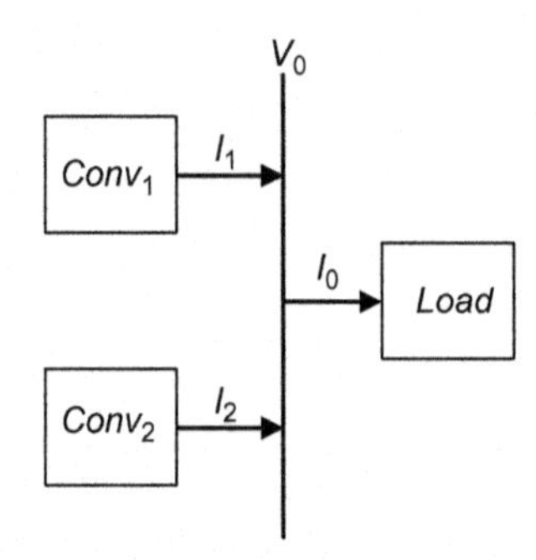

Figure 3.6 Two voltage sources connected in parallel.

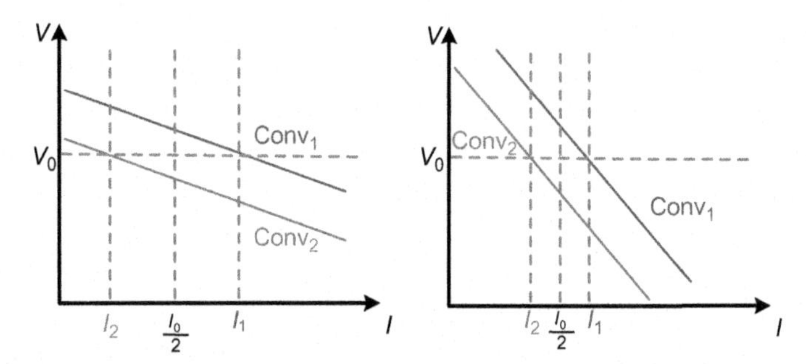

Figure 3.7 Two parallel voltage sources with different operating points at different droop steepness characteristics.

Fig. 3.8 shows the output characteristics of two converters connected in parallel, without employing a current sharing loop. The common output voltage is V_0 while V_i and I_i denote the voltages and currents of each converter with its corresponding output resistance R_i. Then the current sharing error between the two converters can be calculated as:

$$\Delta I = \frac{I_1 - I_2 = R_2 V_1 - R_1 V_2 + 2(V_1 - V_2)R_{Load}}{R_1 R_2 + R_1 R_{Load} + R_2 R_{Load}} \tag{3.24}$$

This error can only be zero if $V_1 = V_2$ and $R_1 = R_2$.

As stated above creating such a characteristic with a physical resistance is only of interest for low power ratings, as the additional resistance leads to large losses [7]. Better is the use of power electronics in the form of a DC/DC converter and a control loop shaping the output voltage as a function of the delivered current.

A mathematical description of the droop control must be derived for this purpose. This can be found with:

$$V = V_0 - I \cdot R \tag{3.25}$$

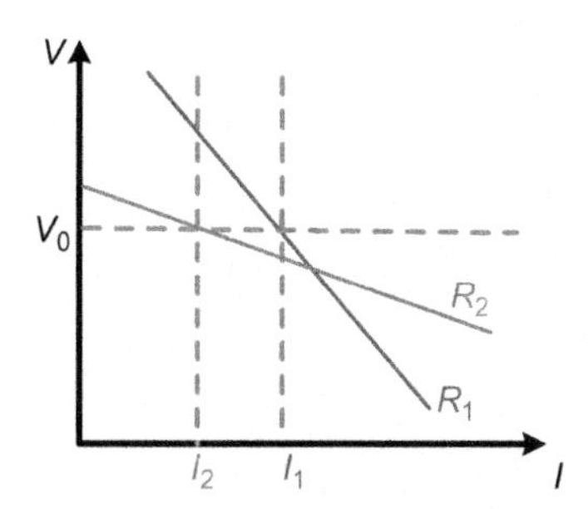

Figure 3.8 Two parallel voltage sources with different operating points at different droop steepness characteristics.

The voltage V and current I are the output voltage and the output current, V_0 is the nominal voltage and R is the gradient of the droop characteristic, which is physically equivalent to a resistor and is therefore referred to as such.

There are different possibilities to realize the droop control with a control system [7–9]. Here the current I is defined as the output variable and voltage V as the input variable of the control system, since the current can be easily controlled with DC/DC converters.

The deviation of the output voltage from the nominal voltage $V_0 - V$ means that the transfer function of the controller can be written as:

$$\frac{I}{V_0 - V} = \frac{1}{R} \tag{3.26}$$

This corresponds to a proportional controller with the gain factor $K = \frac{1}{R}$.

Since a proportional controller always has a permanent control deviation, a proportional integral controller (PI controller) can also be used. However, it must be kept in mind that with PI controllers connected in parallel, the start value of the control deviation has a large influence on the control due to the integrating part. As a result, the stationary state of the control system depends on the control parameters, the line resistances and the noise. This is reflected by unevenly distributed load flows. To avoid this problem, only one PI controller should be implemented in the control system, the others should be kept as P controllers. It should also be noted that the combined voltage and current control systems have defined stationary states, even when using PI controllers, as these control methods regulate both the current (or power) and the voltage, as the name suggests. The block diagram of a generator with voltage droop control is shown in Fig. 3.9.

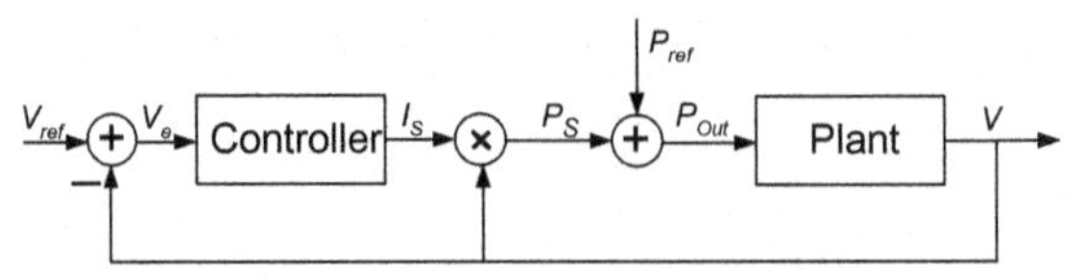

Figure 3.9 Block diagram of the droop control of converter.

The control deviation $V_e = V_{ref} - V$ is passed on to the controller, which generates the control variable I_s as a response. This variable is represented here as current, since the gain factor of the controller (regardless of whether proportional or PI controller) corresponds to a division with a resistor. This current is converted to the output power P_{out} using the output voltage V and the reference power P_{ref} (set point of the converters' power output). This step is carried out in the actuator, in this case a DC/DC converter, which has not been further specified here as it is threatened as a controlled voltage source. The voltage at the load is now determined by the line parameters. It is fed back to calculate the control deviation.

With the voltage droop regulation, the individual network participants do not have to communicate separately with each other. The control is carried out exclusively via the supply voltage. In this case, from the point of view of the overall system, it is an open-loop control [6].

3.6.2 Influence of Droop Coefficients on the Overall System Dynamics

While the droop relation describes an algebraic relation, it has vice versa an effect on the overall system dynamics. It a nutshell, because of the involvement of instantaneous values of system quantities, the droop coefficients are able to move the position of the eigenvalues of the overall grid. Such a consideration can be proved with reference to a very simple system. Let us consider the system of Fig. 3.10 depicting a simple system with two DC sources and a load.

Each source is controlled by a voltage control loop that can be described by its dominant time constant. The reference of each voltage source is then calculated according to a droop law. For both sources it holds:

$$\begin{aligned} V_{1ref} &= V_{1n} - R_{d1} I_1 \\ V_{2ref} &= V_{2n} - R_{d2} I_2 \end{aligned} \tag{3.27}$$

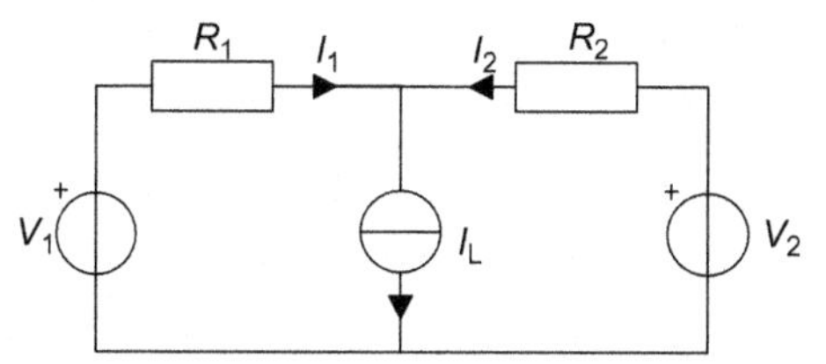

Figure 3.10 A simple circuit controlled by a droop control approach.

It is then necessary to express the line current as a function of the forcing functions. By applying simple circuit solution methods, the following relation is extracted:

$$\begin{bmatrix} I_1 \\ I_2 \end{bmatrix} = \frac{1}{R_1 + R_2} \begin{bmatrix} R_2 & -1 \\ R_1 & 1 \end{bmatrix} \begin{bmatrix} I_L \\ V_2 - V_1 \end{bmatrix} \tag{3.28}$$

Substituting the expression of the current in the droop relation, the overall system equations are obtained

$$\begin{aligned} V_{1ref} &= V_{1n} - \frac{R_{d1} R_2}{R_1 + R_2} I_L + \frac{R_{d1}}{R_1 + R_2}(V_2 - V_1) \\ V_{2ref} &= V_{2n} - \frac{R_{d2} R_1}{R_1 + R_2} I_L + \frac{R_{d2}}{R_1 + R_2}(V_1 - V_2) \end{aligned} \tag{3.29}$$

By applying the hypothesis that the relation between the reference and the actual voltage can be described by means of a simple dominant pole, the following relations hold:

$$\begin{aligned} V_1 + \tau_1 \frac{dV_1}{dt} &= V_{1ref} \\ V_2 + \tau_2 \frac{dV_2}{dt} &= V_{2ref} \end{aligned} \tag{3.30}$$

By substituting these equations in the two main equations the final form of the dynamic model is obtained. It is then possible to extract the state matrix:

$$A = \begin{bmatrix} \left(-1 - \frac{R_{d1}}{R_1 + R_2}\right) \frac{1}{\tau_1} & \left(\frac{R_{d1}}{R_1 + R_2}\right) \frac{1}{\tau_1} \\ \left(\frac{R_{d2}}{R_1 + R_2}\right) \frac{1}{\tau_2} & \left(-1 - \frac{R_{d2}}{R_1 + R_2}\right) \frac{1}{\tau_2} \end{bmatrix} \tag{3.31}$$

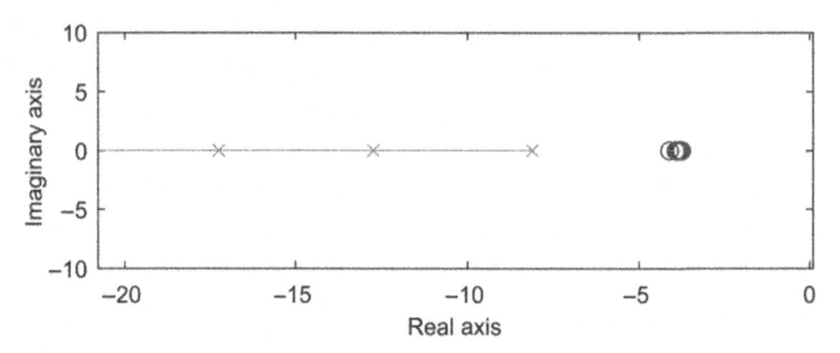

Figure 3.11 Pole positions as function of the value of the droop coefficients.

As an example Fig. 3.11 describes how the eigenvalues of this matrix and then the eigenvalues of the system move by changing the value of the droop coefficients. The following parameters have been used for the calculation: ($R_1 = 0.1\Omega$; $R_2 = 0.2\Omega$; $\tau_1 = 0.2s$; $\tau_2 = 0.3s$;$R_{d1} = 0.1 \div 40\Omega$; $R_{d2} = 0.2 \div 20\Omega$).

Looking at the results, it is possible to see that the system always presents two real poles. While one pole moves in the direction of stability, the second one moves closer and closer to the unstable region.

REFERENCES

[1] K. Ogata, Modern Control Engineering, fifth ed., Prentice-Hall, Boston, 2010.

[2] V.J. Thottuvelil, G.C. Verghese, Analysis and control of paralleled dc/dc converters with current sharing, IEEE Trans. Power Electron. 13 (4) (July 1998) 635−644.

[3] J. Rajagopalan, K. Xing, Y. Guo, F.C. Lee, B. Manners, Modeling and dynamic analysis of paralleled dc−dc converters with master−slave current sharing control, in Proc. IEEE Appl. Power Electron. Conf. Exp., March 1996, pp. 678−684.

[4] Y. Huang, C.K. Tse, Circuit Theoretic Classification of Parallel Connected DC−DC Converters, in IEEE Transactions on Circuits and Systems I: Regular Papers, vol. 54, no. 5, pp. 1099−1108, May 2007.

[5] C.K. Tse, Linear Circuit Analysis, Addison Wesley, London, UK, 1998.

[6] S. Luo, Z. Ye, R.-L. Lin, F.C. Lee, A classification and evaluation of paralleling methods for power supply modules, in 30th Annual IEEE Power Electronics Specialists Conference. Record. (Cat. No.99CH36321), Charleston, SC, 1999, vol. 2, pp. 901−908.

[7] B.T. Irving, M.M. Jovanovic, Analysis, design, and performance evaluation of droop current-sharing method, in APEC 2000. Fifteenth Annual IEEE Applied Power Electronics Conference and Exposition (Cat. No.00CH37058), New Orleans, LA, 2000, vol. 1, pp. 235−241.

[8] Z. Moussaoui, I. Batarseh, H. Lee, C. Kennedy, An overview of the control scheme for distributed power systems, Southcon/96 Conference Record, Orlando, FL, 1996, pp. 584−591.

[9] X. Lu, J.M. Guerrero, K. Sun, J.C. Vasquez, An Improved Droop Control Method for DC Microgrids Based on Low Bandwidth Communication With DC Bus Voltage Restoration and Enhanced Current Sharing Accuracy, in IEEE Transactions on Power Electronics, April 2014, vol. 29, no. 4, pp. 1800−1812.

CHAPTER FOUR

Generation Side Control

As presented in Section 1.3, broad research has been concluded on load side control. One of the main drawbacks of load side control is that each load needs a special converter which has to guarantee stability and this makes the usage of common off-the-shelf components not possible. The complex internal dynamics of parallel connected converter components (e.g., power sources) limit the application of classical control theory as was shown in Ref. [1], where linear-controlled converters exhibited chaotic behavior.

Therefore, the goal is to design a generation control system that meets the operational objectives of the medium voltage direct current (MVDC) power system in presence of wide variations of load or generation capacity. The idea of this approach is inspired by the adaptive control of the generator side interface converters, which compensates for the negative resistance behavior of the loads at every operating point to maintain system-wide stability [2]. To achieve this goal in a multiconverter environment several approaches are proposed and described mathematically in this chapter. Their classification can be seen in Fig. 4.1. Those algorithms will be later described throughout Chapter 7, Hardware in the Loop Implementation and Challenges. But before coming to the control algorithms a system description of the Integrated Shipboard Power Systems (ISPSs) including the circuit basis will be given.

4.1 MVDC SHIPBOARD POWER SYSTEMS

In Chapter 1 the transition from load side control to generation side control in MVDC ISPS was previously highlighted. Therefore, before introducing any generation side control scheme a more detailed look into the application case and its characteristics is necessary.

The generating devices are connected to the MVDC bus via a multistage solution consisting of a rectifier and a DC−DC converter. Recently, the usage of a single stage solution implemented via modular multilevel

Modern Control of DC-Based Power Systems.
DOI: https://doi.org/10.1016/B978-0-12-813220-3.00004-1

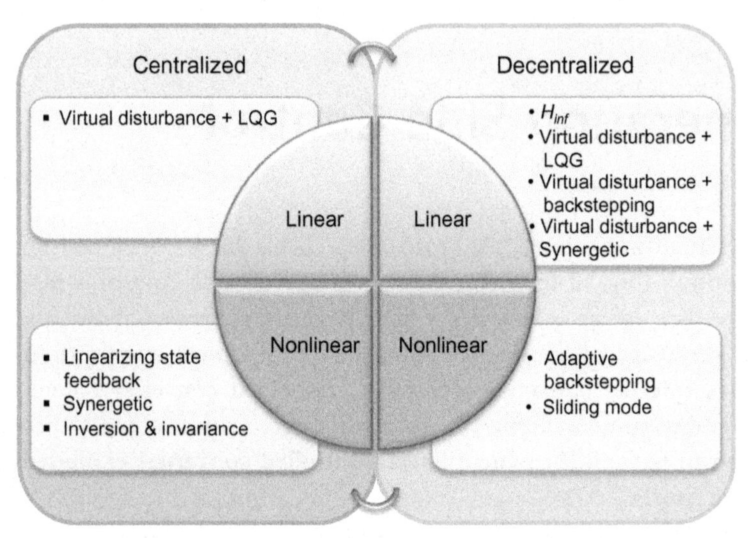

Figure 4.1 Control classification.

converter (MMC) designs in ISPSs applications received considerable attention [3] due to their capability of providing short circuit protection [4]. Another option for implementing the single stage solution in ISPSs could be the usage of dual active bridges (DABs); however, to the author's best knowledge, such an application has been presented only for the connection of storage devices (like batteries, supercapacitors) to the MVDC bus [5], and not as a single stage solution. The work presented in this book does not cover the analysis of a single stage solution interfacing AC generation towards the MVDC bus, since this would result in a very complex model that would likely not provide additional insightful information for the control design and its effect on the control performance.

When taking into consideration a multistage solution with dynamic modeling of two converters altogether, at least four state variables are necessary to develop a fourth-order open-loop model for control. The order of the model will increase for closed-loop control.

Since the focus of the present work lies in the stabilizing control, a model order reduction would be reasonable. Thus, some simplifications were performed in order to decouple the rectification action from the stabilizing action at the device level. It is considered that the rectifier stage has been designed adequately to provide the desired voltage; hence, the rectifier could be replaced by a constant voltage source, and the DC–DC conversion stage would display the effect of the control algorithm for maintaining the bus voltage under disturbances.

For the implementation of the DC−DC conversion stage, various technologies could be used, e.g., buck converter, DAB, or MMC. The buck converter is selected, considering that unidirectional power flow is necessary.

Van der Broeck et al. stated about buck converters [6]:

- They provide a low-cost and robust topology, exhibit well known steady state and dynamic properties and the design procedures for these converters have been studied intensively.
- However, the control of a one quadrant buck converter over a wide load range and varying voltage transfer ratios is challenging.

4.2 STATE-SPACE MODEL

The modeling of the ISPS in this book is based on Ref. [7], which is a radial concept topology. It features a DC bus and the loads are connected either through a DC−DC converter, which throughout this book has been referred to as a point of load converter directly bus or via a load zone. The generation devices are connected via a rectifier and a following DC−DC converter to the MVDC bus, this DC−DC converter will be abbreviated as line regulating converter (LRC). A more detailed representation is depicted in Fig. 4.2 [8]. It will be presumed that each load zone can be represented by a lumped load.

In the model, four generating systems ($G1-G4$) are considered, which are connected through rectifiers ($C1-C4$) in cascade with buck converters, i.e., LRC ($C5-C8$) to the MVDC bus. The MVDC bus feeds all the shipboard loads, through service power converters ($C9-C15$). In Fig. 4.2 the loads $M1$ and $M2$ are ship propulsion motors, each fed by inverters. The loads and propulsion motors are connected to the MVDC bus through cables ($CA1-CA6$). Loads $L2$ and $L3$ are LVDC (low voltage DC utility) and LVAC (low voltage AC utility) loads and are respectively fed by buck converters and inverters. It has to be highlighted that the storage components which are mentioned in Ref. [7], such as supercapacitor, batteries, or flywheel, are omitted from the analysis in this book.

From Fig. 4.2 the general circuit model representation of the system can be derived. It is depicted in Fig. 4.3. The representation shows n DC generating systems in parallel with m loads. In this circuit model the DC

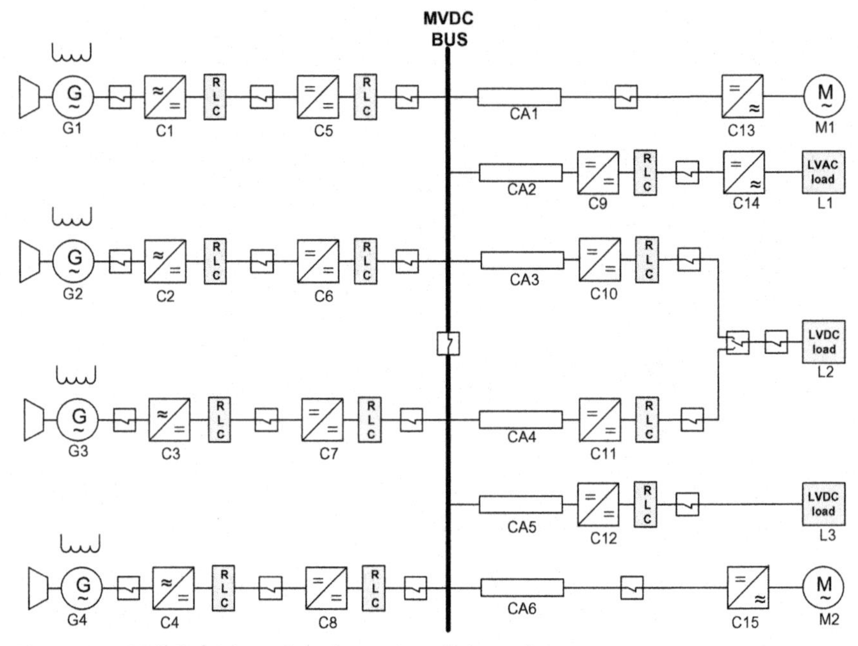

Figure 4.2 MVDC shipboard—schematic radial topology.

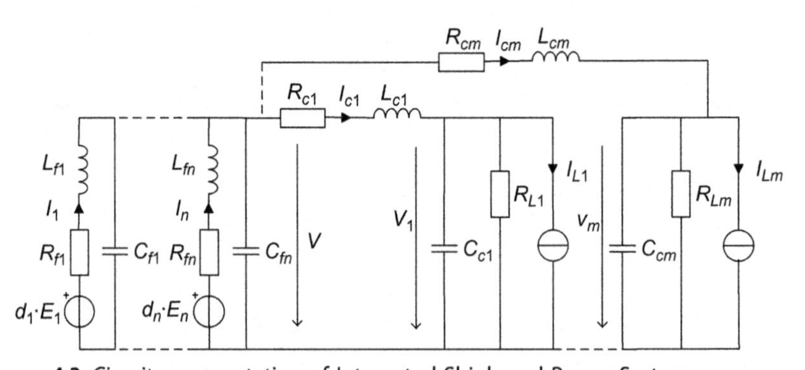

Figure 4.3 Circuit representation of Integrated Shipboard Power System.

voltage generators E_n represent the DC signal-averaged output of the DC–DC "buck" converters.

The circuit model of Fig. 4.3 has been obtained under the following assumptions:

- The averaging method has been adopted;
- The timescale of the electromechanical generation system and its associated control is very slow compared to the voltage transients on the

MVDC bus and therefore the output of this system can be considered constant during the period of observation;

- The generic generating system has been represented by DC ideal voltage DC sources E_n (representing the buck converter output) with the duty cycle d connected to a second-order RLC filter, whose parameters are R_{fn}, L_{fn} and C_{fn};
- The zonal loads and single component loads and their characteristics can be lumped into an equivalent load which exhibits CPL characteristic;
- The generic load has been represented by a load branch resistor R_{Lm} and a controlled current source $I_{Lm} = P_m/V_m$; these two branches represent the linear and the nonlinear (assumed as infinite bandwidth CPL) parts of the load;
- The generic lines which connect the loads to the MVDC bus are characterized by the lumped cable parameters R_{cm}, L_{cm} and by a possible input filtering capacitor C_{cm} of the load converter.

Based on Fig. 4.3 it is possible to derive a state-space average model of a multimachine MVDC system. The Kirchhoff current and voltage laws enable us to express the mathematical model by $n + 2m + 1$ nonlinear differential equations with the following state variables:

- MVDC bus voltage, V (1 equation);
- Generators currents, I_n (n equations);
- Line currents, I_{ch} (m equations);
- Load voltages, V_m (m equations).

The state-space model of the system presented in Fig. 4.3 is therefore defined in (4.1) by the $n + 2m + 1$ equations of the system, where a total capacitor C_{eq} has been defined as the sum of all filter capacitors C_{fn}:

$$\begin{cases} \dfrac{dV}{dt} = \dfrac{1}{C_{eq}}\left(\displaystyle\sum_{k=1}^{n} I_k - \sum_{h=1}^{m} I_{ch}\right) \\ \dfrac{dI_k}{dt} = \dfrac{1}{L_{fk}}(-R_{fk}I_k + V + E_k)\forall k = 1, 2, \ldots, n \\ \dfrac{dI_{ch}}{dt} = \dfrac{1}{L_{ch}}(-R_{ch}I_{ch} + V - V_h)\forall h = 1, 2, \ldots, m \\ \dfrac{dV_h}{dt} = \dfrac{1}{C_{ch}}\left(I_{ch} - \dfrac{P_h}{V_h}\right)\forall h = 1, 2, \ldots, m \end{cases} \tag{4.1}$$

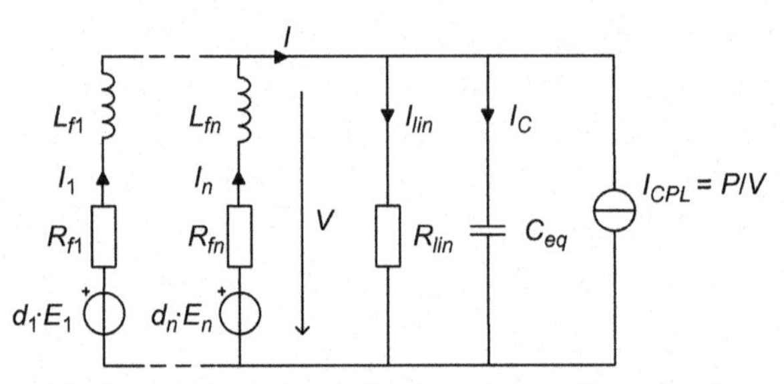

Figure 4.4 Simplified model including a linear and a nonlinear load.

According to [8–11] the series cable parameters are usually neglected in shipboard power systems, which results in the CPLs being connected in parallel to the bus. Since all capacitors and CPLs are in parallel now it is therefore possible to sum them up into an equivalent capacitor, e.g., equivalent constant power load.

For the sake of generality, the network is extended by a linear load, which results in the model depicted in Fig. 4.4. In this way, different kinds of loads schemes can be simulated where the resistance represents the linear part and the current source the nonlinear part. The linear load serves also as a base load to drive the generators in continuous conduction mode before connecting the CPLs.

This simplified model can now be now be described by the following equations.

$$\begin{aligned}\frac{dV}{dt} &= \frac{1}{C_{eq}}(I_1 + I_2 + I_3) - \frac{V}{R_{lin}C_{eq}} - \frac{P_{eq}}{C_{eq}V} \\ \frac{dI_k}{dt} &= -\frac{R_{fk}}{L_{fk}}I_k - \frac{V}{L_{fk}} + \frac{d_k E_k}{L_{fk}} fork = 1, 2, 3\end{aligned} \tag{4.2}$$

In the next section the first generation side control technique is presented.

REFERENCES

[1] I. Batarseh, K. Siri, H. Lee, Investigation of the output droop characteristics of parallel-connected DC-DC converters, in 25th Annual IEEE Power Electronics Specialists Conference, PESC '94 Record, June 20–25, 1994, vol. 2, pp. 1342, 1351.

[2] V. Arcidiacono, A. Monti, G. Sulligoi, Generation control system for improving design and stability of medium-voltage DC power systems on ships, IET Electr. Syst. Transport. 2 (3) (2012) 158–167.

[3] M. Steurer; F. Bogdan, M. Bosworth, O. Faruque, J. Hauer, K. Schoder, et al., Multifunctional megawatt scale medium voltage DC test bed based on modular multilevel converter (MMC) technology, in 2015 International Conference on Electrical Systems for Aircraft, Railway, Ship Propulsion and Road Vehicles (ESARS), March 3–5, 2015, pp. 1, 6.
[4] V. Staudt, M.K. Jager, A. Rothstein, A. Steimel, D. Meyer, R. Bartelt, et al., Short-circuit protection in DC ship grids based on MMC with full-bridge modules, in 2015 International Conference on Electrical Systems for Aircraft, Railway, Ship Propulsion and Road Vehicles (ESARS), March 3–5, 2015, pp. 1, 5.
[5] Y. Tang, A. Khaligh, On the feasibility of hybrid Battery/Ultracapacitor Energy Storage Systems for next generation shipboard power systems, in 2010 IEEE Vehicle Power and Propulsion Conference (VPPC), September 1–3, 2010, pp. 1, 6.
[6] C.H. van der Broeck, R.W. De Doncker, S.A. Richter, J. von Bloh, Unified control of a buck converter for wide load range applications, in IEEE Transactions on Industry Applications, in press.
[7] IEEE recommended practice for 1 kV to 35 kV medium-voltage DC power systems on ships, IEEE Std. 1709–2010, Nov. 2, 2010, pp. 1, 54.
[8] G. Sulligoi, D. Bosich, L. Zhu, M. Cupelli, A. Monti, Linearizing control of shipboard multi-machine MVDC power systems feeding Constant Power Loads, in 2012 IEEE Energy Conversion Congress and Exposition (ECCE), 15–20 2012, pp. 691, 697.
[9] S.D. Sudhoff, K.A. Corzine, S.F. Glover, H.J. Hegner, H.N. Robey Jr, DC link stabilized field oriented control of electric propulsion systems, IEEE Trans. Energy Conv. 13 (1) (1998).
[10] C.H. Rivetta, A. Emadi, G.A. Williamson, R. Jayabalan, B. Fahimi, Analysis and control of a buck DC-DC converter operating with constant power load in sea and undersea vehicles, IEEE Trans. Ind. Applicat. 42 (2) (2006) 559–572.
[11] A.M. Rahimi, G.A. Williamson, A. Emadi, Loop-cancellation technique: a novel nonlinear feedback to overcome the destabilizing effect of constant-power loads, IEEE Trans. Vehic. Technol. 59 (2) (2010) 650–661.

FURTHER READING

Ciezki and Ashton, 1998J.G. Ciezki, R.W. Ashton, The application of feedback linearization techniques to the stabilization of DC-to-DC converters with constant power loads, in Proceedings of the 1998 IEEE International Symposium on Circuits and Systems, 1998. ISCAS '98, 31 May-3 Jun 1998, vol. 3, pp. 526, 529.
Doerry, 2015N. Doerry, Naval Power Systems: integrated power systems for the continuity of the electrical power supply, IEEE Electr. Mag. 3 (2) (2015) 12–21.
Emadi et al., 2006A. Emadi, A. Khaligh, C.H. Rivetta, G.A. Williamson, Constant power loads and negative impedance instability in automotive systems: definition, modeling, stability, and control of power electronic converters and motor drives, IEEE Trans. Vehic. Technol. 55 (4) (2006) 1112–1125.

CHAPTER FIVE

Control Approaches for Parallel Source Converter Systems

This chapter is structured in the following way: First we start with centralized nonlinear control schemes. In Section 5.1 we introduce the Linearizing State Feedback (LSF) where the nonlinearity is compensated. Section 5.2 describes the Synergetic Control, which is a centralized model-based control based on a manifold. In Section 5.3 the Immersion and Invariance Control is presented, which designs a nonlinear control law with guaranteed asymptote. In Sections 5.4 and 5.5 decentralized control concepts are presented. Section 5.4 presents an Observer-based control which decouples the power system on the basis of a Kalman filter and results in a linear Two Degree Of Freedom (2DOF) controller. In Section 5.5 a nonlinear Backstepping control law is presented which combines two possible methods of power system decoupling. Method one is based on a virtual current source, while method two relies on power estimation which is based on an Adaptive Control law in Section 5.6. In Section 5.7 we introduce a H_{∞} controller which offers robustness and optimality in respect to uncertainties. In Section 5.8 we explain the Sliding Mode Control (SMC) which has inherent robustness properties and relies on the switching representation of the converters, while the previous approaches use the state-space averaged model. The chapter is summarized in Section 5.9 with a summary of the unique characteristics of the presented control approaches.

5.1 LINEARIZING STATE FEEDBACK

In this section, a centralized nonlinear control method is considered and later applied to a medium voltage DC (MVDC) system. It is called the linearizing control technique [1]; other names often found in literature are linearization via state feedback or loop cancelation [2]. This

Modern Control of DC-Based Power Systems.
DOI: https://doi.org/10.1016/B978-0-12-813220-3.00005-3

control technique has found successful application in the stabilization of Constant Power Loads (CPLs) in [3–5]. In this control the nonlinear part of the controlled system is compensated and it is possible to design a closed-loop control on a linear control system. Strictly speaking this technique is based on the concept of exact linearization [6] and refers to the input–output linearization. In Section 5.1.1 the theoretical framework of this concept is presented; readers familiar with the theory can jump straight to Section 5.1.2 where this concept is applied to the MVDC ISPs.

5.1.1 Procedure of Linearizing State Feedback

The theory of exact linearization can be explained while considering a nonlinear system described by (5.1):

$$\begin{aligned} \dot{x}(t) &= a(x) + b(x)u \\ y &= c(x) \end{aligned} \tag{5.1}$$

Those are systems which are nonlinear in x but linear in u. Those systems are called input linear systems or control-affine systems. The assumption of a linear acting control variable u is not a big restriction, as many technical systems are linear in u or the manipulated variable [6].

Now the question whether a state feedback control exists, is posed:

$$u = -r(x) + v(x)w \tag{5.2}$$

And a change of variables [1], which will render the resulting system linear.

$$z = T(x) \tag{5.3}$$

The system has the order of $n = dim(x)$. The consideration of multiple input–multiple output (MIMO) systems is possible without great effort and in a similar manner as in the single input–single output (SISO) case; this is presented to the reader in detail in [1,7–9]. Here, just the SISO systems are considered with the aim to present the main idea of the method. When performing an exact linearization two different exact linearizations are possible, depending on the goals of the control design:

- Input-to-State Linearization
- Input–output Linearization.

The focus here lies on the input–output Linearization. For the envisaged controller design, the form of the above system description in (5.1) is not favorable. A suitable transformation is therefore needed. For this purpose the Lie derivative is employed, which is defined as the gradient of a scalar function $h(x)$ multiplied with a vector field $f(x)$, i.e.:

$$L_f h(x) = \frac{\partial h(x)}{\partial x} f(x) = grad^T h(x) f(x) \tag{5.4}$$

Considering the system presented in (5.1), and applying the Lie derivative yields:

$$L_a c(x) = \frac{\partial c(x)}{\partial x} a(x) \tag{5.5}$$

Now forming the time derivative of the output y one obtains:

$$\dot{y}(t) = \frac{dc(x)}{dt} = \frac{\partial c(x)}{\partial x} \dot{x}_1 + \ldots + \frac{\partial c(x)}{\partial x_n} \dot{x}_n x = \frac{\partial c(x)}{\partial x} \dot{x} \tag{5.6}$$

Replacing now:

$$\dot{x}(t) = a(x) + bx(u) \tag{5.7}$$

Results in:

$$\dot{y}(t) = \frac{\partial c(x)}{\partial x} a(x) + \frac{\partial c(x)}{\partial x} b(x) u \tag{5.8}$$

Which can also be described as the Lie derivative:

$$\dot{y}(t) = L_a c(x) + L_b c(x) u \tag{5.9}$$

In most technical systems according to [6] is $L_b = 0$, such that:

$$\dot{y}(t) = L_a c(x) \tag{5.10}$$

The next step would be to calculate $\ddot{y}$ while continuing from (5.10):

$$\ddot{y} = \frac{dL_a c(x)}{dt} = \frac{\partial L_a c(x)}{\partial x} \dot{x} = \frac{\partial L_a c(x)}{\partial x} a(x) + \frac{\partial L_a c(x)}{\partial x} b(x)$$
$$= L_a L_a c(x) + L_b L_a c(x) = L_a^2 c(x) \tag{5.11}$$

For the higher derivates the following is valid:

$$
\begin{aligned}
y &= c(x) \\
\dot{y} &= L_a c(x) \\
\ddot{y} &= L_a^2 c(x) \\
&\vdots \\
y^{(\delta-1)} &= L_a^{\delta-1} c(x) \\
y^{(\delta)} &= L_a^{\delta} c(x) + \frac{\partial L_a^{\delta-1} c(x)}{\partial x} b(x) u \\
L_b L_a^i c(x) &= \frac{\partial L_a^i c(x)}{\partial x} b(x) = 0; i = 0, \ldots, \delta - 2
\end{aligned}
\tag{5.12}
$$

Only for the index $\delta - 1$ the term $L_b L_a^i c(x)$ is not 0. δ is also called the relative degree of the system. The relative degree is in linear system equal to the difference between numerator and denominator of the transfer function. If $n = dim(x) = \delta$ the system has therefore no internal dynamics. Those internal dynamics are not observable and thus cannot be linearized.

According to [6] the input signal has to be in the following form, to linearize the system, this corresponds to the nonlinear Ackermann formula, which is only dependent on the original state x:

$$
u = -\frac{L_a^n c(\boldsymbol{x}) + a_{n-1} L_a^{\delta-1} c(x) + \ldots + a_1 L_a c(x) + a_0 c(x)}{L_{\boldsymbol{b}} L_a^{n-1} c(\boldsymbol{x})} + \frac{V}{L_{\boldsymbol{b}} L_a^{n-1} c(\boldsymbol{x})} w
\tag{5.13}
$$

By means of the transformations, the nonlinear system is converted into a linear representation. For this procedure, the name of exact input–output linearization or exact linearization is used. The choice of the coefficients a_i of the control normal form is unrestricted, so that one can imprint any desired eigenvalue configuration on the transformed system. Also, the value V can be freely selected. Comparing Eq. (5.13) with (5.2) it is possible to say that $r(x)$ is the controller, while $v(x)$ has the function of a prefilter [6].

5.1.2 Application to MVDC System

The LSF controller is a centralized control architecture, which means that it collects the measurements of each generator and load. Therefore, the controller has perfect knowledge as all relevant states can be measured. This can be a significant advantage in terms of control performance.

A drawback could be that the centralized model requires a communication infrastructure that has to be protected against faults as well. That could be a reason why the LSF controller might not be the best solution for a real implementation even if its control performance is sufficient. This is already shown in [3] and [10].

The overall schematic of this control approach is depicted in Fig. 5.1, where V_{bus} corresponds to the bus voltage and I_n are the inductor currents presented in Fig. 5.2.

The design of the LSF controller involves exact input to output linearization [6] of the MIMO system and transforms the system using the compensation term. This compensation term injects through the converters a signal on their output voltage, which is capable of compensating the nonlinear part of the system. Obviously, a converter must have enough regulation bandwidth to effectively apply the signal. Therefore the LSF can be successfully applied only to converters with high switching frequency, such as the DC–DC ones based on IGBT/IGCTs. Using other types of converters to interface generators with the MVDC bus, such as diode or thyristor rectifiers, the LSF cannot be applied on the bus controlling converters but must be relegated to the load side for CPL compensation.

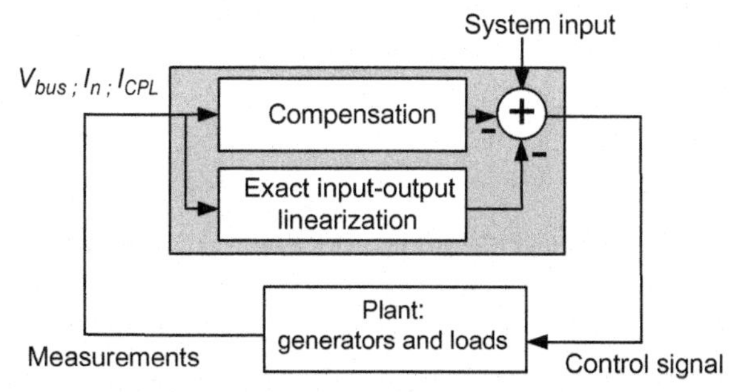

Figure 5.1 Scheme of the linearizing state feedback controller.

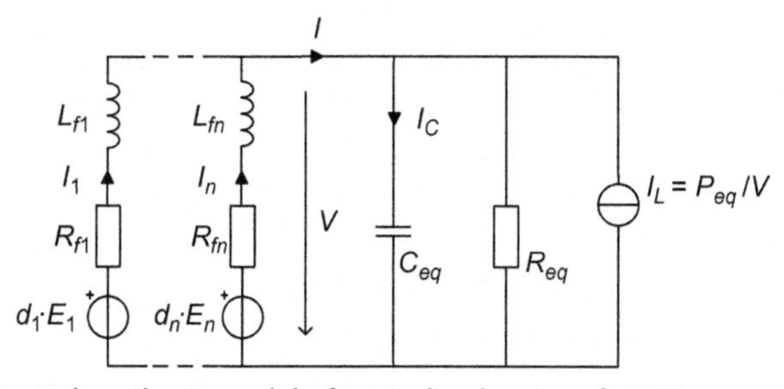

Figure 5.2 Control synthesis model of centralized system for LSF.

After the compensation a linear system is present, where the classical pole placement technique can be performed so that the natural frequency and damping coefficient are shifted to the desired values. Additionally, a PI controller is added in parallel, which drives the system output to the reference value.

It is possible to make an analogy to terrestrial power systems where the fast stabilizing action of the LSF corresponds to the primary control action, while the slower dynamics where the PI control bring the system to the reference value corresponds to a secondary control action in terrestrial power systems.

In the following, it is shown how to achieve LSF using the general approach of [6]. Furthermore, it is explained how the control law can be developed from conditions on the natural frequency and damping ratio as it was done in [10]. Additionally, the derivation of the control law incorporates the Lie derivative.

As explained in Section 5.2, the cable parameters can be neglected and the CPL capacitances and output filter capacitances can be replaced by an equivalent component: $C_{eq} = C_{CPL} + C_{f1} + C_{f2} + C_{f3}$

The system can be derived from the model in Fig. 5.2 as follows:

$$
\begin{aligned}
\dot{V} &= \frac{1}{C_{eq}}(I_1 + I_2 + I_3) - \frac{V}{R_L C_{eq}} - \frac{P_{eq}}{C_{eq} V} \\
\dot{I}_k &= -\frac{R_{fk}}{L_{fk}} I_k - \frac{V}{L_{fk}} + \frac{d_k \cdot E_k}{L_{fk}} fork = 1, 2, 3
\end{aligned}
\tag{5.14}
$$

To shorten the calculations a time constant $T_f = \frac{L_{fk}}{R_{fk}}$ is defined. This is only valid if the ratios are equal for all filters. Besides, the equations are simplified by summing up the inductances $\frac{1}{L_{eq}} = \frac{1}{L_{f1}} + \frac{1}{L_{f2}} + \frac{1}{L_{f3}}$ and currents $i = I_1 + I_2 + I_3$.

$$
\begin{aligned}
\dot{V} &= \frac{I}{C_{eq}} - \frac{V}{R_L C_{eq}} - \frac{P_{eq}}{C_{eq} V} \\
\dot{I}_k &= -\frac{1}{T_f} I - \frac{V}{L_{eq}} + \frac{d_1 E_1}{L_{f1}} + \frac{d_2 E_2}{L_{f2}} + \frac{d_3 E_3}{L_{f3}}
\end{aligned}
\tag{5.15}
$$

Still, the system description (5.15) is not ideal for the LSF design as it is described in [6]. Therefore, the following state-space representation will be used to show the transformation of the system:

$$
\begin{aligned}
\dot{\boldsymbol{x}} &= \boldsymbol{A}(\boldsymbol{x}) + \boldsymbol{B}(\boldsymbol{x}) \cdot \boldsymbol{u} \\
y &= c(\boldsymbol{x})
\end{aligned}
\tag{5.16}
$$

with

$$x = \begin{bmatrix} V & I \end{bmatrix}^T$$

$$A(x) = \begin{bmatrix} \dfrac{1}{C_{eq}} I - \dfrac{V}{R_L C_{eq}} - \dfrac{P_{eq}}{C_{eq} V} \\ -\dfrac{1}{T_f} I - \dfrac{v}{L_{eq}} \end{bmatrix} \quad c(x) = V \tag{5.17}$$

$$B(x) = \begin{bmatrix} 0 & 0 & 0 \\ \dfrac{E_1}{L_{f1}} & \dfrac{E_2}{L_{f2}} & \dfrac{E_3}{L_{f3}} \end{bmatrix} \quad u = \begin{bmatrix} d_1 & d_2 & d_3 \end{bmatrix}^T$$

Moreover, the Lie derivative was already introduced in Section 5.1.1 as a tool to shorten the calculation of the exact linearization:

$$L_f h(x) = \frac{\partial h(x)}{\partial x} f(x) = \text{grad}^T h(x) \cdot f(x) \tag{5.18}$$

Using the Lie derivative it is now possible to obtain the first time derivative of the output signal y.

$$\begin{aligned} \dot{y} &= L_a c(x) + L_b c(x) \cdot u \\ &= \begin{bmatrix} 1 & 0 \end{bmatrix} \cdot A(x) \\ &= \frac{I}{C_{eq}} - \frac{V}{R_L C_{eq}} - \frac{P_{eq}}{C_{eq} v} = \dot{V} \end{aligned} \tag{5.19}$$

The previous step is executed two times since the system is of order two. The second step leads to

$$\begin{aligned} \ddot{y} &= L_a L_a c(x) + L_b L_a c(x) \cdot u \\ &= \begin{bmatrix} -\dfrac{1}{R_L C_{eq}} + \dfrac{P_{eq}}{C_{eq} V^2} & \dfrac{1}{C_{eq}} \end{bmatrix} a(x) + \begin{bmatrix} -\dfrac{1}{R_L C_{eq}} + \dfrac{P_{eq}}{C_{eq} V^2} & \dfrac{1}{C_{eq}} \end{bmatrix} B(x) u \\ &= -\frac{\dot{V}}{R_L C_{eq}} + \frac{P_{eq}}{C_{eq} v^2} \dot{V} - \frac{I}{C_{eq} T_f} - \frac{V}{C_{eq} L_{eq}} + \frac{1}{C_{eq}} \begin{bmatrix} \dfrac{E_1}{L_{f1}} & \dfrac{E_2}{L_{f2}} & \dfrac{E_3}{L_{f3}} \end{bmatrix} \cdot \begin{bmatrix} d_1 & d_2 & d_3 \end{bmatrix}^T \\ &= -\frac{1}{C_{eq} T_f} I - \frac{1}{C_{eq} L_{eq}} V + \frac{1}{C_{eq}} \left(\frac{d_1 E_1}{L_{f1}} + \frac{d_2 E_2}{L_{f2}} + \frac{d_3 E_3}{L_{f3}} \right) + \left(-\frac{1}{C_{eq} R_L} + \frac{P_{eq}}{C_{eq} V^2} \right) \dot{V} \\ &= \ddot{V} \end{aligned} \tag{5.20}$$

In this way the system (5.16) is transformed to be represented in the so-called nonlinear controllable canonical form [6]:

$$\begin{bmatrix} \dot{z}_1 \\ \dot{z}_2 \end{bmatrix} = \begin{bmatrix} z_2 \\ L_a^2 c(\boldsymbol{x}) \end{bmatrix} + \begin{bmatrix} \boldsymbol{0} \\ L_b L_a c(\boldsymbol{x}) \end{bmatrix} \cdot \boldsymbol{u} \tag{5.21}$$

with the following states:

$$\begin{bmatrix} z_1 \\ z_2 \end{bmatrix} = \begin{bmatrix} y \\ \dot{y} \end{bmatrix} = \begin{bmatrix} c(\boldsymbol{x}) \\ L_a c(\boldsymbol{x}) \end{bmatrix} \tag{5.22}$$

A noteworthy detail that should be pointed out is that after the exact input—output linearization the linearized system has the same order as the nonlinear system. This means that there are no internal dynamics and the property of observability remained in this case intact. In order to linearize a SISO system, the input signal has to be of the following form:

$$u = -\frac{L_a^2 c(\boldsymbol{x}) + \boldsymbol{k}^T \boldsymbol{z}}{L_b L_a c(\boldsymbol{x})} + \frac{V}{L_b L_a c(\boldsymbol{x})} w \tag{5.23}$$

with

$$\boldsymbol{k}^T = \begin{bmatrix} a_0 & a_1 \end{bmatrix} \tag{5.24}$$

Then, the resulting system can be written in the linear controllable canonical form:

$$\begin{bmatrix} \dot{z}_1 \\ \dot{z}_2 \end{bmatrix} = \begin{bmatrix} 0 & 1 \\ -a_0 & -a_1 \end{bmatrix} \begin{bmatrix} z_1 \\ z_2 \end{bmatrix} + \begin{bmatrix} 0 \\ V \end{bmatrix} \cdot w \tag{5.25}$$

where w is the new input of the system.

The system considered here, on the contrary, is a MIMO system. That is why the control law, i.e., the duty cycle d_i, has to be slightly modified to divide the linearizing first term between the weighted number of participating LRC (S_i).

$$d_i = -\frac{1}{S_i} \frac{L_a^2 c(\boldsymbol{x}) + \boldsymbol{k}^T \boldsymbol{z}}{L_b L_a c(\boldsymbol{x})_i} + \frac{V_i}{L_b L_a c(\boldsymbol{x})_i} w_i \tag{5.26}$$

The overall control vector is

$$\boldsymbol{u} = \begin{bmatrix} d_1 & d_2 & d_3 \end{bmatrix}^T. \tag{5.27}$$

The share of every generator to the linearizing function is determined by the sharing coefficient S_i. The sum of all sharing coefficients has to be equal to one in order to ensure linearization. Because the system is supposed to have a certain natural frequency ω_0 and damping ratio ξ, the variables of (5.25) and (5.26) are chosen as follows:

$$
\begin{aligned}
a_0 &= \omega_0^2 \\
a_1 &= 2\xi\omega_0 \\
V_i &= \frac{E_i}{C_{eq}L_{fi}} = L_b L_a c(\boldsymbol{x})_i
\end{aligned}
\tag{5.28}
$$

V_i is chosen like this to be compatible to the controller designed in [11]. Inserting the control law (5.27) in system (5.21) yields

$$
\begin{aligned}
\ddot{V} &= L_a^2 c(\boldsymbol{x}) + L_b L_a c(\boldsymbol{x}) \cdot \boldsymbol{u} \\
&= L_a^2 c(\boldsymbol{x}) - L_a^2 c(\boldsymbol{x}) + k^T z + \frac{w_1 E_1}{C_{eq}L_{f1}} + \frac{w_2 E_2}{C_{eq}L_{f2}} + \frac{w_3 E_3}{C_{eq}L_{f3}} \\
&= -a_1 \dot{V} - a_0 V + \frac{w_1 E_1}{C_{eq}L_{f1}} + \frac{w_2 E_2}{C_{eq}L_{f2}} + \frac{w_3 E_3}{C_{eq}L_{f3}} \\
&= -2\xi\omega_0 \dot{V} - \omega_0^2 V + \frac{w_1 E_1}{C_{eq}L_{f1}} + \frac{w_2 E_2}{C_{eq}L_{f2}} + \frac{w_3 E_3}{C_{eq}L_{f3}}
\end{aligned}
\tag{5.29}
$$

The control input is calculated to be

$$
\begin{aligned}
d_i &= -\frac{C_{eq}L_{fi}}{S_i E_i}\left(L_a^2 c(\boldsymbol{x}) + \boldsymbol{k}^T \boldsymbol{z}\right) + w_i \\
&= -\frac{C_{eq}L_{fi}}{S_i E_i}\left(-\frac{\dot{V}}{R_L C_{eq}} + \frac{P_{eq}}{C_{eq}V^2}\dot{V} - \frac{I}{C_{eq}T_f} - \frac{V}{C_{eq}L_{eq}} - a_1 \dot{V} - a_0 V\right) + w_i.
\end{aligned}
\tag{5.30}
$$

Using Kirchhoff's law, the following representation of the overall current is obtained:

$$
I = I_C + I_{R_L} + I_{CPL} = C_{eq}\dot{V} + \frac{V}{R_L} + \frac{P_{eq}}{V}
\tag{5.31}
$$

Inserting (5.31) in (5.30), the control law can be transformed so that the control law is not dependent on the currents anymore. This is done to retrieve a representation that is equivalent to the one in [11].

$$d_i = -\frac{C_{eq}L_{fi}}{S_iE_i}\left(-\frac{\dot{V}}{R_LC_{eq}} + \frac{P_{eq}}{C_{eq}v^2}\dot{V} - \frac{\dot{V}}{T_f} - \frac{V}{C_{eq}T_fR_L}\right.$$

$$\left. - \frac{P_{eq}}{C_{eq}T_fV} - \frac{V}{C_{eq}L_{eq}} - 2\xi\omega_0\dot{V} - \omega_0^2V\right) + w_i$$

$$= -\frac{C_{eq}L_{fi}}{S_iE_i}\left(\left(-\frac{1}{C_{eq}T_fR_L} - \frac{1}{C_{eq}L_{eq}} - \omega_0^2\right)V\right.$$

$$\left. + \left(-\frac{1}{R_LC_{eq}} - \frac{1}{T_f} - 2\xi\omega_0\right)\dot{V} + \frac{P_{eq}}{C_{eq}v^2}\dot{V} - \frac{P_{eq}}{C_{eq}T_fV}\right) + w_i \quad (5.32)$$

Since the system described in [3] does not include a resistive load, the calculations have to be repeated including this load which yields the linearized system (5.33). The detailed derivation of Eq. (5.33) is depicted in the Appendix.

$$\ddot{V} + \left(\frac{1}{R_LC_{eq}} + \frac{1}{T_f} + K_2\right)\dot{V}$$

$$+ \left(\frac{1}{C_{eq}T_fR_L} + \frac{1}{C_{eq}L_{eq}} + K_1\right)V$$

$$= \frac{w_1E_1}{C_{eq}L_{f1}} + \frac{w_2E_2}{C_{eq}L_{f2}} + \frac{w_3E_3}{C_{eq}L_{f3}} + K_1V_0$$

$$K_1 = \omega_0^2 - \frac{1}{C_{eq}T_fR_L} - \frac{1}{C_{eq}L_{eq}}$$

$$K_2 = 2\xi\omega_0 - \frac{1}{T_f} - \frac{1}{R_LC_{eq}} \quad (5.33)$$

Inserting the values of K_1 and K_2, it can be seen that the systems are equal. Summarizing, the LSF can be split up into a linearizing part f_l and a compensating part f_c of (5.34).

$$
\begin{aligned}
f_l &= -\frac{P_{eq}}{C_{eq}T_fV} + \frac{P_{eq}}{C_{eq}V^2}\dot{V} = -\frac{I_{CPL}}{C_{eq}T_f} + \frac{I_{CPL}}{C_{eq}v}\frac{I - I_{R_L} - I_{CPL}}{C_{eq}} \\
f_c &= K_1V + K_2\dot{V} = K_1V + K_2\frac{V - I_{R_L} - I_{CPL}}{C_{eq}}
\end{aligned}
\tag{5.34}
$$

With the power sharing factor of 1/3 as three generation side converters are used.

$$
F_k = (f_l + f_c)\frac{C_{eq}L_{fk}}{3E_k} \tag{5.35}
$$

5.1.3 Simulation Results

5.1.3.1 Cascaded System

A cascaded converter setup has been modeled in Matlab-SIMULINK with the parameters described in Table 6.1. In the simulation, a step variation of the CPL load has been performed. The variation of the bus voltage during the step change of the CPL load is described in Fig. 5.3. At $t = 0.25$ s the load step of 7.5 MW was performed, demonstrating that the LSF control stabilizes the system towards the reference point $V_{ref} = 1$. The dynamic of the inductor current is described in Fig. 5.4.

5.1.3.2 Shipboard Power System

A Shipboard Power System has been modeled in Matlab\SIMULINK, based on the parameters in Table 6.2. The load steps of 22.5 MW have

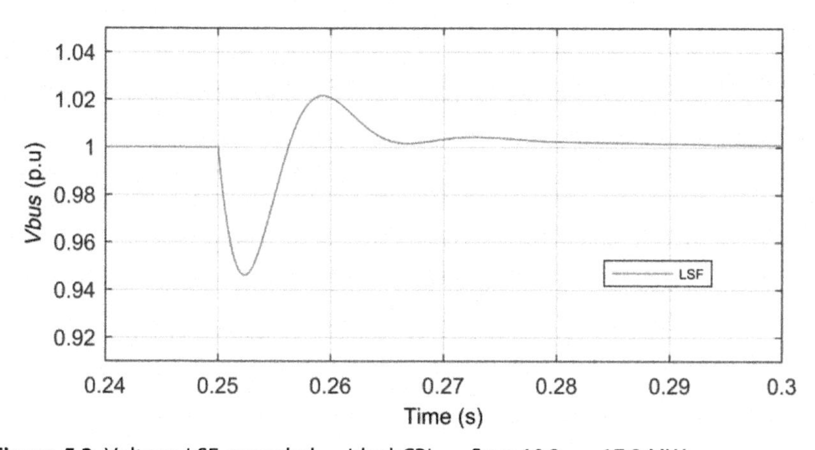

Figure 5.3 Voltage LSF cascaded – Ideal CPL – Step 10.3 → 17.8 MW.

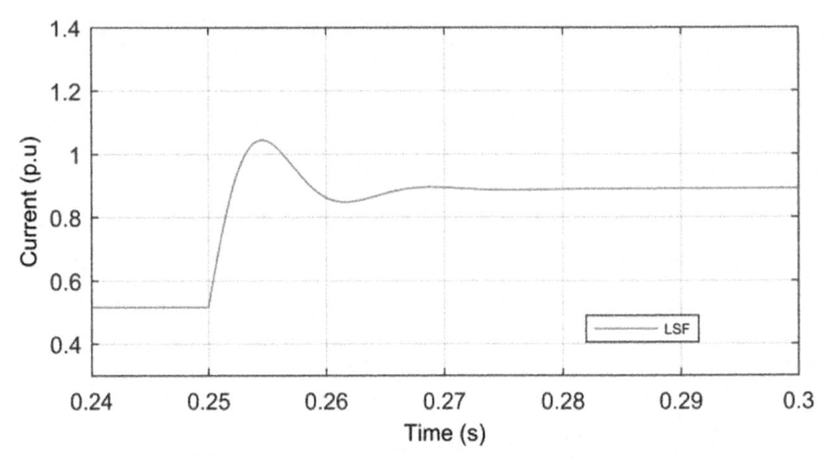

Figure 5.4 Current LSF cascaded – Ideal CPL – Step 10.3 → 17.8 MW.

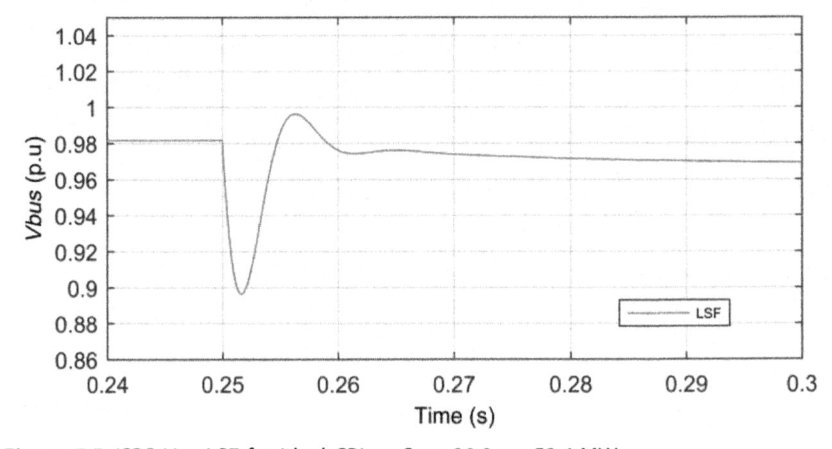

Figure 5.5 ISPS V_{bus} LSF for ideal CPL – Step 30.9 → 53.4 MW.

been performed at $t = 0.25$ s. The coefficients of the droop controllers were chosen to obtain an even current sharing among the converters. All converters equally share the total load power between them.

The effect of the increasing and decreasing of the CPL load on the load voltage is described in Fig. 5.5, demonstrating that the voltage remains stable and reaches the voltage set-points that are gradually modified by the action of the droop control. In Fig. 5.6 one can observe the transient for the current.

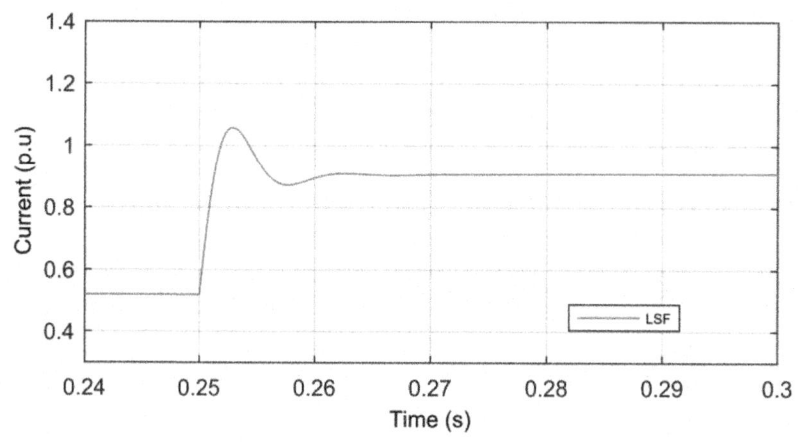

Figure 5.6 ISPS $I_{L1=L2=L3}$ LSF for ideal CPL – Step 30.9 → 53.4 MW Synergetic Control.

5.2 SYNERGETIC CONTROL

In this section, a centralized nonlinear control method is considered and later applied to MVDC system. It was developed by Kolesnikov [12] and called synergetic control. This control technique has found successful application in the stabilization of CPLs in [13–16]. In this control the nonlinear part of the controlled system is not compensated as the LSF was in Section 5.1.1, but a nonlinear control is designed which uses a model of the system and forces the system onto a manifold. Once the system is on the manifold the order is reduced [17]. For switching converters, the control variable is the state δ of the active switch, which has two states, on (or 1) and off (or 0); they are therefore variable-structure systems. The synergetic control applies to the converter the duty cycle d which is the averaging of the two states over time. It uses constant switching frequency; which makes it appealing to implement in power converters. The Synergetic control uses a model of the system to be controlled in order to synthesize the control law. This approach follows the internal model principle which says that an accurate control can only be achieved if the control system contains some representation of the process to be controlled [18]. In this chapter, the procedure of synergetic control design with the resulting synergetic control law is presented in Section 5.2.1. The application of the technique to a multiconverter test system based on the MVDC shipboard power system is presented in Section 5.2.2 for highlighting the actual implementation steps.

5.2.1 Procedure of Synergetic Control

Synergetic control was developed by Kolesnikov and his coworkers [12]; it is a state-space approach based on dynamic control of the interaction of energy and information within the system [15]. The authors of [19] often refer to this approach as the stabilization and control via system restriction and manifold invariance. The major idea behind this concept is to restrict the motion or trajectories of the system to a manifold or hyperplane. This includes forcing the system onto the manifold when it is not on the manifold already. The control problem addressed in this work is voltage stability (i.e., damping any voltage oscillation and transient behavior that may arise due to disturbances in the system). The synergetic controller forces the system to take the characteristics of the manifold. Therefore, the manifold has to be constructed in such a way that the closed-loop system behavior is stable.

The design approach consists of two major steps: the construction of a stable manifold or hyperplane and the synthesis or design of a controller such that the trajectories of the system are forced onto, and subsequently remain on, the manifold. The objective of the controller is responsible that the manifold is invariant and attractive.

The technique of synergetic control is similar to that of SMC in the way the manifold is being constructed [20,21]. The difference lies in the fact of how the system is forced to reach the manifold. In the case of synergetic control, the system is caused to reach the manifold in an exponential manner and therefore not showing a chattering effect which appears in SMC due to the fact that the system is forced to reach the manifold in a finite period of time which introduces some form of discontinuities in the control action and creates chattering on the manifold [22].

The synergetic control design procedure follows the Analytical Design of Aggregated Regulators (ADAR) method. The steps involved in the control design can summarized according to [17]. Assume the system to be controlled can be described by a set of nonlinear equation of the form:

$$\dot{x} = f(x, u) \tag{5.36}$$

where x and u are the system state variable vector and the input vector respectively.

During operation the system is influenced by a dynamic disturbance $M_i(t)$:

$$\frac{dM_i(t)}{dt} = f_{mi}(x) \tag{5.37}$$

The first step of the procedure is defining a macrovariable (ψ) as a function of the state variables:

$$\psi = \psi(x) \tag{5.38}$$

The characteristic of the macrovariable can be selected in accordance to the control specifications which can be either regulating or stabilizing the output of the system. The definition of the macrovariable defines how the system behaves once it reaches the manifold. For each input channel a macrovariable is needed.

It has been shown in [19], that (5.39) is a solution of the optimum problem whose objective function J is given in (5.40).

$$T\dot{\psi} + \psi = 0 \tag{5.39}$$

$$J = \int_{t_0}^{t_1} L(t, \psi, \dot{\psi}) \tag{5.40}$$

where $\{t_0, t_1\}$ are the initial and final time, and L is defined as:

$$L(t, \psi, \psi) = \psi K^T K \psi + \psi^T \psi \tag{5.41}$$

T is a design parameter that specifies the time it takes the macrovariable to converge to zero or evolve into the manifold, which can also be viewed as the speed at which the system variable reaches the manifold. Substituting (5.36) and (5.38) into (5.39) leads to:

$$T\frac{\partial\psi}{\partial x}\dot{\psi} + \psi = T\frac{\partial\psi}{\partial x}f(x, u) + \psi = 0 \tag{5.42}$$

By defining an appropriate macrovariable and choosing the parameter T, the control output can be derived from (5.42).

There is no unique way of constructing the manifold. The manner in which the manifold is constructed depends on the type of problem (regulation or tracking) and determines the quality of the performance of the controller [23].

Three conditions have to be fulfilled by the manifold [15]:

- Reachability of the manifold;
- A control that keeps the system on the manifold;
- Stability of the system has to be guaranteed while being on the manifold.

To guarantee stability a locus of steady-state operating points has to be identified. In the case of power converters this characterizes a curve $g(x) = 0$ in the state plane which is characterized by the duty cycle d. Afterwards a function F has to be identified whose directional derivative along its trajectories corresponds to (5.43), where $b = const.$

$$\frac{dF}{dt} = \frac{\partial F}{\partial x}\dot{x} = bg(x) \tag{5.43}$$

After the synergetic control design is completed an analytical control law is the result which ensures stable motion to the equilibrium point of the closed-loop system. The trajectory depends on the structure of the manifold. Subsequently the stability conditions on the manifold have to be analyzed.

The fact that synergetic control uses a model of the system for control synthesis can be considered both an advantage and a disadvantage. It appears desirable that the control uses all available information on the system for control purposes, but on the other hand it makes the control more sensitive to model and parameter errors [17].

5.2.2 Application to MVDC System

The Synergetic control is a centralized nonlinear control architecture which includes a model of the controlled system. In contrast to linear control, synergetic control is capable of handling system nonlinear dynamics and ensuring global system stability [15], while systems consisting of parallel-connected power converters under linear control may not show proper behavior and may lead to system collapse [24].

In this chapter the synergetic control is applied to the MVDC System. The overall procedure can be summarized as following (Fig. 5.7):

1. Derivation of state-space model and state variable definition;
2. Definition of macrovariable as function of state variables;
3. Calculation of control law.

Fig. 5.8 shows a system with parallel-connected converters. This system is under several assumptions: The system operates in continuous conduction mode (CCM). The switching occurs at a high-frequency (*1* kHz) relative to the filter dynamics. The parasitic effects are ignored. The state variables are the voltage over the capacitor v and the inductor currents i_{Li}

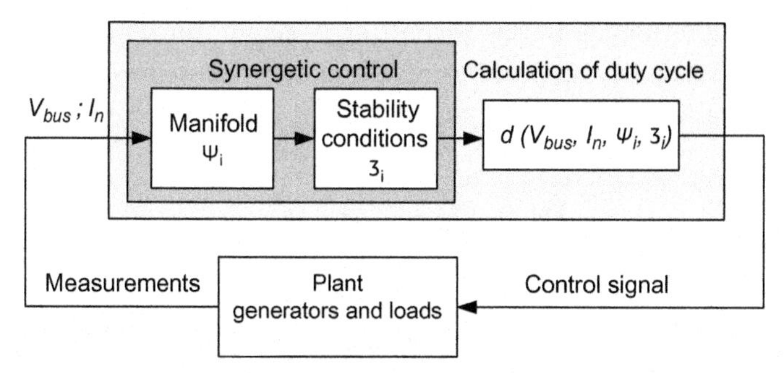

Figure 5.7 Overview of synergetic control scheme.

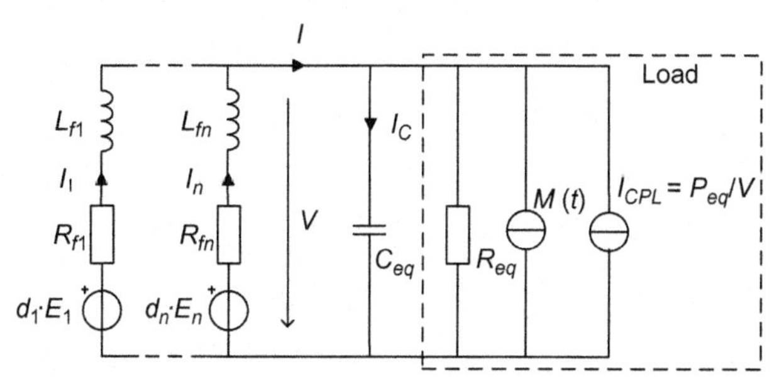

Figure 5.8 Control synthesis model of centralized system for synergetic control containing disturbance *M(t)*.

inductance currents. The CPLs are lumped again into one equivalent load and are represented as a dependent current source [13]. The number of controlled buck converters corresponds to $m = 3$.

When designing this system-wide control, the following objective has to be fulfilled: The MVDC bus voltage has to be kept to $V_{c,ref}$ while achieving the desired current.

The state-space averaged model of the system, which includes $m = 3$ generation side buck converters, is given in (5.44) according to [15], where V, I_{Li} are state variables:

$$\begin{cases} \dot{V} = \dfrac{1}{C_{eq}}\left(\displaystyle\sum_{i=1}^{3} V_{Li}\right) - \dfrac{V + V_{ref}}{R_{eq}C_{eq}} - \dfrac{P_{eq}}{\left(V + V_{ref}\right)C_{eq}} + \dfrac{\delta M}{C_{eq}} \\ \dot{I}_{Li} = -\dfrac{1}{L_i}\left(V + V_{ref}\right) + u_i \\ \dot{M} = \eta V \\ u_i = d_i \dfrac{E_i}{L_i} \\ V = V_c - V_{c,ref} \end{cases} \tag{5.44}$$

Special attention should be given to the newly introduced term $M(t)$, which is a dependent currents source and represents the changes of the system load [25]. This is approximated by a piecewise constant function which acts as an integral action:

$$\frac{dM}{dt} = \frac{\left(V - V_{ref}\right)}{T_I R_d} = \frac{V}{T_I R_d} \tag{5.45}$$

To design the synergetic controller, it is necessary to define macrovariables which are based on system states. Recall that the number of macrovariables is equal to number of control channels [25]. The result of the synergetic control design is a control law which ensures stable movement to and along the manifolds towards the equilibrium point [15].

$$\psi_i(I_{L1}, I_{L2}, \ldots, V) = 0; \quad i = 1 \ldots 3 \tag{5.46}$$

The following macrovariables were chosen to obtain the desired dynamic properties.

$$T_i.\dot{\psi}_i = -\psi_i; T_i > 0 \tag{5.47}$$

The convergence speed is defined by the vector T_i whose size corresponds to the number of participating converters. There are three control channels, therefore three macrovariables $\psi_1 \ldots \psi_3$.

$$\begin{aligned}
\psi_1 &= a_{11}M + a_{12}V + a_{13}I_{L1} + a_{14}I_{L2} + a_{15}I_{L3} + \frac{a_{16}}{\left(V + V_{ref}\right)C_{eq}} \\
\psi_2 &= a_{21}M + a_{22}V + a_{23}I_{L1} + a_{24}I_{L2} + a_{25}I_{L3} + \frac{a_{26}}{\left(V + V_{ref}\right)C_{eq}} \\
\psi_3 &= a_{31}M + a_{32}V + a_{33}I_{L1} + a_{34}I_{L2} + a_{35}I_{L3} + \frac{a_{36}}{\left(V + V_{ref}\right)C_{eq}}
\end{aligned} \tag{5.48}$$

An interesting finding according to [25] of the synergetic control law in parallel power converter system is that coefficients a_{x3}, a_{x4}, and a_{x5} allocate the current sharing ratings. The coefficients a_{x6} provide the ability to compensate the impact of the CPL. A detailed explanation of the properties of the coefficients can be found in [15] and [25]. Defining ψ as a vector enables us to generalize the system:

$$\psi = AX - B \tag{5.49}$$

with $\Psi = (\psi_1 \ldots \psi_3)^T$ and $X = (I_{L1} \ldots I_{L3})^T$

$$A = \begin{pmatrix} a_{13} & a_{14} & a_{15} \\ a_{23} & a_{24} & a_{25} \\ a_{33} & a_{34} & a_{35} \end{pmatrix} \qquad B = \begin{pmatrix} b_1 \\ \vdots \\ b_m \end{pmatrix}$$
$$b_i = -a_{i1}M - a_{i2}V - \frac{a_{i,n+1}}{\left(V + V_{ref}\right)} \qquad i = 1, m \tag{5.50}$$

The control law can be calculated with $U = A^{-1}G$ and results in:

$$\begin{aligned} u_1 &= A^{-1}(1,1)g_1 + A^{-1}(1,2)g_2 + A^{-1}(1,3)g_3 \\ u_2 &= A^{-1}(2,1)g_1 + A^{-1}(2,2)g_2 + A^{-1}(2,3)g_3 \\ u_3 &= A^{-1}(3,1)g_1 + A^{-1}(3,2)g_2 + A^{-1}(3,3)g_3 \end{aligned} \tag{5.51}$$

Where G is a vector $G = (g_1 g_2 \ldots g_m)^T$, and each element corresponds to:

$$\begin{aligned} g_i = &-\frac{\psi_i}{T_i} + \left(\sum_{j=1}^{3} \frac{a_{ij+2}}{L_j} - \frac{a_{i2}}{R_{eq}C_{eq}} + \frac{a_{in+1}}{\left(V + V_{ref}\right)^2 R_{eq}.C_{eq}}\right)\left(V + V_{ref}\right) \\ &-\eta a_{i1}V - \left(a_{i2} - \frac{a_{in+1}}{\left(V + V_{ref}\right)^2}\right)\left(\sum_{j=1}^{3} \frac{i_j}{C_{eq}} + \delta M - \frac{P_{eq}}{\left(V + V_{ref}\right)C_{eq}}\right) \end{aligned} \tag{5.52}$$

Substituting (5.52) in (5.51)

$$u_i = \sum_{j=1}^{m} A^{-1}(i,j)d_j i = 1, m \tag{5.53}$$

As it is mentioned in [12], control laws (5.53) ensure stability towards the manifolds (5.46). After reaching the manifold, the coefficients in (5.48) define the current sharing. In other words, after passing the transient time and reaching steady state, by choosing the proper coefficients in (5.48), the current sharing is provided [25]. As the system was forced by the control laws (5.53) to hit the manifold, the stability on the manifold has to be guaranteed as part of the synergetic control design. The stability conditions (F_1, F_2, F_3) can be analyzed by (5.54), including (5.55)–(5.57).

$$\ddot{V}_{\psi} + F_1 V_{\psi} + F_2 V_{\psi} + \frac{F_3}{\left(V_{\psi} + V_{ref}\right)} = 0 \tag{5.54}$$

$$F_1 = \frac{1}{R_{eq} C_{eq}} + \sum_{j=1}^{3} \sum_{i=1}^{3} A^{-1}(j, i) a_{i2} \tag{5.55}$$

$$F_2 = \eta \left(\sum_{j=1}^{3} \sum_{i=1}^{3} A^{-1}(j, i) a_{i1} - \delta \right) \tag{5.56}$$

$$F_3 = \left(\frac{1}{C_{eq}} + \sum_{j=1}^{3} \sum_{i=1}^{m3} A^{-1}(j, i) a_{in+1} \right) P_{eq} \tag{5.57}$$

Eq. (5.54) can also be rewritten to investigate the system's internal behavior.

$$\dot{V}_{\psi} = -F_1 V_{\psi} - F_2 \int V_{\psi} dt = 0 \tag{5.58}$$

This corresponds to a PI control law inside the manifold, which has as a consequence that asymptotic stability can be achieved and a constant disturbance can be successfully rejected.

The system of (5.54) can also be expressed by its characteristic equation and therefore the stability conditions of (5.60) apply.

$$s^2 + F_1 s + F_2 s + \frac{F_3}{\left(s + V_{c,ref}\right)} = 0 \tag{5.59}$$

$$\det(A) \neq 0, F_3 = 0, F_1 > 0, F_2 > 0, T_i > 0, i = 1, m \tag{5.60}$$

Without restricting the generality $F_3 = 0$ can be set in order to determine the stability condition easily. By selecting a specific choice of coefficients of (5.61), system asymptotic stability can be ensured and the nonlinearity is compensated [25].

$$F_3 = 0 \rightarrow \left(\frac{1}{C_{eq}} + A^{-1}(1,1)a_{16} + A^{-1}(1,2)a_{16} + A^{-1}(1,3)a_{16} \right.$$
$$+ A^{-1}(2,1)a_{26} + A^{-1}(2,2)a_{26} + A^{-1}(2,3)a_{26} + A^{-1}(3,1)a_{36}$$
$$\left. + A^{-1}(3,2)a_{36} + A^{-1}(3,3)a_{36} \right) = 0$$

$$F_1 > 0 \rightarrow \frac{1}{R_{eq}C_{eq}} + A^{-1}(1,1)a_{12} + A^{-1}(1,2)a_{12} + A^{-1}(1,3)a_{12}$$
$$+ A^{-1}(2,1)a_{22} + A^{-1}(2,2)a_{22} + A^{-1}(2,3)a_{22}$$
$$+ A^{-1}(3,1)a_{32} + A^{-1}(3,2)a_{32} + A^{-1}(3,3)a_{32} > 0$$
$$F_2 > 0 \rightarrow A^{-1}(1,1)a_{11} + A^{-1}(1,2)a_{11} + A^{-1}(1,3)a_{11} + A^{-1}(2,1)a_{21}$$
$$+ A^{-1}(2,2)a_{21} + A^{-1}(2,3)a_{21} + A^{-1}(3,1)a_{31} + A^{-1}(3,2)a_{31}$$
$$+ A^{-1}(3,3)a_{31} - \delta > 0 \tag{5.61}$$

After fulfilling the stability conditions and the desired current sharing methodology of the synergetic control law [25], the resulting duty cycles (5.62)–(5.64) are:

$$d_1 = -\frac{\psi_1}{T_1} + \left(\frac{a_{13}}{L_1} + \frac{a_{14}}{L_2} + \frac{a_{15}}{L_3} - \frac{a_{12}}{R_{eq}C_{eq}} + \frac{a_{16}}{\left(V + V_{c,ref}\right)^2 R_{eq}C_{eq}}\right)\left(V + V_{c,ref}\right)$$
$$-\eta a_{11}V - \left(a_{12} - \frac{a_{16}}{\left(V + V_{c,ref}\right)^2}\right)\left(\frac{I_1 + I_2 + I_3}{C_{eq}} + \delta M - \frac{P_{eq}}{\left(V + V_{c,ref}\right)C_{eq}}\right) \tag{5.62}$$

$$d_2 = -\frac{\psi_2}{T_2} + \left(\frac{a_{23}}{L_1} + \frac{a_{24}}{L_2} + \frac{a_{25}}{L_3} - \frac{a_{22}}{R_{eq}C_{eq}} + \frac{a_{26}}{\left(v + V_{c,ref}\right)^2 R_{eq}C_{eq}}\right)\left(V + V_{c,ref}\right)$$
$$-\eta a_{21}V - \left(a_{22} - \frac{a_{26}}{\left(V + V_{c,ref}\right)^2}\right)\left(\frac{I_1 + I_2 + I_3}{C_{eq}} + \delta M - \frac{P_{eq}}{\left(v + V_{c,ref}\right)C_{eq}}\right) \tag{5.63}$$

$$d_3 = -\frac{\psi_2}{T_2} + \left(\frac{a_{33}}{L_1} + \frac{a_{34}}{L_2} + \frac{a_{35}}{L_3} - \frac{a_{32}}{R_{eq}C_{eq}} + \frac{a_{36}}{\left(V + V_{c,ref}\right)^2 R_{eq}C_{eq}}\right)\left(V + V_{c,ref}\right)$$
$$-\eta a_{31}V - \left(a_{32} - \frac{a_{36}}{\left(V + V_{c,ref}\right)^2}\right)\left(\frac{I_1 + I_2 + I_3}{C_{eq}} + \delta M - \frac{P_{eq}}{\left(V + V_{c,ref}\right)C_{eq}}\right) \tag{5.64}$$

5.2.3 Simulation Results

5.2.3.1 Cascaded System

A cascaded converter setup has been modeled in Matlab-SIMULINK with the parameters described in Table 6.1. In the simulation, a step variation of the CPL load has been performed. The variation of the bus voltage during the step change of the CPL load is described in Fig. 5.9. At $t = 0.25$ s the load step of 7.5 MW was performed, demonstrating that the Synergetic control stabilizes the system towards the reference point $V_{ref} = 1$. The dynamic of the inductor current is described in Fig. 5.10.

5.2.3.2 Shipboard Power System

A Shipboard Power System has been modeled in Matlab\SIMULINK, based on the parameters in Table 6.2. The load steps of 22.5 MW have been performed at $t = 0.25$ s. The coefficients of the droop controllers were chosen to obtain an even current sharing among the converters. All converters equally share the total load power between them.

The effect of the increasing and decreasing of the CPL load on the load voltage is described in Fig. 5.11, demonstrating that the voltage remains stable and reaches the voltage set-points that are gradually modified by the action of the droop control. In Fig. 5.12 one can observe the transient for the current.

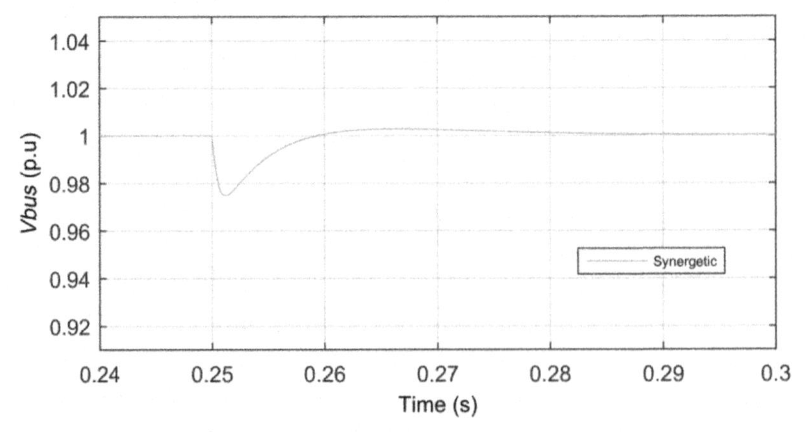

Figure 5.9 Voltage synergetic cascaded – Ideal CPL – Step 10.3 → 17.8 MW.

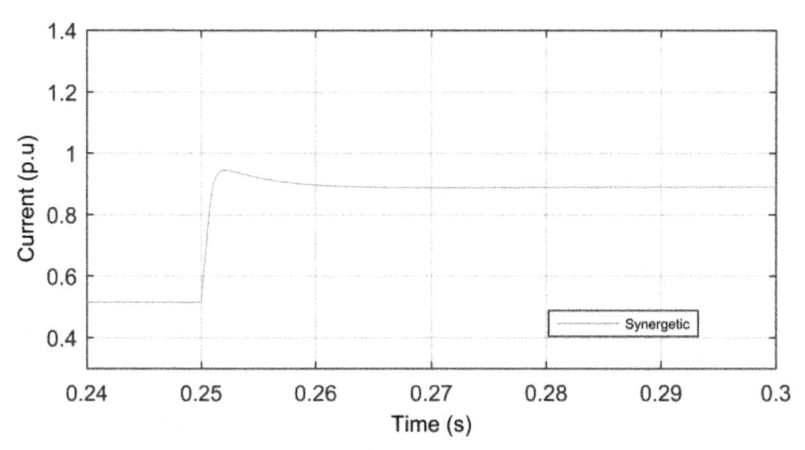

Figure 5.10 Current synergetic cascaded – Ideal CPL – Step 10.3 → 17.8 MW.

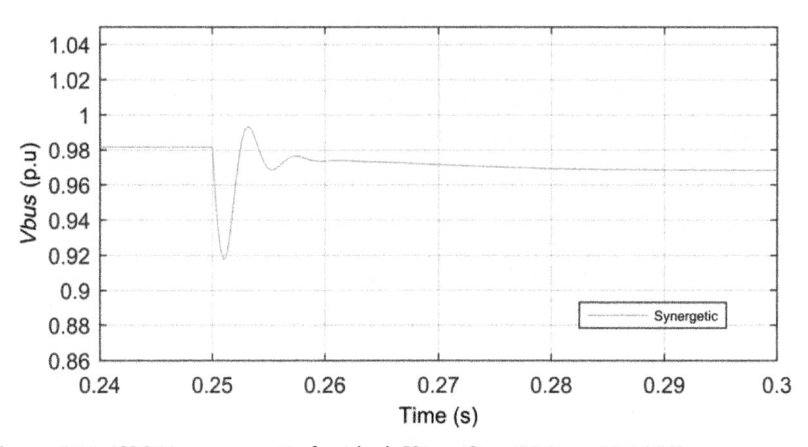

Figure 5.11 ISPS V_{bus} synergetic for ideal CPL – Step 30.9 → 53.4 MW.

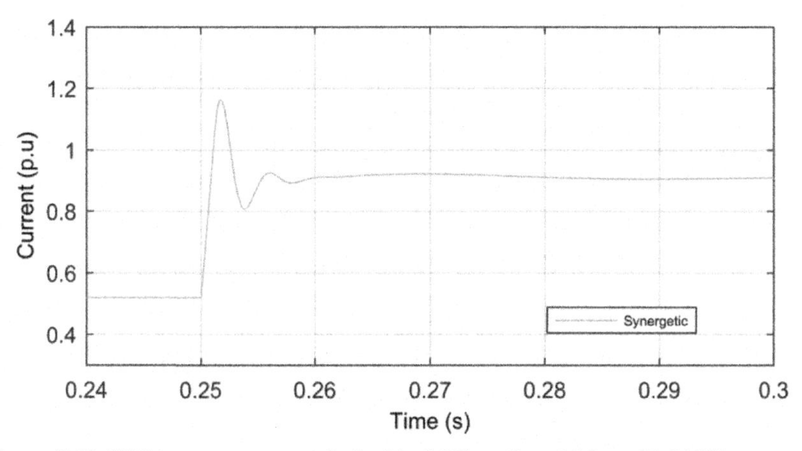

Figure 5.12 ISPS $I_{L1=L2=L3}$ synergetic for ideal CPL – Step 30.9 → 53.4 MW.

5.3 IMMERSION AND INVARIANCE CONTROL

The chapter introduces the Immersion and Invariance (I&I) control method for the design of nonlinear controllers for nonlinear systems. The method has been introduced by A. Astolfi and R. Ortega with the purpose of combining the notions of system immersion and manifold invariance [26] into a new method for the asymptotic stabilization of nonlinear systems [26].

The I&I control realizes the immersion of the systems trajectories into a low-order target-system with a desired dynamic. In its nonadaptive formulation, the control theory does not require the definition of a Lyapunov function, unlike many other methodologies, and, for its geometrical interpretation as a manifold-based control, can be considered a more general theory that includes the sliding mode [27] and synergetic controllers [17].

The control method is based on the theorem, introduced in [28], that contains the definition of the theory and describes the steps for defining the stabilizing control output.

The adaptive formulation of the I&I, which is not part of this chapter, has been described in [28] as a novel approach for the adaptive stabilization of nonlinear system.

The chapter introduces the general theory of the nonlinear stabilizing controller, describing the theorem and giving some interpretation to it. Therefore, the application of the theory to a simple two-dimensional system and to a buck converter in the presence of nonlinear CPL load are performed.

5.3.1 The Immersion and Invariance Stabilization

The section defines the theory of the Immersion and Invariance control for the stabilization of nonlinear systems towards an equilibrium point. In the basic formulation, the control uses the state feedback for generating the control output of a system in the following form:

$$\dot{x} = f(x) + g(x)u \tag{5.65}$$

where $x \in R^n$, $u \in R^m$ and $f(x), g(x)$ are two nonlinear functions. The role of the controller is to define a control output $u = c(x)$ that makes the closed-loop system asymptotically stable.

The calculation of the control law starts with the definition of the target system to map into the nonlinear system to be controlled. The target system has typically a low-order dimension, and it is described by the following expression:

$$\dot{\xi} = \alpha(\xi) \tag{5.66}$$

$$\xi^* = 0 \tag{5.67}$$

As described in (5.67), the target system has an equilibrium point at the origin that must be asymptotically stable.

The target system is usually defined a priori by the designer, by specifying the desired dynamic. This allows the target system to comply with the theorem in [28], because it can be defined such as (5.67) is satisfied. In the I&I approach, the target system defines also the dynamic of the closed-loop system, since the system controlled by the feedback loop contains the copy of the dynamic of $\dot{\xi} = \alpha(\xi)$.

Supposing that there exists a mapping $\pi{:}R^p \to R^n$ such as $x = \pi(\xi)$, with $\xi \in R^p$, the following equations must hold:

$$\pi(\xi^*) = x^* \tag{5.68}$$

$$\pi(0) = 0 \tag{5.69}$$

The application of the mapping $x = \pi(\xi)$ and of the control law to the nonlinear system (5.65) results in the partial differential Eq. (5.70).

$$f(\pi(\xi)) + g(\pi(\xi))c(\pi(\xi)) = \frac{\partial \pi}{\partial \xi}\alpha(\xi) \tag{5.70}$$

Once the conditions (5.68)–(5.70) are satisfied, the manifold M, which is the geometric interpretation of the mapping procedure, can be defined implicitly by:

$$\{x \in R^n | \Phi(x) = 0 = \{x \in R^n | x = \pi(\xi) for\ \ some\ \ \xi \in R^p \tag{5.71}$$

The Eq. (5.71) shows that the aim of the controller is to drive to zero all the off-the-manifold trajectories the manifold, described by $z = \Phi(x)$. The implicit definition of the manifold demonstrates that the only trajectories permitted are the ones that satisfy Eq. (5.71), which are strictly connected to the target system.

Moreover, the equation demonstrates an equivalence between the map $\pi{:}R^p \to R^n$ and the manifold, defining a correlation between the mathematical and geometrical interpretation of the stabilization procedure.

The solution of the Eq. (5.72) allows the calculation of the final control law. The equation must guarantee that the trajectories of z converge to $z = 0$.

$$\dot{z} = f(z) \tag{5.72}$$

If the application of the control law is able to drive to zero the off-the-manifold trajectories and it guarantees that the trajectories of the closed loop are bounded, then x^* is the asymptotically stable equilibrium point of the closed-loop system. In calculation of the control law, the Lyapunov function is not used, but the authors in [28] have demonstrated that the mapping procedure represents the dual of the I&I approach.

The correlation between the target system and the nonlinear system is described in Fig. 5.13. The trajectory $\xi(t)$ on the one-dimension target system is mapped on the two-dimension space. The mapping procedure results in a trajectory that converges to the manifold that contains the equilibrium point.

5.3.2 Example

The I&I control previously described is applied to a two-dimension nonlinear system as an example of the application of the control theory. The calculation of the control output follows the steps previously defined and it is in line with other examples presented in [28].

The two dimensions system is characterized by two nonlinear state equations with one control input, defined as:

$$\begin{cases} \dot{x}_1 = x_1 x_2^3 \\ \dot{x}_2 = x_2^2 + x_1 + u \end{cases} \tag{5.73}$$

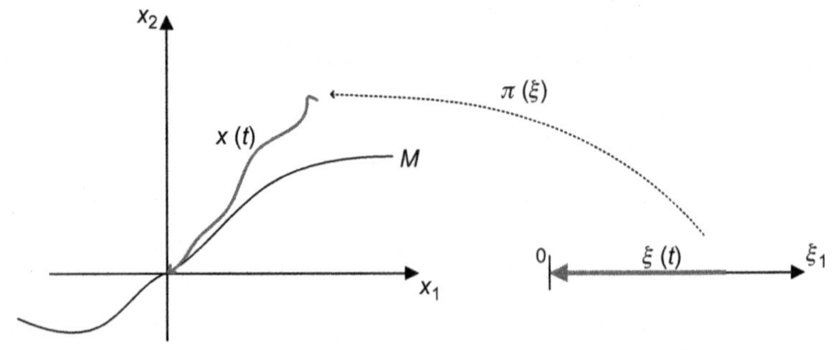

Figure 5.13 Graphical illustration of the mapping between the target system and the system to be controlled.

The aim of the controller is to bring the system to the set-points values $[x_1^*, 0]$. Therefore, the system variables are modified in order to comply with the Eq. (5.68) as follows:

$$\begin{cases} v_1 = x_1 - x_1^* \\ v_2 = x_2 \end{cases} \tag{5.74}$$

The nonlinear target system is given by:

$$\dot{\xi} = -K_1 \xi^3 \tag{5.75}$$

which has an asymptotically stable equilibrium point in $\xi^* = 0$.

The mapping function is π: $R \to R^2$, meaning that $(v_1, v_2) = (\pi_1(\xi), \pi_2(\xi))$. By setting $\pi_1(\xi) = \xi = v_1$, the equilibrium point of $\pi_1(\xi^*) = \xi^* = 0$, which is equal to the equilibrium point of $v_1^* = 0$ and satisfies the hypothesis (5.68).

The application of the map π to the system (5.73) in the new variables (v_1, v_2) results in a nonlinear system with the partial differential Eq. (5.70):

$$\begin{cases} -K_1 \dot{\xi}_3 = \xi \pi_2^3 \\ -\dfrac{\partial \pi_2}{\partial \xi} K_1 \xi = \pi_2^2 + \xi + c(\pi(\xi)) \end{cases} \tag{5.76}$$

where $u = c(\pi(\xi))$ is the control output that drives to zero the off-the-manifold trajectories and $\pi_1(\xi)$ has been already substituted with ξ. The solution of the first equation of (5.76) results in $\pi_2 = -K_1 \xi^{2/3}$.

The implicit definition of the manifold, which forces the trajectories of the off-the-manifold coordinate $z = \Phi(x)$ to zero, can be described as:

$$\Phi(x) = v_2 - \pi_2(\xi_1)_{|\xi_1 = v_1} = v_2 + K_1 {v_1}^{2/3} = 0 \tag{5.77}$$

The dynamic of z must have bounded trajectories to satisfy the theorem in [28], therefore the time derivative of (5.77) is defined by the following equation:

$$\dot{z} = -K_2 z \tag{5.78}$$

The result of (5.78) defines the control output that drives to zero the coordinate z and renders the manifold invariant is given by:

$$u = c(\pi) = -v_2^2 - v_1 - \frac{2}{3} K_1 v_1^{2/3} v_2^3 - K_2 \left(v_2 + K_1 v_1^{2/3} \right) \tag{5.79}$$

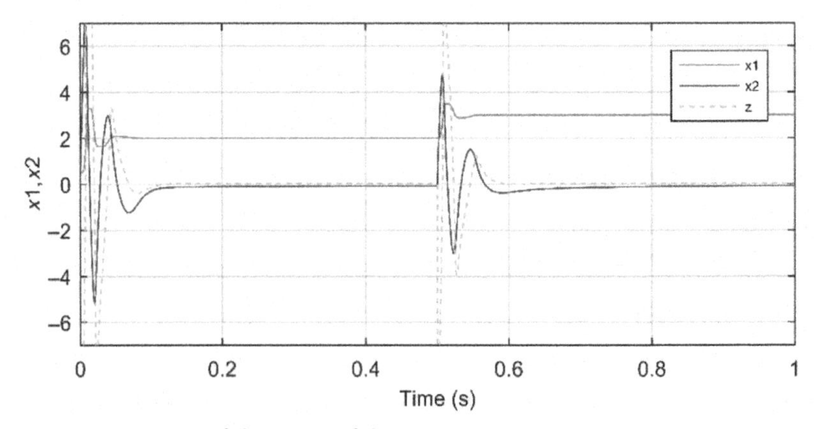

Figure 5.14 Dynamic of the states of the system.

Finally, the variables ν_1 and ν_2 are substituted with original variables of the system.

The system has been simulated with Matlab/SIMULINK to demonstrate the validity of the control law calculated. The parameters of the controller are $K_1 = 10$ and $K_2 = 100$ and $x_1(0) = 0.5$ and $x_2(0) = 0$. At time t = 0.5 s, the reference is varied from $x^*_{1,1} = 2$ to $x^*_{1,2} = 3$, meaning that the manifold varies consequently.

Fig. 5.14 shows the dynamic evolution of the states, where it is clear that x_1 and x_2 are controlled to follow their references. Moreover, the evolution of the off-the-manifold variable z is described, showing that the controller is able to drive to zero the trajectory.

By considering (5.71), the Eq. (5.77) for the different set-points results in:

$$\begin{aligned} \Phi_1 &= \left\{ \left(\nu_1^1, \nu_2^1\right) = \left(\pi_1(\xi), \pi_2(\xi)\right) \right\} \\ x_2 &= -K_1\left(x_1 - x^*_{1,1}\right)^{2/3} \end{aligned} \tag{5.80}$$

$$\begin{aligned} \Phi_2 &= \left\{ \left(\nu_1^2, \nu_2^2\right) = \left(\pi_1(\xi), \pi_2(\xi)\right) \right\} \\ x_2 &= -K_1\left(x_1 - x^*_{1,2}\right)^{2/3} \end{aligned} \tag{5.81}$$

The trajectories of the states and of the implicit definition of the manifolds are therefore shown in Fig. 5.15, which highlights that the control is able to drive the trajectories to the reference set-points.

5.3.3 Application to MVDC System

In this section, the I&I control is applied to the buck converter connected to a resistive load and a CPL, which represents the nonlinear element of

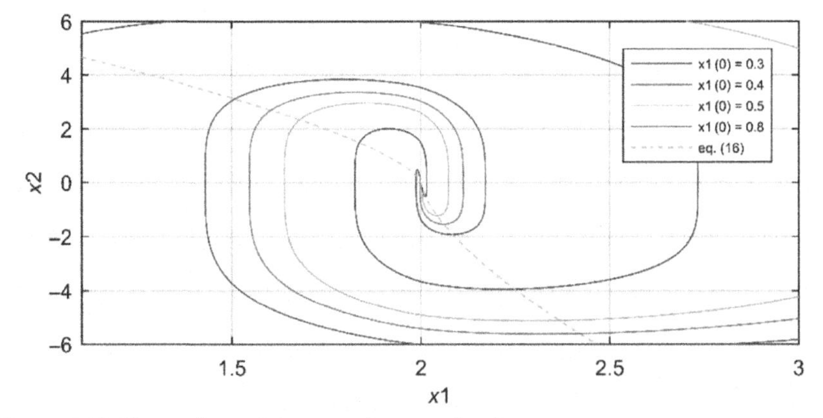

Figure 5.15 Phase plane of the generic second-order system.

the system [29]. The Eq. (5.82) represents the buck converter system are defined in section. The system is rewritten with a new set of variables $x_1 = V_{ref} - V$ and $x_2 = I_L$, adding a new variable x_3, that does not represent any physical element but integrates the voltage error. The system in the new set of variables is described by the following set of equations:

$$\begin{cases} \dot{x}_1 = \dfrac{x_2}{C_f} - \dfrac{x_1}{R_L C_f} - \dfrac{P_{eq}}{C_f x_1} + x_3 \\ \dot{x}_2 = -\dfrac{x_1}{L_f} + \dfrac{d \cdot E}{L_f} \\ \dot{x}_3 = -k_g x_1 \end{cases} \tag{5.82}$$

The nonlinear target system is a two-dimension system defined by the equations describing its dynamic:

$$\begin{cases} \dot{\xi}_1 = -\xi_1^5 \\ \dot{\xi}_2 = -\xi_2 \end{cases} \tag{5.83}$$

The mapping $\pi : R^2 \to R^3$ is applied to the system (5.82) such that $\pi_1(\xi) = \xi_1 = x_1$ and $\pi_2(\xi) = \xi_2 = x_3$, resulting in the following nonlinear system:

$$\begin{cases} -\xi_1^5 = \dfrac{\pi_3(\xi)}{C_f} - \dfrac{\xi_1}{R_L C_f} - \dfrac{P_{eq}}{C_f \xi_1} + \xi_2 \\ -\dfrac{\partial \pi_3(\xi)}{\partial \xi} \xi_1^5 - \dfrac{\partial \pi_3(\xi)}{\partial \xi} \xi_2 = -\dfrac{\xi_1}{L_f} + \dfrac{c(\pi(\xi)) \cdot E}{L_f} \\ -\xi_2 = -k_g \xi_1 \end{cases} \tag{5.84}$$

The solution of the first equation of the system (5.84) results in:

$$\pi_3(\xi) = -C_f\xi_1^5 + \frac{\xi_1}{R_L} + \frac{P_{eq}}{\xi_1} - \xi_2 C_f \tag{5.85}$$

An implicit definition of the manifold can be obtained by defining:

$$\Phi(x) = x_2 - \pi_3(\xi)_{|\xi_1=x_1,\xi_2=x_3} = x_2 + C_f x_1^5 - \frac{x_1}{R_L} - \frac{P_{eq}}{x_1} + x_3 C_f = 0 \tag{5.86}$$

As shown before, the dynamic of the off-the-manifold trajectories is defined by:

$$\dot{z} = -K_2 z \tag{5.87}$$

The result of the Eq. (5.87) is the control output that stabilizes the system:

$$d = c(\pi) = \frac{L_f}{E}\left(\frac{x_1}{L_f} + \left(\frac{x_2}{C_f} - \frac{x_1}{R_L C_f} - \frac{P_{eq}}{C_f x_1} + x_3\right)\right.$$
$$\left.\left(-5x_1^4 C_f + \frac{1}{R_L} - \frac{P_{eq}}{x_1^2}\right) + k_g C_f x_1 - K_2\left(x_2 + C_f x_1^5 - \frac{x_1}{R_L} - \frac{P_{eq}}{x_1} + x_3 C_f\right)\right) \tag{5.88}$$

Since the term P_{eq}/x_1 represents the current absorbed by the CPL load, the measurement of the current I_{CPL} can be used for calculating the control law, resulting in:

$$d = \frac{L_f}{E}\left(\frac{x_1}{L_f} + \left(\frac{x_2}{C_f} - \frac{x_1}{R_L C_f} - \frac{I_{CPL}}{C_f} + x_3\right)\right.$$
$$\left.\left(-5x_1^4 C_f + \frac{1}{R_L} - \frac{I_{CPL}}{x_1}\right) + k_g C_f x_1 - K_2\left(x_2 + C_f x_1^5 - \frac{x_1}{R_L} - I_{CPL} + x_3 C_f\right)\right) \tag{5.89}$$

The overall structure of the I&I control applied to the buck converter is described in Fig. 5.16.

The application of the I&I control on a system characterized by parallel-connected buck converters consists of replicating the procedure

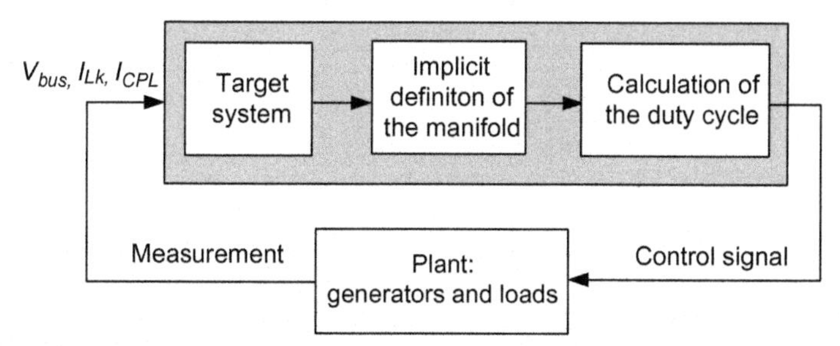

Figure 5.16 Structure of the proposed I&I control.

previously explained for all the generation-side converters. The load is characterized by a resistance and an ideal CPL and it is connected to the same bus where all the converters are connected [11].

The buck converter is controlled by means of the control law described in (5.89) where $x_1 = V^*_{ref} - V$ and $x_2 = I_{L,k}$, where $k = 1, \ldots, 3$ represents the measurement of the inductor current of the converter k.

The current sharing among the buck converters is controlled by the droop characteristic described by:

$$V^*_{ref} = V_{ref} - r_d \cdot I_{L,k} \tag{5.90}$$

The control of the sharing is obtained by changing the value of the fictitious resistance [30].

The current measurement of the CPL is shared among the four buck converters and multiplied by a factor that reduces the impact of the CPL current on the calculation of the control output, given that the values in p.u. of the CPL current are very high compared to the other variables.

The overall control structure of the parallel-connected system is described by Fig. 5.17.

Fig. 5.17 demonstrates that the proposed control approach for the parallel-connected converters exhibits a centralized configuration, given that the CPL current must be shared among the different controllers.

5.3.4 Simulation Results

5.3.4.1 Cascaded System

A cascaded converter setup has been modeled in Matlab-SIMULINK with the parameters described in Table 6.1. In the simulation, a step variation of the CPL load has been performed. The variation of the bus voltage during the step change of the CPL load is described in Fig. 5.18. At $t = 0.25$ s the load step of 7.5 MW was performed, demonstrating that

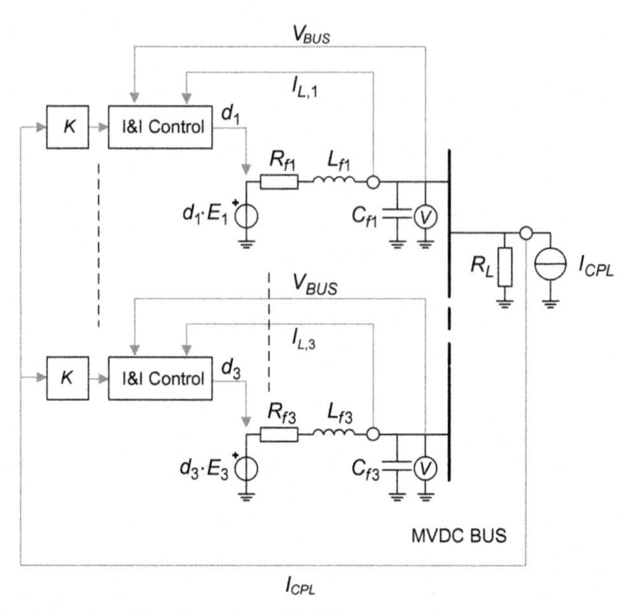

Figure 5.17 Centralized I&I Control for MVDC System.

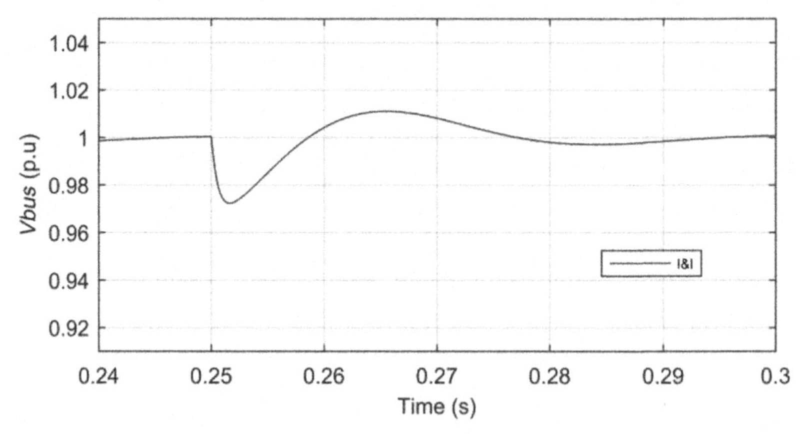

Figure 5.18 Voltage I&I cascaded – Ideal CPL – Step 10.3 → 17.8 MW.

the I&I control stabilizes the system towards the reference point $V_{ref} = 1$. The dynamic of the inductor current is described in Fig. 5.19.

5.3.4.2 Shipboard Power System

The Shipboard Power System has been modeled in Matlab\SIMULINK, based on the parameters in Table 6.2. The load steps of 22.5 MW have

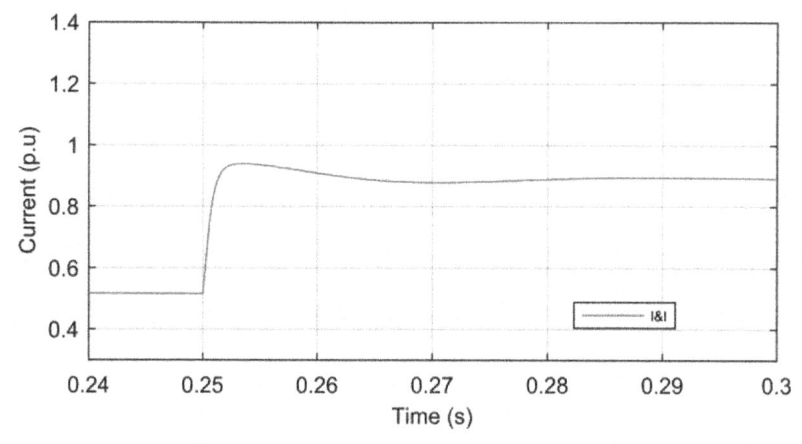

Figure 5.19 Current I&I cascaded – Ideal CPL – Step 10.3 → 17.8 MW.

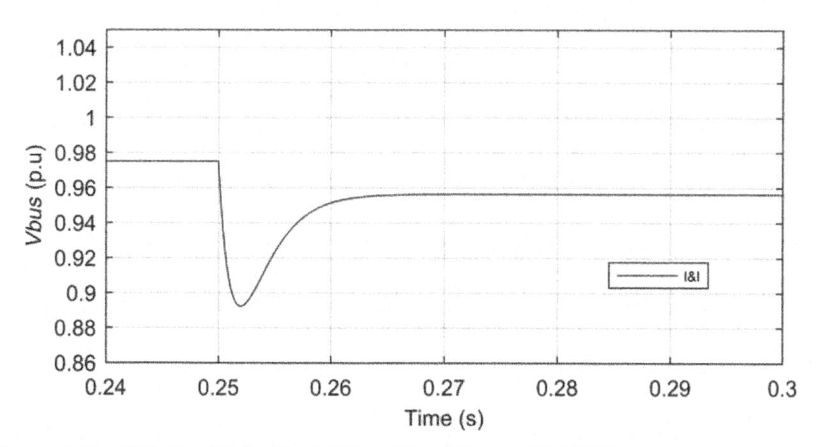

Figure 5.20 ISPS V_{bus} I&I for ideal CPL – Step 30.9 → 53.4 MW.

been performed at $t = 0.25$ s. The coefficients of the droop controllers were chosen to obtain an even current sharing among the converters. All converters equally share the total load power between them.

The effect of the increasing and decreasing of the CPL load on the load voltage is described in Fig. 5.20, demonstrating that the voltage remains stable and reaches the voltage set-points that are gradually modified by the action of the droop control. In Fig. 5.21 one can observe the transient for the current.

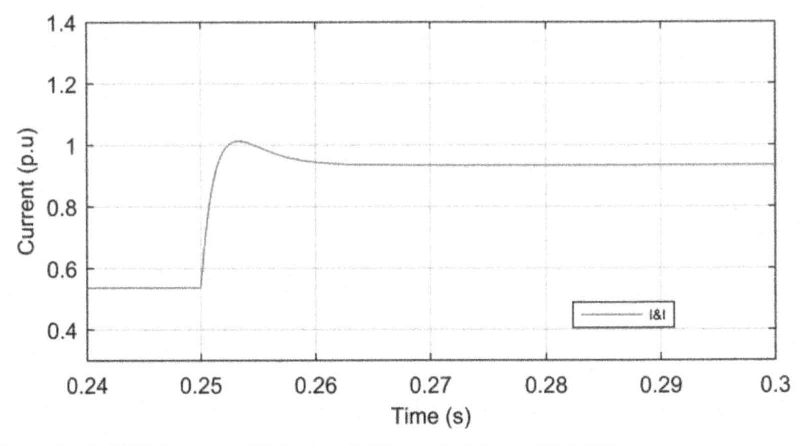

Figure 5.21 ISPS $I_{L1=L2=L3}$ I&I for ideal CPL and 30.9 → 53.4 MW.

5.4 DECENTRALIZED CONTROLS

The previous two sections presented nonlinear schemes, which are based on a centralized architecture. In this section, an approach is introduced that is based on a decentralized control architecture and in addition is also linear.

To enable a decentralized control approach the ISPS has to be decoupled. Logical decoupling points are every generation side interfacing converter as depicted in Fig. 5.22. In [31] and [32] the concept of the "virtual disturbance" for power system decoupling was introduced, according to which, under proper modeling and the usage of an augmented Kalman filter, it is possible to estimate simultaneously local states and interactions with other subsystems.

It consists of two parts. The first part is a linear controller which is based on the principle of the Linear Quadratic Regulator (LQR) [33], which successfully stabilized a CPL in [34]. The second part is based on an observer—Kalman filter [33]. A Kalman filter is similar to Leunberger observer, except that the basic framework is stochastic rather than deterministic. The concept was proposed in [35] and [36]. Both together form the Linear Quadratic Gaussian (LQG) control.

This control approach has found successful application on the stabilization of CPLs in [37,38]. The optimal control was chosen because it has the following properties which the other small-signal approaches do not have. On a small-signal basis the results obtained by linear optimal control

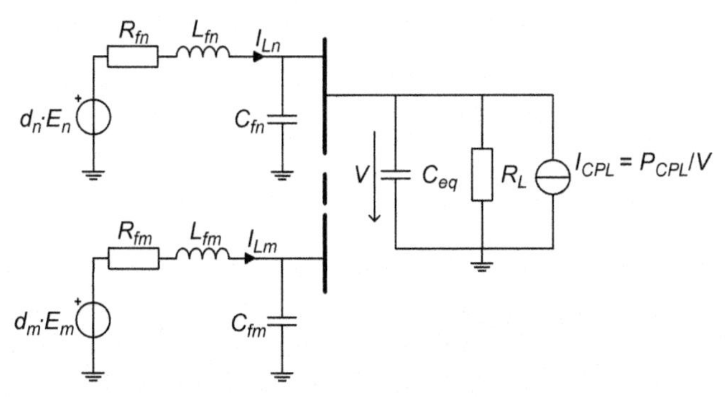

Figure 5.22 Decoupled Shipboard Power System.

results may be applied to nonlinear systems. An optimal control is designed for a nonlinear system with the assumption that the system will start in a defined initial state. For each state there exists an optimal control. If the system starts in a slightly different initial state, then a first approximation to the difference between the two optimal controls may normally be derived, if desired, by solving a linear optimal control problem. This holds independently of the criterion for optimality for the nonlinear system [39–41]. This difference between the different operating points is precompensated in this research work with the set-point trajectory generator.

In Sections 5.4.1.1 and 5.4.1.2 the theoretical framework of this concept is presented. The reader who is familiar with these concepts may jump straight to Section 5.4.2 where the application of the decentralized LQG on the MVDC system is presented, while in Section 5.4.3 a centralized formulation of this control concept is shown.

5.4.1 Procedure of Observer-Based Control

5.4.1.1 Linear Quadratic Regulator

Linear optimal control is a special sort of optimal control. The plant that is controlled is assumed linear, and the controller, the device that generates the optimal control, is constrained to be linear. Linear controllers are achieved by working with quadratic performance indices. These are quadratic in the control and regulation/tracking error variables. Such methods that achieve linear optimal control are termed Linear Quadratic (LQ) methods [33,41].

The LQ controller, also called Riccati controller, is a state controller for a linear dynamic system whose feedback matrix is determined by the minimization of a quadratic cost functional. Its synthesis is thus a subproblem of optimal control.

A common method for the design of a state controller is the pole placement. The eigenvalues of the closed loop and thus its dynamics are defined selectively as it was performed with LSF in Section 5.1.1. The disadvantages of this method are that the performances of individual states are not placed in the foreground and actuator saturation or actuation effort can be considered only indirectly. Both are, however, often desired in the practical application and are enabled by the LQR controller. For this purpose, a cost function J with the following form is assumed for a finite time horizon T.

$$J(u, x_0) = \frac{1}{2}x(T)^T Sx(T) + \frac{1}{2}\int_0^T (x(T)^T Qx(T) + u(t)^T Ru(t))dt \quad (5.91)$$

The states and actuating variables are factored in quadratically. With the weighting matrices S, Q, R the end values of the states, the state trajectories, and the actuating variable trajectories are prioritized. With the diagonal elements of Q the velocity of the single states can be driven to zero. If the system has only one input R will be a scalar. The bigger the value of R chosen, the slower the control would be. S is used when considering a finite time horizon to minimize the final values of the states, if the time is not sufficient to drive the states to zero. When considering an infinite time horizon, this weighting does not apply because the states for $t \to \infty$ have to trend to zero as otherwise the integral would not converge.

The synthesis of this control is highlighted for the infinite time horizon care in the following steps; for more in depth explanation the reader is referred to [33]. A continuous time linear system, defined by:

$$\dot{x} = Ax(t) + Bu(t) \quad (5.92)$$

with a quadratic cost function as defined in (5.93)

$$J(u, x_0) = \int_0^\infty (x(T)^T Qx(T) + u(t)^T Ru(t))dt \quad (5.93)$$

The feedback control law that minimizes the value of the cost is given by:

$$u = -Kx \quad (5.94)$$

The optimal gain is determined by the optimality principle, stating that, if the closed-loop control is optimal over an interval, it will be optimal over any subinterval within it. Where K is defined by:

$$K = R^{-1}B^{T}P(t) \tag{5.95}$$

And P is found by solving the continuous algebraic Riccati equation.

$$PA = A^{T}P - PBR^{-1}B^{T}P + Q = 0 \tag{5.96}$$

The performance index is therefore minimized for the infinite time horizon, thus obtaining a constant optimal state feedback gain, by assuming that the controller will operate for a longer period that the transient time of the optimal gains [33].

5.4.1.2 Kalman Filter

The ISPS is an application where possible transients happen in a very short time and it is desirable to estimate those transients exactly, therefore it is not reasonable to neglect them. Thereby applying a steady-state Kalman filter with constant gain will not lead to optimal results, but it yields to sufficient estimation performance. The advantage of the steady-state Kalman filter in comparison with its more advanced versions [42,43] is that it provides fast computation, as the covariance matrixes and the Kalman gain are not updated. According to [33] it brings savings in filter complexity and computational effort while sacrificing only small amounts of performance.

The Kalman filter fulfills in this context the role of a linear optimal dynamic state estimator. The Kalman filter has two main steps: the prediction step, where the next state of the system is projected given the previous measurements, and the update step, where the current state of the system is estimated given the measurement at that time step [44,45].

A key structural result that allows optimal state estimation to be considered as a dual to optimal control is the separation principle [46]. The paper shows that the optimal feedback control of a process whose state is not available may be separated, without loss of optimality, into two linked tasks:

- Estimating optimally the state x by a state estimator to produce a best estimate $\hat{x}$.
- Providing optimal state feedback based on the estimated state $\hat{x}$, rather than the unavailable true state x.

The Kalman filter estimates the state of a plant based on the input signals and measurements. However, the plant is not just driven by a known control signal $u(t)$ but also by an unknown plant noise $w(t)$. Furthermore, the measurements $y(t)$ include a noise signal $v(t)$ which is called measurement noise. This leads to the following continuous linear time invariant plant model:

$$\begin{aligned} \dot{\boldsymbol{x}}(t) &= \boldsymbol{A}\boldsymbol{x}(t) + \boldsymbol{B}\boldsymbol{u}(t) + \boldsymbol{F}\boldsymbol{w}(t) \\ y(t) &= \boldsymbol{C}\boldsymbol{x}(t) + \boldsymbol{H}\boldsymbol{v}(t) \end{aligned} \tag{5.97}$$

Typically, the noise signals are assumed to be uncorrelated white Gaussian noise with zero mean and the following covariance matrices S_w and S_v.

For an infinite time horizon the optimal state estimation problem can be formulated such that the cost J of the mean square estimation error is minimized:

$$J = \lim_{T \to \infty} E\left\{ \frac{1}{T} \int_0^T \left[(\boldsymbol{x}(t) - \tilde{\boldsymbol{x}}(t))\boldsymbol{C}^T \boldsymbol{C}(\boldsymbol{x}(t) - \tilde{\boldsymbol{x}}(t) \right] dt \right\} \tag{5.98}$$

Where $E\ldots$ is the expected value. Under the assumption that $\boldsymbol{CA}$ is observable the solution of (5.99) applied to the system of (5.98) will lead to the Kalman filter [33]. The structure of the Kalman filter can be depicted in various forms of which one of them is the predictor/corrector form [45]:

$$\dot{\hat{\boldsymbol{x}}}(t) = \boldsymbol{A}\hat{\boldsymbol{x}}(t) + \boldsymbol{B}u(t) + \boldsymbol{K}[y(t) - \boldsymbol{C}\hat{\boldsymbol{x}}(t)] \tag{5.99}$$

This terminology appeared first for discrete Kalman filters. It describes how the derivative of the estimation is first predicted from old data and then updated using new measurements [33]. Eq. (5.99) clarifies that the only unknown variable is the Kalman gain $\boldsymbol{K}$, which is calculated as follows:

$$\boldsymbol{K} = \boldsymbol{P}\boldsymbol{C}^T(\boldsymbol{H}\boldsymbol{S}_v^{-1}\boldsymbol{H}^T)^{-1} \tag{5.100}$$

Since the time-varying solution of the differential equation is close to the steady-state solution when the system is running for a long time, the derivatives in the Riccati equation are set to zero and where $\boldsymbol{P}$ is the solution of the algebraic Riccati Eq. (5.101):

$$\boldsymbol{A}\boldsymbol{P} + \boldsymbol{A}^T\boldsymbol{P} - \boldsymbol{P}\boldsymbol{C}^T(\boldsymbol{H}\boldsymbol{S}_v^{-1}\boldsymbol{H}^T)^{-1}\boldsymbol{C}\boldsymbol{P} + \boldsymbol{F}\boldsymbol{S}_w\boldsymbol{F}^T = \boldsymbol{0} \tag{5.101}$$

5.4.1.3 Augmented Kalman Filter

Typically ISPSs can be divided in local subsystems with local states $\boldsymbol{x}_i$ which are dynamically coupled such as in Fig. 5.7. In the context of a decentralized control architecture which is able to accommodate sudden changes in topology and operates without the need of additional measurements, each subsystem assumes to lump the nonlocal part into an equivalent virtual source. Additionally, this source is considered to be a disturbance with unknown dynamics as it is illustrated in Fig. 5.23.

It is considered now that the ISPS can be represented by a LTI system like (5.97). This allows expressing any subsystem as [32]:

$$\dot{\boldsymbol{x}}_i = \boldsymbol{A}_i\boldsymbol{x}_i + \boldsymbol{B}_i\boldsymbol{u}_i + \sum_{j=k,j\neq 1}^{m} \boldsymbol{F}_i\boldsymbol{x}_j \tag{5.102}$$

$$\dot{\boldsymbol{y}}_i = \boldsymbol{C}_i\boldsymbol{x}_i + \sum_{j=k,j\neq 1}^{m} \mathbf{G}_i x_j + \boldsymbol{H}_i\boldsymbol{v}_i \tag{5.103}$$

Where the local state vectors are $\boldsymbol{x}_i$ and $\boldsymbol{u}_i$ and $\boldsymbol{y}_i$ represent the local measurements which are degraded by the noise $\boldsymbol{v}_i$. The remote j-th subsystem is represented by its state $\boldsymbol{x}_j$. The interaction between subsystems are represented by the terms $\sum_{j=k,j\neq 1}^{m} \boldsymbol{F}_i\boldsymbol{x}_j$, $\sum_{j=k,j\neq 1}^{m} \mathbf{G}_i x_j$.

Incorporating now the concept of a virtual disturbance source, defined in [32], into the Eqs. (5.117) and (5.118) and denoting it with $\boldsymbol{x}_{di}$ permits to define the following system model:

$$\dot{\boldsymbol{x}}_i = \boldsymbol{A}_i\boldsymbol{x}_i + \boldsymbol{B}_i\boldsymbol{u}_i + \boldsymbol{D}_i\boldsymbol{x}_{di} \tag{5.104}$$

$$\dot{\boldsymbol{y}}_i = \boldsymbol{C}_i\boldsymbol{x}_i + \boldsymbol{E}_i\boldsymbol{x}_{di} + \boldsymbol{H}_i\boldsymbol{v}_i \tag{5.105}$$

Where the terms related to $\boldsymbol{x}_{di}$ have now substituted the interconnection terms, and therefore the local subsystem dynamics depend only on these states and the dynamics of the virtual disturbance. As the virtual

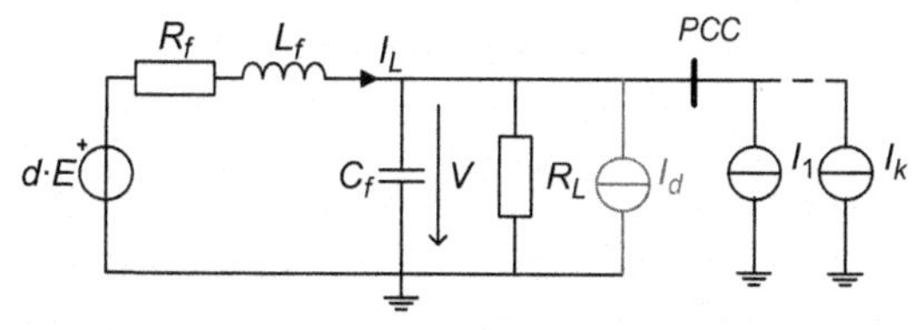

Figure 5.23 Decentralized model including the virtual disturbance.

source is considered as an unknown dynamic disturbance, the local dynamic states are augmented by $\boldsymbol{x}_{di}$.

$$\boldsymbol{x}_{i,ext} = \begin{bmatrix} \boldsymbol{x}_i \\ \boldsymbol{x}_{di} \end{bmatrix} \tag{5.106}$$

Applying (5.106) on (5.119), (5.120) enables to write the local state-space model:

$$\dot{\boldsymbol{x}}_{i,ext} = \boldsymbol{A}_{i,ext}\boldsymbol{x}_{i,ext} + \boldsymbol{B}_{i,ext}\boldsymbol{u}_i \tag{5.107}$$

$$\dot{\boldsymbol{y}}_i = \boldsymbol{C}_{i,ext}\boldsymbol{x}_{i,ext} + \boldsymbol{H}_{i,ext}\boldsymbol{x}_{di} \tag{5.108}$$

On this local state-space model the Kalman filter is designed, where the matrices $\boldsymbol{A}_{i,ext}, \boldsymbol{B}_{i,ext}, \boldsymbol{C}_{i,ext}, \boldsymbol{H}_{i,ext}$ include the local states and the virtual disturbance source [32]. Applying the augmented Kalman filter will lead to (5.109) where $\boldsymbol{K}_{i,ext}$ is augmented Kalman gain computed by solving the algebraic Riccati equation, and $\hat{\boldsymbol{x}}_{i,ext}$ is the estimation of $\boldsymbol{x}_{i,ext}$.

$$\dot{\hat{\boldsymbol{x}}}_{i,ext} = \boldsymbol{A}_{i,ext}\hat{\boldsymbol{x}}_{i,ext} + \boldsymbol{B}_{i,ext}\boldsymbol{u}_i + \boldsymbol{K}_{i,ext}(\boldsymbol{x}_i - \boldsymbol{C}_{i,ext}\hat{\boldsymbol{x}}_{i,ext}) \tag{5.109}$$

As the observer estimates an additional state, the 2DOF of the LQR is augmented with L_{ext}, a term expressing the online deviation of the system set-point due to the virtual disturbance.

5.4.1.4 Shaping Filter

The virtual disturbance was presented in the previous section. To include the virtual disturbance inside the state estimation of the Kalman filter an appropriate model is needed. In control and estimation theory unknown inputs are assumed to be white noise. However, this model is not sufficient to describe the influences of the rest of the network since external currents can have a nonzero mean value and correlation time. Therefore, a non-Gaussian signal (say, $U(s)$) can be treated by synthesizing a filter transfer function ($G(s)$) such that:

$$G(s) = \frac{W(s)}{U(s)} = \frac{b_d}{s + a_d} \tag{5.110}$$

Where $W(s)$ is a white noise signal. Thus, by adding a new element G to the process model, the requirement that w shall be a Gaussian signal of zero mean may be met [47]. The element G used in this way is sometimes referred to as a coloring or shaping filter.

Therefore, the white noise is transformed by a linear filter which is proposed in [48]. The disturbance model is then:

$$\dot{u}_d = -a_d u_d + b_d w \tag{5.111}$$

With the filter parameters according to [33], τ_c is the correlation time and σ_w is the variance of the noise that has to be estimated:

$$a_d = \frac{1}{\tau_c}; b_d = \sqrt{\frac{2\sigma_w^2}{\tau_c}} \tag{5.112}$$

5.4.2 Application to Decentralized Controlled MVDC System (LQG + Virtual Disturbance)

The LQG design is composed of a LQR and a linear Kalman filter and is applied to the decentralized system which is depicted in Fig. 5.22. The LQR is implemented with static feedback of the estimated state variables of the controlled system and the feed-forward of the control variables and estimated. The Kalman filter is an optimal linear observer which estimates the state and disturbance from the measured variables. The achievable control performance is dependent on the speed of the estimation of the disturbance. The LQR guarantees optimal state feedback given that the state estimation by the Kalman filter is sufficiently accurate. This structure is shown in Fig. 5.24. The modifications are explained in the following.

In order to facilitate the understanding the overall system model is presented again in Fig. 5.25 while the decentralized system model is depicted in Fig. 5.26.

In [36] the authors introduced a concept to handle those unknown disturbances on the bus. They proposed to sum up all unknown currents

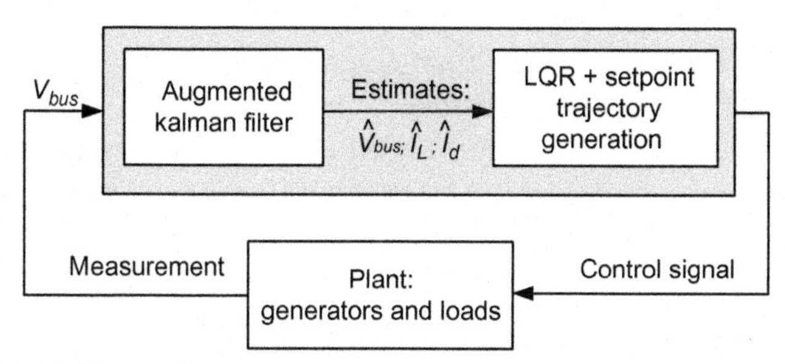

Figure 5.24 LQG controller structure using an augmented Kalman filter and a LQR with set-point trajectory generation.

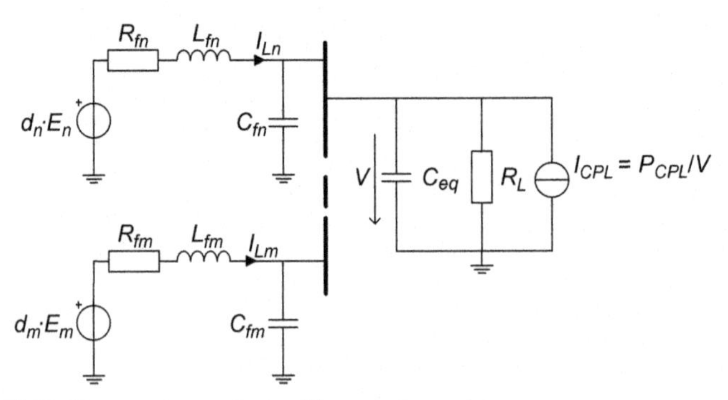

Figure 5.25 Multiconverter system with main bus.

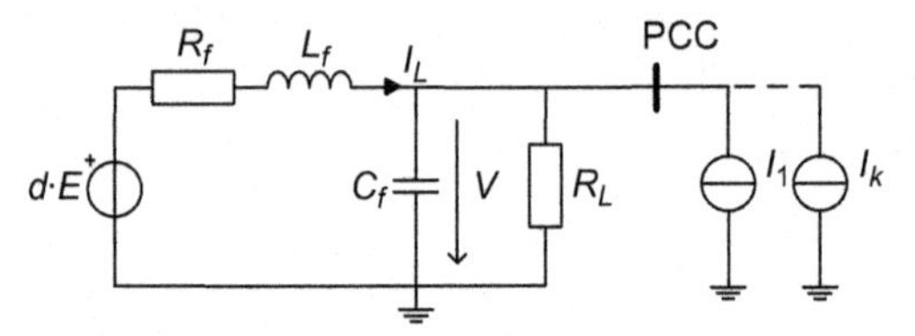

Figure 5.26 Decentralized model.

which flow into the Point of Common Coupling (PCC), in a virtual disturbance as it is shown in Eq. (5.113). This disturbance current I_d is then estimated by a Kalman filter.

$$I_d = -\sum_{i=1}^{k} I_i \tag{5.113}$$

This leads to the circuit depicted in Fig. 5.27. The current sources $I_1 \dots I_k$ are now substituted by the current source I_d, in accordance with Eq. (5.113). Where Fig. 5.27 represents the local converter model which is used for the subsequent control and estimator design.

5.4.2.1 LQG – Set-point Trajectory

The LQG controller used here is based on the design proposed in [31] and [34] which is an application of the 2DOF structure that is explained in detail in [49]. The 2DOF structure combines the feed-forward control and feedback control to achieve performance. The standard scheme is extended by a trajectory generation which is depicted in Fig. 5.28. Thereby, the disturbance caused by the part of the network which is not considered in the decentralized model can be taken into account.

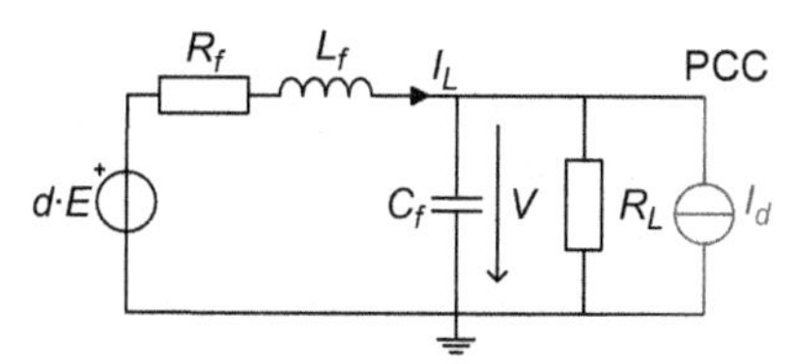

Figure 5.27 Decentralized model including the virtual disturbance.

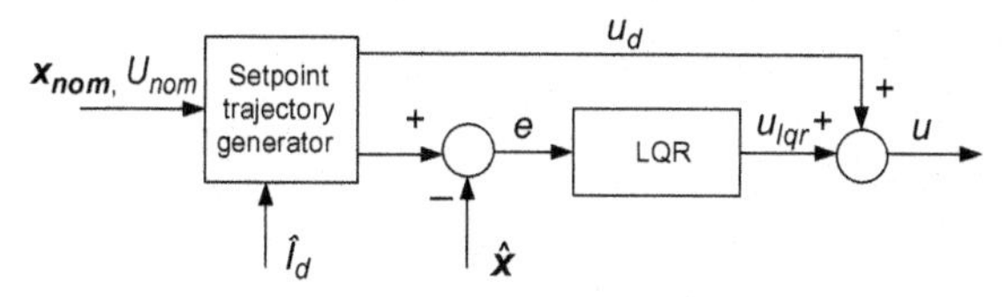

Figure 5.28 LQG controller with two degree of freedom structure.

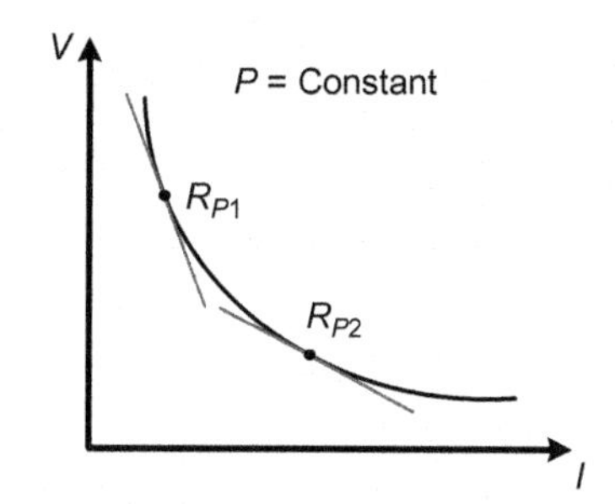

Figure 5.29 Illustration of the set-point generation.

Essentially, the trajectory generation is responsible for the linearization of the system because the nonlinearities of the loads that generate the disturbance current are not included in the model but hidden in the disturbance variable I_d. Hence, the system is handled by definition as a linear system, where the trajectory generation will handle the deviations of the "linear assumed operating point" from the actual operating point.

Strictly speaking the LQR controller is unaware, from a global point of view, where the system operating point is at the moment.

This global point of view, the entire CPL curve, compared to the linearized small-signal region (e.g., R_{P1}) is illustrated in Fig. 5.29. The set-point trajectory generation is the component which determines in which operating point the system is. Then, the LQR controller is responding to disturbances around this point.

The focus lies now on the calculation of the trajectory generation and the optimal state feedback. For the decentralized model depicted in Fig. 5.27 the state equations are given in (5.114):

$$\begin{aligned}\dot{x}_1 &= \frac{x_2}{C_f} - \frac{x_1}{R_L C_f} - \frac{I_d}{C_f} \\ \dot{x}_2 &= \frac{dE}{L_f} - \frac{x_2 R_f}{L_f} - \frac{x_1}{L_f}\end{aligned} \tag{5.114}$$

Where the states are $x_1 = V$ and $x_2 = I_L$.

The system can be written in matrix form, where d in (5.115) corresponds to the duty cycle:

$$\begin{aligned}\dot{\boldsymbol{x}} &= \boldsymbol{A}\boldsymbol{x} + \boldsymbol{B}d + \boldsymbol{N}I_d \\ y &= \boldsymbol{C}\boldsymbol{x}\end{aligned} \tag{5.115}$$

$$\begin{aligned}\boldsymbol{x} = \begin{bmatrix} V \\ I_L \end{bmatrix} \boldsymbol{A} &= \begin{bmatrix} -\frac{1}{R_L C_f} & \frac{1}{C_f} \\ -\frac{1}{L_f} & -\frac{R_f}{L_f} \end{bmatrix} \\ \boldsymbol{B} = \begin{bmatrix} 0 & \frac{E}{L_f} \end{bmatrix}^T & \boldsymbol{C} = \begin{bmatrix} 1 & 0 \end{bmatrix} \\ \boldsymbol{N} &= \begin{bmatrix} -\frac{1}{C_f} & 0 \end{bmatrix}^T\end{aligned} \tag{5.116}$$

The nominal set-point of the system where V_{nom} is equal to the reference bus voltage while the external disturbance is zero can be determined:

$$\begin{aligned}\boldsymbol{0} &= \boldsymbol{A}\boldsymbol{x}_{nom} + \boldsymbol{B}d_{nom} \\ V_{nom} &= \boldsymbol{C}\boldsymbol{x}_{nom}\end{aligned} \tag{5.117}$$

The equilibrium point of the system is described by Eq. (5.117) as all derivates are set to zero. The decentralized model does not take into account the effects of the whole system; the disturbance has to be considered additionally when generating the set-point. This disturbance I_d is estimated and the actual set-point is adapted.

$$\begin{aligned}\boldsymbol{0} &= \boldsymbol{A}\boldsymbol{x}_e + \boldsymbol{B}d_e + \boldsymbol{N}I_d \\ V_{nom} &= \boldsymbol{N}\boldsymbol{x}_e\end{aligned} \tag{5.118}$$

The system would be stable if there were no modeling errors or disturbances are present and the virtual disturbance estimate is perfect. This would mean that the set-point trajectory generator alone could output the control signal which drives the system to its equilibrium point.

Since in real applications this is not the case, an additional LQR controller is added for optimal state feedback. The goal of this control strategy is the minimization of a cost function that includes the states as well as the control signal. By assuming that the controller will operate for a longer period than the transient time of the optimal gains [50], the LQG controller is supposed to minimize the cost function for an infinite time interval.

The steady-state solution of the following cost function (J_i) has to be found:

$$J_i = E\left[\int_0^\infty \left(\mathbf{Q}_y(V - V_{nom})^2 + \mathbf{R}(d - d_e)^2\right)\right] dt \tag{5.119}$$

Eq. (5.119) can be transformed into the general cost function form in [33]:

$$J = E\left[\int_0^\infty \left((\boldsymbol{x} - \boldsymbol{x}_e)^T \mathbf{Q}(\boldsymbol{x} - \boldsymbol{x}_e) + (d - d_e)^T \mathbf{R}(d - d_e)\right)\right] dt \tag{5.120}$$

Where $\mathbf{Q} = \boldsymbol{C}^T \boldsymbol{Q} \boldsymbol{C}$ is a symmetric, positive-semidefinite weighting matrix and $\boldsymbol{R}$ becomes a positive weighting scalar.

By using the feedback gain matrix of the LQR on the state estimates of the Kalman filter a state feedback which minimizes the optimization problem can be obtained:

$$d = -\boldsymbol{R}^{-1}\boldsymbol{B}^T\boldsymbol{P}x = -\boldsymbol{K}\hat{x} \tag{5.121}$$

$\boldsymbol{P}$ is found from the steady-state solution of the Riccati differential equation which satisfies the algebraic Riccati equation:

$$\boldsymbol{PA} + \boldsymbol{A}^T\boldsymbol{P} - \boldsymbol{PBR}^{-1}\boldsymbol{B}^T\boldsymbol{P} + \mathbf{Q} = \mathbf{0} \tag{5.122}$$

5.4.2.2 Augmented Local Kalman Filter

The design will be performed in two consecutive steps. First, the disturbance I_d in the system is simply included as noise rather than as a state and the Kalman filter is introduced to estimate the inductor current of the buck converter. Second, this Kalman filter is augmented to include the disturbance current estimation as a state based on a model of white noise as input to a linear filter.

The disturbance estimation is handled by the augmented Kalman filter. The design of the augmented Kalman filter for decentralized controlled MVDC systems is performed in two consecutive steps.

1. The disturbance I_d in the system is simply included as noise rather than a state and the Kalman filter is introduced to estimate the inductor current of the buck converter.
2. This Kalman filter is augmented to include the disturbance current estimation as a state based on a model of white noise as input to a linear shaping filter for representing nonstationary colored noise [47] and [34].

At first, the system model looks exactly like the one for the LQR controller:

$$\begin{aligned} \dot{\boldsymbol{x}} &= \boldsymbol{A}\boldsymbol{x} + \boldsymbol{B}d + \boldsymbol{N}I_d \\ y &= \boldsymbol{C}\boldsymbol{x} \end{aligned} \tag{5.123}$$

$$\begin{gathered} \boldsymbol{x} = \begin{bmatrix} V \\ I_L \end{bmatrix} \boldsymbol{A} = \begin{bmatrix} -\frac{1}{R_L C_f} & \frac{1}{C_f} \\ -\frac{1}{L_f} & -\frac{R_f}{L_f} \end{bmatrix} \\ \boldsymbol{B} = \begin{bmatrix} 0 & \frac{E}{L_f} \end{bmatrix}^T \boldsymbol{C} = \begin{bmatrix} 1 & 0 \end{bmatrix} \\ \boldsymbol{N} = \begin{bmatrix} -\frac{1}{C_f} & 0 \end{bmatrix}^T \end{gathered} \tag{5.124}$$

According to Section 5.4.1.4, the disturbance current model for nonwhite noise is given in (5.125):

$$\dot{I}_d = -a_d I_d + b_d w \tag{5.125}$$

with the filter parameters:

$$a_d = \frac{1}{\tau_c}; b_d = \sqrt{\frac{2\sigma_w^2}{\tau_c}} \tag{5.126}$$

where τ_c is the correlation time and σ_w is the variance of the noise that has to be estimated. Extending the system (5.115) by the linear filter leads to, with v being the measurement noise in the local available measurement (e.g., the bus voltage):

$$\begin{aligned} \dot{\boldsymbol{x}}_{ext} &= \boldsymbol{A}_{ext}\boldsymbol{x}_{ext} + \boldsymbol{B}_{ext}d + \boldsymbol{N}_{ext}w \\ y &= \boldsymbol{C}_{ext}\boldsymbol{x}_{ext} + v \end{aligned} \tag{5.127}$$

$$\boldsymbol{x}_{ext} = \begin{bmatrix} V \\ I_L \\ I_d \end{bmatrix} \qquad \boldsymbol{A}_{ext} = \begin{bmatrix} -\frac{1}{R_L C_f} & \frac{1}{C_f} \\ -\frac{1}{L_f} & -\frac{R_f}{L_f} \\ 0 & -a_d \end{bmatrix}$$

$$\boldsymbol{B}_{ext} = \begin{bmatrix} 0 \\ \frac{E}{L_f} \\ 0 \end{bmatrix} \qquad \boldsymbol{C}_{ext} = \begin{bmatrix} 1 & 0 & 0 \end{bmatrix} \tag{5.128}$$

$$\boldsymbol{N}_{ext} = \begin{bmatrix} -\frac{1}{C_f} & 0 & b_d \end{bmatrix}^T$$

The difference between the augmented version and the Kalman filter from Section 5.4.1.2 is that the state-space matrices incorporate the additional state which will lead to the following augmented state:

$$\dot{\hat{\boldsymbol{x}}}_{ext} = \boldsymbol{A}_{ext}\hat{\boldsymbol{x}}_{ext} + \boldsymbol{B}_{ext}\boldsymbol{d}_i + \boldsymbol{K}_{KF}(\boldsymbol{x}_i - \boldsymbol{C}_{ext}\hat{\boldsymbol{x}}_{ext}) \tag{5.129}$$

where $\boldsymbol{K}_{KF}$ corresponds to the steady-state Kalman gain which can be determined by the equations of (5.100) and (5.101), while assuming that the observability condition of the augmented system is kept valid.

As the observer estimates one additional state, the 2DOF structure is augmented with L_{ext}, a term expressing the online deviation of the system set-point due to the virtual disturbance. This set-point trajectory generator is also used to mitigate the difference between the linearized model values of the states and the actual nonlinear values. In steady-state the derivatives of the differential equations of the system are equal to zero, yielding at the equilibrium point of (5.118). The nominal set-point of the system without the presence of I_d, thus based on the nonaugmented system, results in a set-point only dependent on V_{nom}; adding the adaption term corresponding to I_d and substituting it by its estimate $\hat{I}_d$ leads to a set-point dependent on $\hat{I}_d$ and V_{nom}.

$$\begin{bmatrix} \hat{x} \\ \hat{u} \end{bmatrix} = \begin{bmatrix} \boldsymbol{A} & \boldsymbol{B} \\ c & 0 \end{bmatrix}^{-1} \left(\begin{bmatrix} 0 \\ V_{nom} \end{bmatrix} + \begin{bmatrix} -n\hat{I}_d \\ 0 \end{bmatrix} \right) \tag{5.130}$$

Correspondingly the estimate of $\hat{I}_d$ needs to be multiplied by the factor L_{ext} (6) in order to correct the set-point variations induced by the disturbance.

$$L_{ext} = \begin{bmatrix} \boldsymbol{A} & \boldsymbol{B} \\ c & 0 \end{bmatrix}^{-1} \begin{bmatrix} \boldsymbol{0} \\ \frac{1}{c_k} \end{bmatrix} \tag{5.131}$$

The set-point adaptation term is added to the nominal state and added to the nominal set-point values to form the control variables of the two degree of freedom structure. The final control structure is depicted in Fig. 5.30.

5.4.3 Application to Centralized Controlled MVDC System (LQG + Virtual Disturbance)

The control approach highlighted in Sections 5.4–5.4.2 can also be extended to the centralized system that was depicted in Fig. 5.2 and was used as the basis for the LSF control of Section 5.1. The design is still based on the 2DOF structure [49]. The simplification which takes place here is that due to the centralized control architecture all currents are now available, but they are still being estimated. The difference from the LSF control is that a linear control is employed and this control assumes that it is always in the operating point. Therefore, the Kalman filter is still necessary for adjusting the assumed linear operating point to the actual operating point which is defined by the nonlinear CPL. An augmentation is not necessary (Fig. 5.31).

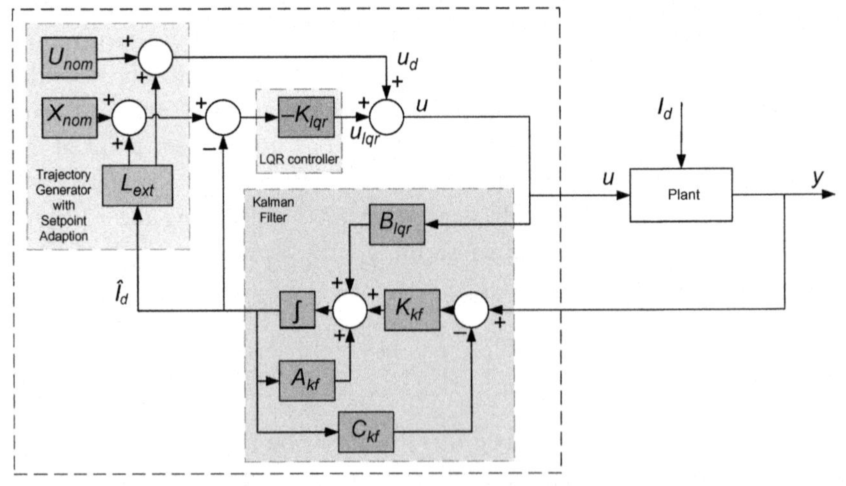

Figure 5.30 Final structure of the observer-based controller.

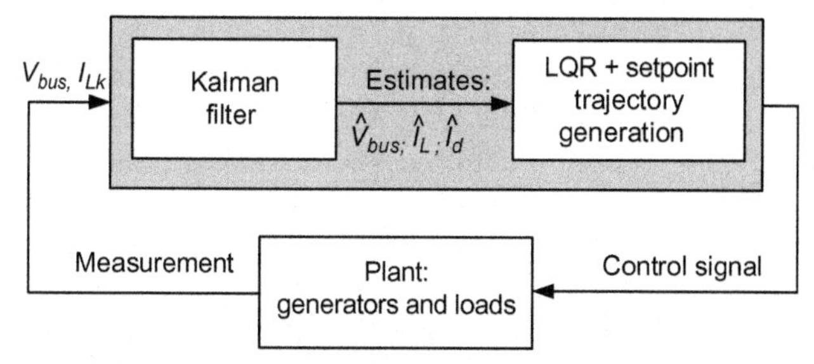

Figure 5.31 Centralized LQG controller with set-point trajectory.

This leads to the system description, where I_d is the noise introduced, by the CPL nonlinearity:

$$\begin{aligned} \dot{\boldsymbol{x}} &= \boldsymbol{A}\boldsymbol{x} + \boldsymbol{B}d + \boldsymbol{N}I_d \\ y &= \boldsymbol{C}\boldsymbol{x} \end{aligned} \tag{5.132}$$

where the state vector $\boldsymbol{x}$ includes all inductor currents. Those currents are still estimated by a Kalman filter.

$$\boldsymbol{x} = \begin{bmatrix} V \\ I_{L1} \\ I_{L2} \\ I_{L3} \end{bmatrix} \boldsymbol{A} = \begin{bmatrix} -\dfrac{1}{R_L C_{eq}} & \dfrac{1}{C_{eq}} & \dfrac{1}{C_{eq}} & \dfrac{1}{C_{eq}} \\ -\dfrac{1}{L_f} & -\dfrac{R_f}{L_f} & 0 & 0 \\ -\dfrac{1}{L_f} & 0 & -\dfrac{R_f}{L_f} & 0 \\ -\dfrac{1}{L_f} & 0 & 0 & -\dfrac{R_f}{L_f} \end{bmatrix} \tag{5.133}$$

$$\begin{gathered} \boldsymbol{C} = \begin{bmatrix} 1 & 0 & 0 & 0 \end{bmatrix} \boldsymbol{B} = \begin{bmatrix} 0 & 0 & 0 \\ E/L_f & 0 & 0 \\ 0 & E/L_f & 0 \\ 0 & 0 & E/L_f \end{bmatrix} \\ \boldsymbol{N} = \begin{bmatrix} -\frac{1}{C_f} & 0 & 0 & 0 \end{bmatrix}^T \end{gathered} \tag{5.134}$$

5.4.4 Simulation Results

5.4.4.1 Cascaded System

A cascaded converter setup has been modeled in Matlab-SIMULINK with the parameters described in Table 6.1. In the simulation, a step variation of the CPL load has been performed. The variation of the bus voltage during the step change of the CPL load is described in Fig. 5.32. At $t = 0.25$ s the load step of 7.5 MW was performed, demonstrating that the *Virt Dist LQG* control stabilizes the system towards the reference point $V_{ref} = 1$. The dynamic of the inductor current is described in Fig. 5.33.

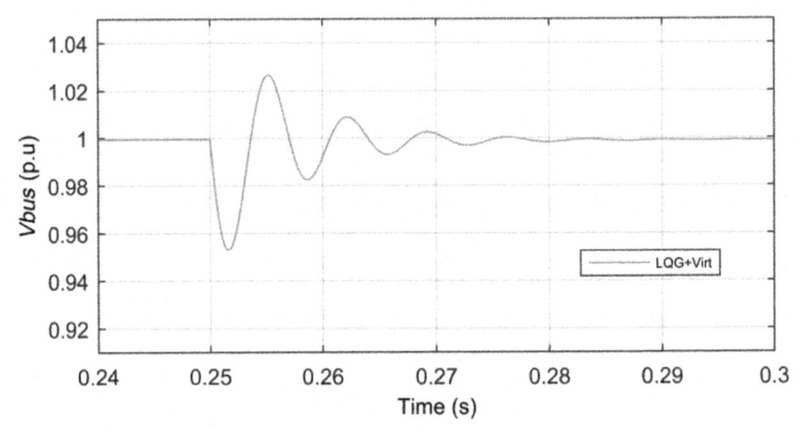

Figure 5.32 Voltage virtual distribution LQG cascaded – Ideal CPL – Step 10.3 → 17.8 MW.

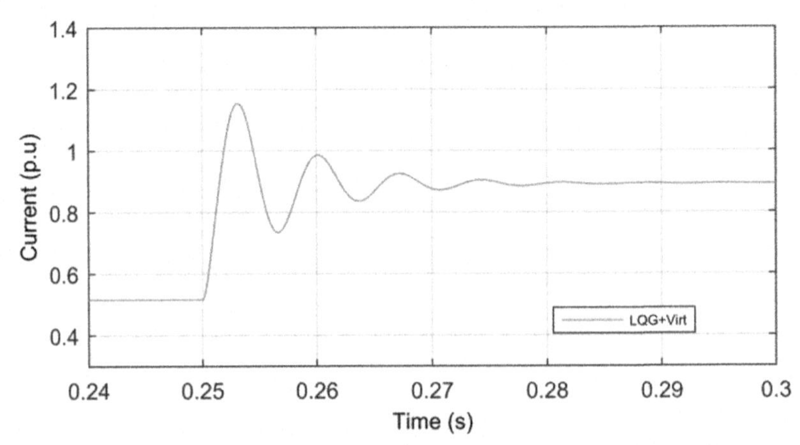

Figure 5.33 Current virtual distribution LQG cascaded – Ideal CPL – Step 10.3 → 17.8 MW.

5.4.4.2 Shipboard Power System

A Shipboard Power System has been modeled in Matlab\SIMULINK, based on the parameters in Table 6.2. The load steps of 22.5 MW have been performed at $t = 0.25$ s. The coefficients of the droop controllers were chosen to obtain an even current sharing among the converters. All converters equally share the total load power between them.

The effect of the increasing and decreasing of the CPL load on the load voltage is described in Fig. 5.34, demonstrating that the voltage remains stable and reaches the voltage set-points that are gradually modified by the action of the droop control. In Fig. 5.35 one can observe the transient for the current.

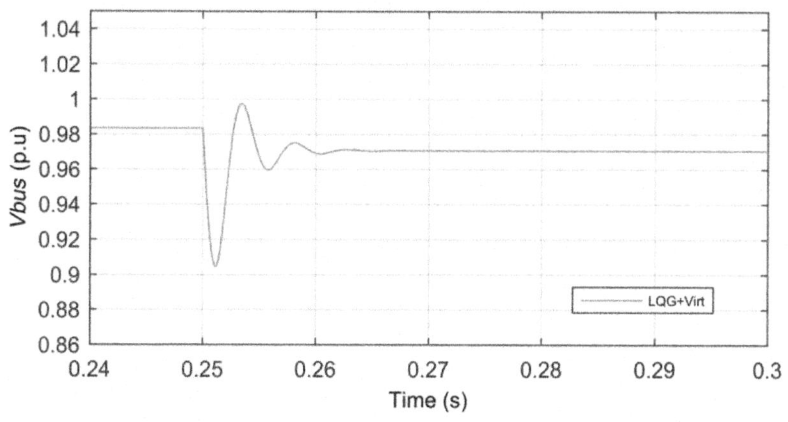

Figure 5.34 ISPS V_{bus} virtual distribution LQG for ideal CPL – Step 30.9 → 53.4 MW.

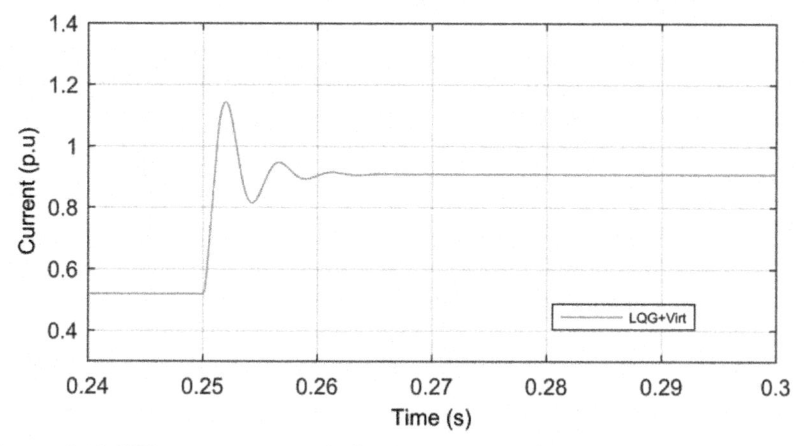

Figure 5.35 ISPS $I_{L1=L2=L3}$ virtual distribution LQG for ideal CPL and 30.9 → 53.4 MW.

5.5 BACKSTEPPING BASED CONTROL

In the following subsections the reader is introduced to the theory behind Backstepping [51]. In a nutshell it can be described as a systematic way to construct a control law which is based on Lyapunov function while partitioning the system and extending the controlled system size with each step. The reader will encounter a short description of what Backstepping is, which will be followed by an overview of the stability concept according to Lyapunov, which is the basis for the Control Lyapunov Function (CLF). The Backstepping concept was successfully applied to the control of DC–DC converters in [52] supplying a resistive load. In [53] and [54] an adaptive design with time-varying loads is presented, although with the restriction that load change is slow compared to the converter dynamics.

To the authors' best knowledge, the concept of Backstepping was applied for the first time on the stabilization of CPLs and multiple parallel LRCs in [55] and [56].

In Section 5.5.1 the theory behind Backstepping is explained; this includes the stability concept of Lyapunov in 5.5.1.1 and CLFs in 5.5.1.2. In Section 5.5.2 the procedure of Backstepping is explained. The reader who is familiar with those theoretical concepts will find in Section 5.5.3 the application of the Backstepping theory in combination with the augmented Kalman filter of Section 5.4.1.3 for decoupling the network on the ISPS circuit.

5.5.1 Theory Behind Backstepping

This section gives an overview of Lyapunov theory in general and the integrator Backstepping technique. Furthermore, this basic principle will be just referred to as Backstepping throughout this book.

The peculiarity with nonlinear systems is that stability is no longer a global property; it is restricted to certain trajectories of the system. Therefore, many different stability criteria exist for specific classes of nonlinear systems but the theory is not as developed as for linear systems yet. A very often used stability concept is based on the theory introduced by Lyapunov.

According to Lyapunov, the stability of the system can be ensured by finding a so-called Lyapunov function which fulfills the criteria defined in the direct method of Lyapunov. However, in practice there are cases where it proves to be very difficult to find such a solution for complex systems.

Backstepping is a concept that allows the user to follow a scheme when trying to find a suitable Lyapunov function candidate for the given system. Hence, the challenge of finding a Lyapunov function candidate while using Lyapunovs's direct method is tackled systematically.

5.5.1.1 Lyapunov

Lyapunov proposed an indirect method and a direct method to prove system stability. The indirect method of Lyapunov, also called first method, uses the linearization of a system to determine the local stability of the original system. Lyapunov's direct method, also called the second method of Lyapunov, allows us to determine the stability of a system without explicitly integrating the differential Eq. (5.135).

$$\dot{x}(t) = f(x(t)) \tag{5.135}$$

This method is a generalization of the idea that if there is a "measure of energy" in a system; then, the rate of change of the system's energy can be studied to ascertain stability [57]. The second method will be used further on as it is the foundation of the Backstepping theory.

The following definition from [6] describes the properties of a valid Lyapunov function:

Direct method of Lyapunov: Let the differential equation $\dot{x} = f(x)$ with equilibrium point $\mathrm{x_E} = 0$ have a continuous and unique solution for every initial condition in a neighborhood $\mathrm{D_1}(0)$ of the equilibrium point. If a function $V(x)$ exists which is positive definite and therefore fulfills:

$$\begin{gathered} V(0) = 0 \\ V(x) < 0, x \neq 0 \end{gathered} \tag{5.136}$$

and is continuous with continuous derivatives:

$$\dot{V}(x) \leq 0 \tag{5.137}$$

Then, the equilibrium point $x_E = 0$ is stable in the sense of Lyapunov in the neighborhood. In addition, the equilibrium point is asymptotically stable if the inequality of eq. (5.137) holds true for all x.

The derivative of a Lyapunov function is calculated with gradients as follows

$$\dot{V}(\mathbf{x}) = \dot{x}^T \mathrm{grad}\, V(\mathbf{x}) = \sum_{i=1}^{n} \dot{x}_i \frac{\partial V}{\partial x_i} \tag{5.138}$$

where $\dot{x} = f(x)$. Finally one can check if Eq. (5.137) is fulfilled.

The fact that the differential equations are used directly without having to calculate their solution gave the method its name.

An interpretation of the definition given above is displayed in Figs. 5.36 and 5.37. While Fig. 5.36 describes the case where $\dot{V} = 0$, which is sufficient for Lyapunov stability, Fig. 5.37 shows a system that has a tendency to move to the equilibrium point in the neighborhood of the equilibrium. According to that, $\dot{V}$ is always smaller than zero and the function decreases for every trajectory.

5.5.1.2 Control Lyapunov Function

The previous chapter introduced the direct method of Lyapunov. Consequently, the usage of this theorem is explained in employing CLFs according to [51].

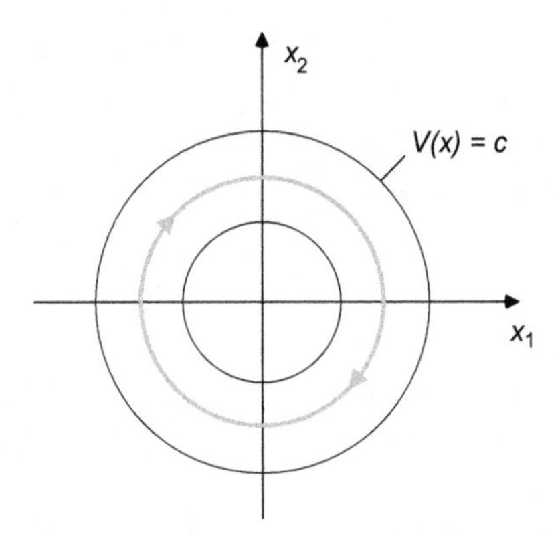

Figure 5.36 Lyapunov stability [6].

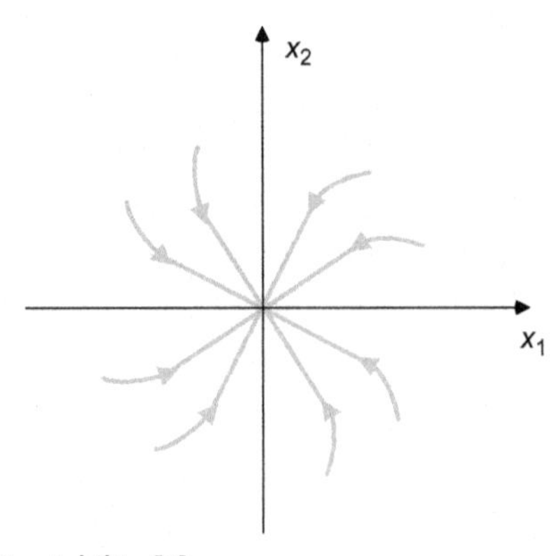

Figure 5.37 Asymptotic stability [6].

At first, a general dynamic system is considered, where x is the system state and u is the control input.

$$\dot{x} = f(x, u), \quad f(0, 0) = 0 \tag{5.139}$$

The goal is to achieve global asymptotic stability for the equilibrium point $x = 0$. Therefore, one needs to find a feedback law $u = a(x)$ that guarantees for all $x \neq 0$.

$$\dot{V}(x) \leq 0 \tag{5.140}$$

It is possible to choose u in such a way to attain the infimum, which is the greatest lower bound of the term in brackets. This will lead to minimal response time see Eq. (5.141).

$$\inf_{u} \{\dot{V}(x) = f^T(x, u) \cdot \text{grad} \ V(x)\} \tag{5.141}$$

The Lyapunov function $V(x)$ is used to find a stabilizing control law $u = k(x)$; hence, it is named CLF. The existence of a globally stabilizing control law is equivalent to the existence of a CLF which is known as Artstein's theorem [58]. This approach can be illustrated by considering the system, which is affine (i.e., linear) in the control input:

$$\dot{x} = f(x) + h(x)u \tag{5.142}$$

Under the assumption that a CLF for the system is known, Sontag proposes a particular choice of control law, which is commonly referred to as Sontag's formula [59]:

$$\begin{aligned} u = k(x) &= -\frac{(a - \sqrt{a^2 + b^2}}{b} \\ a &= V_x(x) f(x) \\ b &= V_x(x) h(x) \end{aligned} \tag{5.143}$$

This control law often uses parts of the system to help stabilize it. This reduces the magnitude of the control input u.

$$\dot{V} = V_x(x)(f(x) + h(x)u) = a + b\left(-\frac{(a - \sqrt{a^2 + b^2}}{b}\right) = -\sqrt{a^2 + b^2} \tag{5.144}$$

An approach was proposed in [60] where u is chosen for minimizing the control effort necessary to satisfy:

$$\dot{V} \leq -W(x) \tag{5.145}$$

Which leads to:

$$\dot{V}\Big|_{u=0} = V_x(x)\dot{f}(x) < -W(x) \tag{5.146}$$

It was mentioned in the beginning of Section 5.5.1.1, that it can sometimes be very difficult to find a Lyapunov function for a given system since by now there is no general approach to this problem.

5.5.2 Procedure of Backstepping

As already mentioned Backstepping or more specifically the integrator Backstepping technique provides a schematic way of finding Lyapunov control functions for systems that can be expressed in strict feedback form, which corresponds to a lower triangular system matrix. In the following, the integrator Backstepping is explained for an example which was taken from [6] and [61].

A general structure for a system that fulfills these requirements is given in eq. (5.147). For strict feedback form it is important that nonlinearities f_i and h_i of the derivative of state x_i are only dependent on the previous states which are fed back.

$$\begin{aligned}\dot{x}_1 &= f_1(x_1) + h_1(x_1)\cdot x_2\\ \dot{x}_2 &= f_2(x_1,x_2) + h_2(x_1,x_2)\cdot x_3\\ &\vdots\\ \dot{x}_k &= f_k(x_1,x_2,\ldots,x_3) + h_k(x_1,x_2,\ldots,x_k)\cdot u\end{aligned} \tag{5.147}$$

Compared to the LSF (Section 5.1) which is suitable for a similar subclass of nonlinear systems, Backstepping is advantageous because nonlinearities, which might be useful to keep for the system dynamics, are not canceled out. Therefore, one could avoid the use of big control inputs which lead to a high energy consumption or might be even impossible to generate. Furthermore, the LSF requires a detailed knowledge of the system to be controlled, whereas Backstepping allows attaining robustness against parameter variation to some extent.

The Backstepping approach will be now explained for a general system including two integrators. This two integrator system has similarities with the buck converter system including the LC-output filter which corresponds to a second-order system.

$$\begin{aligned}\dot{x}_1 &= f_1(x_1) + h_1(x_1)\cdot x_2 \\ \dot{x}_2 &= u\end{aligned} \tag{5.148}$$

A graphical presentation of the system is depicted in Fig. 5.38 to illustrate better the Backstepping concept. The graphical representation will be transformed along with the equations.

The output is defined as $y = x_1$ which should track a reference signal $y_{ref}(t)$. This tracking control problem can be transformed to a regulation problem by introducing the tracking error variable. The error signal is defined as $z_i = x_i - x_{i,d}$ (e.g., $z_1 = x_1 - y_{ref}$). Afterwards the first system equation is rewritten as:

$$\dot{z}_1 = \dot{x}_1 - \dot{y}_{ref} = f_1(x_1) + h_1(x_1)\cdot x_2 - \dot{y}_{ref} \tag{5.149}$$

Since the system is in strict feedback form, the state x_2 can be used as a virtual control input for the z_1-input subsystem. The idea of Backstepping is to set the state which acts as virtual input to a value that stabilizes the previous state and makes it globally asymptotically stable. Since x_2 is a state variable and not a real control input it is called *virtual control* and its desired value is referred to as a stabilizing function. The tracking error variable represents the difference between the virtual control x_2 and its desired value $\alpha(x_1, y_{ref}, \dot{y}_{ref})$

$$z_2 = x_2 - x_2^{des} = x_2 - \alpha\left(x_1, y_{ref}, \dot{y}_{ref}\right) \tag{5.150}$$

So far this is identical to the system (5.148) since the first equation was just extended by $\alpha(x_1) - \alpha(x_1) = 0$. The same transformation is represented in Fig. 5.39.

The goal is now to select a CLF in such way that the stabilizing virtual control law renders its time derivative along the solutions of z_1 subsystem (5.149) negative definite, where $W(z_1)$ is positive definite. The Lyapunov

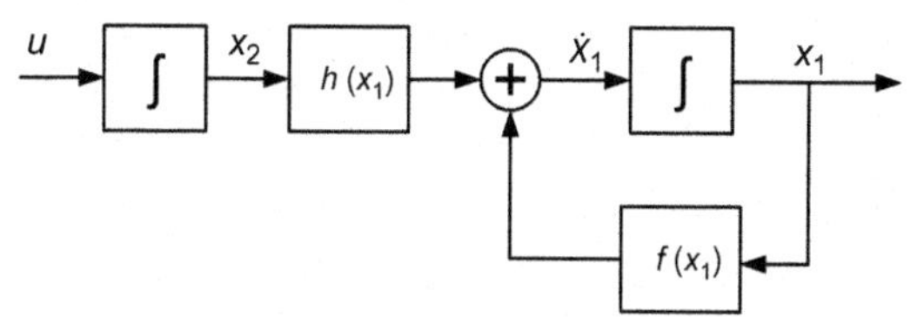

Figure 5.38 Graphical representation of system model (5.148).

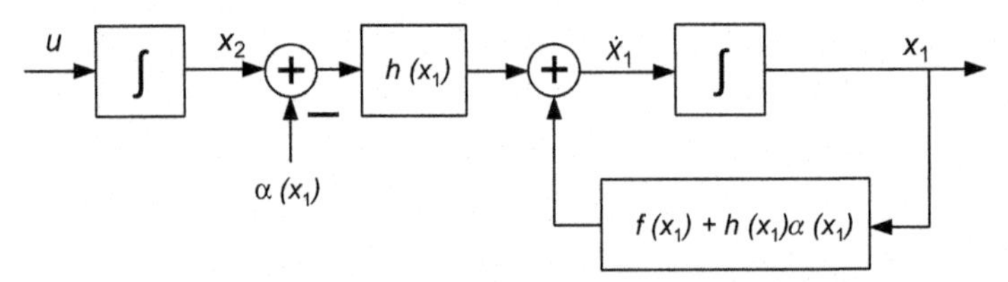

Figure 5.39 Graphical representation of new system including the control function α.

function in Eq. (5.151) is proposed for the stability analysis. The direct method of Lyapunov was described previously in Section 5.5.1.1.

$$\dot{V}_1 = \frac{\partial V_1}{\partial z_1}\left[f(x_1) + h(x_1)\cdot \alpha\left(x_1, y_{ref}, \dot{y}_{ref}\right) - \dot{y}_{ref}\right] \leq \mathrm{W}(z_1) \qquad (5.151)$$

Inserting the control function $\alpha(x_1)$ (5.150) in (5.148) leads to a new system representation (5.152) where the system is rewritten in the terms of the new state z_2:

$$\begin{aligned} \dot{z}_1 &= f + h(z_2 + \alpha) - \dot{y}_{ref} \\ \dot{z}_2 &= u - \dot{\alpha} = u - \frac{\partial \alpha}{\partial x_1}\left[f + h(z_2 + \alpha)\right] + \frac{\partial \alpha}{\partial y_{ref}}\dot{y}_{ref} + \frac{\partial \alpha}{\partial \dot{y}_{ref}}\ddot{y}_{ref} \end{aligned} \qquad (5.152)$$

The error between x_2 and the desired control function α was defined in (5.150) and according to (5.151) the system representation changed consequently again.

The previous step in Eq. (5.152) is the reason why this technique is called Integrator Backstepping. When comparing Fig. 5.39 with Fig. 5.40, it becomes apparent that the control function $\alpha(x_1)$ is shifted in front of the first integrator.

In conclusion, the original system is transformed to be represented in a form where all state variables have to be rendered to zero. The task is now to find a control law for u that ensures that z_2 converges to zero, i.e., x_2 converges to its desired value α. Therefore a CLF for the z_1, z_2 system is needed. A natural assumption is to use the CLF in (5.151) and augment it by a quadratic term, which penalizes the error z_2. Therefore, an extension of the previous Lyapunov function that includes both states is defined and by using Eq. (5.141) it is possible to derive $\dot{V}_2$:

$$V_2(z_1, z_2) = V(z_1) + \frac{1}{2}z^2 \qquad (5.153)$$

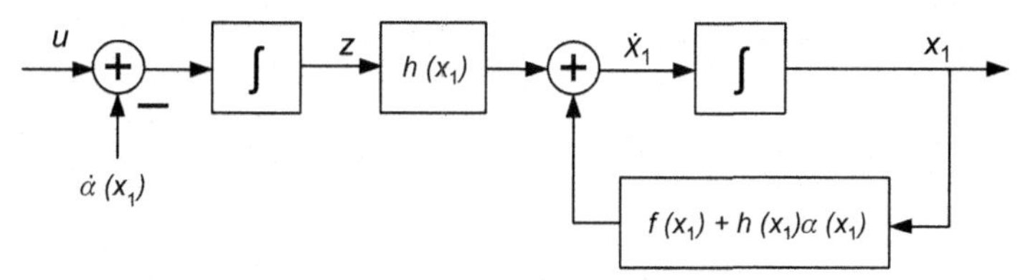

Figure 5.40 Control function αshifted in front of the integrator.

$$\begin{aligned}
\dot{V}_2(x_1,x_2) &= \dot{V}_1 + z_2\dot{z}_2 \\
&= \dot{V}_1 + z_2\left[u - \frac{\partial\alpha}{\partial x_1}\left[f + h(z_2+\alpha)\right] + \frac{\partial\alpha}{\partial y_{ref}}\dot{y}_{ref} + \frac{\partial\alpha}{\partial \dot{y}_{ref}}\ddot{y}_{ref}\right] \\
&= \frac{\partial V_1}{\partial z_1}\left[f + h(z_2+\alpha) - \dot{y}_{ref}\right] \\
&\quad + z_2\left[u - \frac{\partial\alpha}{\partial x_1}\left[f + h(z_2+\alpha)\right] + \frac{\partial\alpha}{\partial y_{ref}}\dot{y}_{ref} + \frac{\partial\alpha}{\partial \dot{y}_{ref}}\ddot{y}_{ref}\right] \\
&= \frac{\partial V_1}{\partial x_1}\left[f + h(z_2+\alpha) - \dot{y}_{ref}\right] \\
&\quad + z_2\left[\frac{\partial V_1}{\partial z_1}h + u - \frac{\partial\alpha}{\partial x_1}\left[f + h(z_2+\alpha)\right] - \frac{\partial\alpha}{\partial y_{ref}}\dot{y}_{ref} - \frac{\partial\alpha}{\partial \dot{y}_{ref}}\ddot{y}_{ref}\right] \\
&\le W(z_1) + z_2\left[\boldsymbol{\frac{\partial V_1}{\partial z_1}h + u - \frac{\partial\alpha}{\partial x_1}\left[f(x_1) + h(z_2+\alpha)\right]}\right. \\
&\quad \left.\boldsymbol{- \frac{\partial\alpha}{\partial y_{ref}}\dot{y}_{ref} - \frac{\partial\alpha}{\partial \dot{y}_{ref}}\ddot{y}_{ref}}\right]
\end{aligned} \tag{5.154}$$

Since the first part of the sum was already stated to be less than zero, the same has to be achieved for the two last parts (in blue) to ensure stability. This can be done by selecting the control law accordingly:

$$u = -cz_2 - \frac{\partial V}{\partial z_1}h(x_1) + \frac{\partial\alpha}{\partial x_1}\left[f + h(z_2+\alpha)\right] - \frac{\partial\alpha}{\partial y_{ref}}\dot{y}_{ref} - \frac{\partial\alpha}{\partial \dot{y}_{ref}}\ddot{y}_{ref};\ c>0 \tag{5.155}$$

Consequently, the derivative of Eq. (5.153) results in:

$$\dot{V}_2 = -W(z_1) - cz_2^2 < 0 \tag{5.156}$$

Therefore, the equilibrium point of $(z_1, z_2) = 0$ is globally stable. With (5.156) the tracking problem is also solved as for $t \rightarrow \infty, x_1 \rightarrow y_{ref}$. It should be mentioned this is not restricted only to the quadratic CLF and may in certain conditions render better results as certain nonlinearities are not canceled out [62].

5.5.3 Application to MVDC System – Backstepping With Virtual Disturbance

The controller which is synthesized here is based on similar principles to the controller presented in Section 5.4.2. It uses also the virtual disturbance current as a decoupling approach. This simplifies the Backstepping approach, as now only a linear system must be controlled but requires the design of a linear Kalman filter for the estimation of the disturbance. Fig. 5.41 depicts the overall control scheme. The design is composed of two independent blocks. The Kalman filter receives only the bus voltage measurement and estimates the inductor current and disturbance current which are fed to the controller.

The design of the controller is carried out in two steps:

1. The Backstepping procedure is applied including the estimated disturbance current as if it was already known.
2. The design of the Kalman filter is performed afterwards.

The circuit which is presented in Fig. 5.42, leads to the following system description with the following two states, $x_1 = V$ and $x_2 = I_L$.:

$$
\begin{aligned}
\dot{x}_1 &= \frac{x_2}{C_f} - \frac{x_1}{R_L C_f} - \frac{I_d}{C_f} \\
\dot{x}_2 &= \frac{dE}{L_f} - \frac{x_2 R_f}{L_f} - \frac{x_1}{L_f}
\end{aligned}
\tag{5.157}
$$

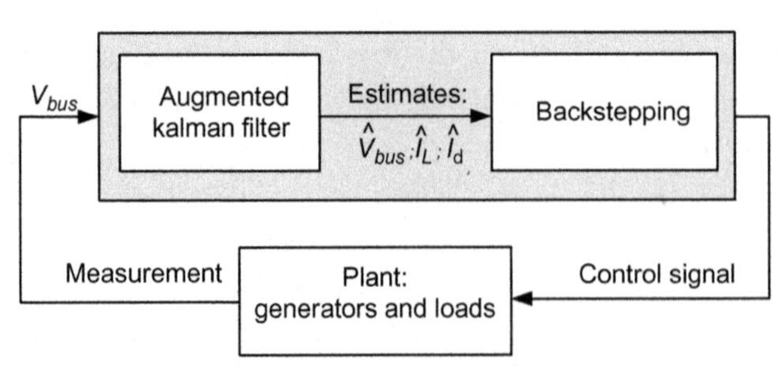

Figure 5.41 Structure of Backstepping controller with a Kalman filter for disturbance estimation.

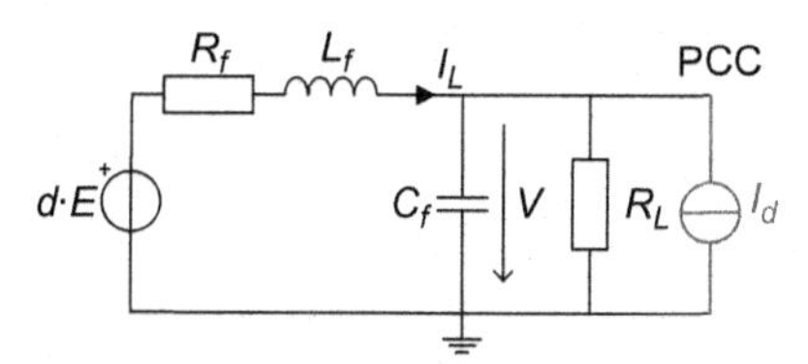

Figure 5.42 Decentralized model with disturbance current I_d.

Those equations are based on the averaged model in CCM, where v is the voltage across the capacitor and and I_L the inductor current of the converter output filter. E is the input voltage of the buck converter, d the duty cycle, R_L the value of the resistive load, and L_f, R_f, and C_f the converter output filter values. It has to be highlighted that the disturbance current I_d is included in the system description and it is assumed to be a known quantity at this stage of the design process.

The aim is to track a certain reference voltage V_{ref} rather than forcing V to zero. A change of coordinates is introduced, where z_1 is the new state error variable: The error variable z_1 is defined as:

$$z_1 = x_1 - V_{ref} \tag{5.158}$$

Deriving Eq. (5.158) and setting $\dot{V}_{ref} = 0$, Eq. (5.159) is obtained:

$$\dot{z}_1 = \frac{x_2}{C_f} - \frac{x_1}{R_L C_f} - \frac{i_d}{C_f} \tag{5.159}$$

If the reference voltage V_{ref} was not constant, it would have to be sufficiently smooth. Otherwise, step changes in V_{ref} would cause strong perturbations in the output voltage since the first and second derivatives of V_{ref} are introduced in the control signal.

The first step consists of finding a virtual control input which renders z_1 Lyapunov stable. The only variable that can be influenced in Eq. (5.159) is x_2. This is the virtual control input for $\dot{z}_1$. The intention is to set:

$$\dot{z}_1 = -c_1 z_1 \tag{5.160}$$

The estimation is performed by the Kalman filter. Therefore, the model includes a known disturbance estimation that can be canceled. The design parameter is $c_1 > 0$. To eliminate the term $-\frac{x_1}{R_L C_f} - \frac{I_d}{C_f}$ and ensure that Eq. (5.160) is satisfied:

$$\alpha(x_1) = \frac{x_2}{C_f} = -c_1 z_1 + \frac{x_1}{R_L C_f} + \frac{I_d}{C_f} \tag{5.161}$$

Consequently, Eq. (5.159) can be written as follows:

$$\dot{z}_1 = -c_1 z_1 + z_2 \tag{5.162}$$

Differentiating the frequently used Lyapunov function equation with respect to time, and inserting $V_1(x_1) = \frac{1}{2} z_1^2$ results in:

$$\dot{V}_1 = z_1 \dot{z}_1 = -c_1 z_1^2 + z_1 z_2 \tag{5.163}$$

The second error variable is now defined as:

$$z_2 = \frac{x_2}{C_f} - \alpha(x_1) = \frac{x_2}{C_f} + c_1 z_1 - \frac{x_1}{R_L C_f} - \frac{I_d}{C_f} \tag{5.164}$$

For which the derivative corresponds to:

$$\dot{z}_2 = \frac{\dot{x}_2}{C_f} - c_1^2 z_1 + c_1 z_1 - \frac{x_2}{R_L C_f^2} + \frac{x_1}{R_L^2 C_f^2} + \frac{I_d}{R_L C_f^2} - \frac{\dot{I}_d}{C_f} \tag{5.165}$$

Inserting now (5.164) and (5.165) in the Lyapunov function:

$$V_2(x_1, x_2) = \dot{V}_1 + \frac{1}{2} z_2^2 \tag{5.166}$$

Enables us now to calculate the derivative in the following way:

$$\begin{aligned}
\dot{V}_2 &= \dot{V}_1 + z_2 \dot{z}_2 \\
&= -c_1 z_1^2 + z_1 z_2 + z_2 \left(\frac{\dot{x}_2}{C_f} - c_1^2 z_1 + c_1 z_2 - \frac{x_2}{R_L C_f^2} \right. \\
&\qquad \left. + \frac{x_1}{R_L^2 C_f^2} + \frac{I_d}{R_L C_f^2} - \frac{\dddot{i}_d}{C_f} \right) \\
&= -c_1 z_1^2 - c_2 z_2^2 + z_2 \left(\frac{\dot{x}_2}{C_f} - (c_1^2 - 1) z_1 + (c_1 + c_2) z_2 \right. \\
&\qquad \left. - \frac{x_2}{R_L C_f^2} + \frac{x_1}{R_L^2 C_f^2} + \frac{I_d}{R_L C_f^2} - \frac{\dot{I}_d}{C_f} \right)
\end{aligned} \tag{5.167}$$

After assuring the stability of V_1, it has to be ensured that:

$$\dot{z}_2 = -c_2 z_2 - z_1 \tag{5.168}$$

To guarantee Lyapunov stability the inequality $\dot{V}_2 \leq 0$ has to be valid. Again, $c_2 > 0$ is a design parameter. This can be achieved by setting the term inside of the brackets to zero. Combining Eqs. (5.167) and (5.168) leads to a duty cycle:

$$d = \frac{L_f C_f}{E}\left(\frac{x_1}{L_f C_f} + \frac{x_2 R_f}{L_f C_f} + (c_1^2 - 1)z_1 - (c_1 + c_2)z_2 + \frac{x_2}{R_L C_f^2} - \frac{x_1}{R_L^2 C_f^2} - \frac{I_d}{R_L C_f^2} + \frac{\dot{I}_d}{C_f}\right) \tag{5.169}$$

The duty cycle is also the final control law for the virtual disturbance Backstepping procedure. This leads to:

$$\begin{aligned} \dot{V}_2 &= \dot{V}_1 + z_2\dot{z}_2 \\ &= -c_1 z_1^2 + z_1 z_2 + z_2(-c_2 z_2 - z_1) \\ &= -c_1 z_1^2 - c_2 z_2^2 < 0, \quad z_1, z_2 \neq 0. \end{aligned} \tag{5.170}$$

Therefore, ensuring Eq. (5.169) guarantees that the derivative of the Lyapunov function V_2 is less than zero. According to the Lyapunov direct method, this means that the system is asymptotically stable around the point where the voltage equals the reference voltage. It is noteworthy that the tuning of the design parameters c_1 and c_2 influences how much each state overshoots. An example can be given by decreasing the design parameter c_1, which has as a consequence a slower convergence of the voltage and less overshoot in the inductor current, since the weight of error z_1 is also decreased.

A very interesting conclusion can be taken when rewriting (5.170):

$$-c_1 z_1^2 - c_2 z_2^2 = \dot{V}_2 \leq -2\theta V_2; \quad \theta = \min(c_1, c_2) \tag{5.171}$$

Therefore (5.171) is described by an exponential function, which means that V converges exponentially fast to zero, and as $z_{1,2}$ are the error variables consequently the error converges exponentially fast to zero. This has consequences for (5.158), which represents the deviation of the bus voltage from V_{ref}, where the steady-state error disappears. This implies that the DC bus is tightly controlled. Extending this thought to (5.164), which represents the current, implies that the disturbance current I_d, which is the reference, is perfectly tracked. Taking now into account (5.163), (5.166), and (5.170), where V is positive definite and $\dot{V}$ is negative definite, has as a consequence for the error system that it is globally asymptotical stable around the origin.

5.5.4 Simulation Results

5.5.4.1 Cascaded System

A cascaded converter setup has been modeled in Matlab-SIMULINK with the parameters described in Table 6.1. In the simulation, a step variation of the CPL load has been performed. The variation of the bus voltage during the step change of the CPL load is described in Fig. 5.43. At $t = 0.25$ s the load step of 7.5 MW was performed, demonstrating that the *Virt Dist BS* control stabilizes the system towards the reference point $V_{ref} = 1$. The dynamic of the inductor current is described in Fig. 5.44.

5.5.4.2 Shipboard Power System

A Shipboard Power System has been modeled in Matlab\SIMULINK, based on the parameters in Table 6.2. The load steps of 22.5 MW have been performed at $t = 0.25$ s. The coefficients of the droop controllers were chosen to obtain an even current sharing among the converters. All converters equally share the total load power between them.

The effect of the increasing and decreasing of the CPL load on the load voltage is described in Fig. 5.45, demonstrating that the voltage remains stable and reaches the voltage set-points that are gradually modified by the action of the droop control. In Fig. 5.46 one can observe the transient for the current.

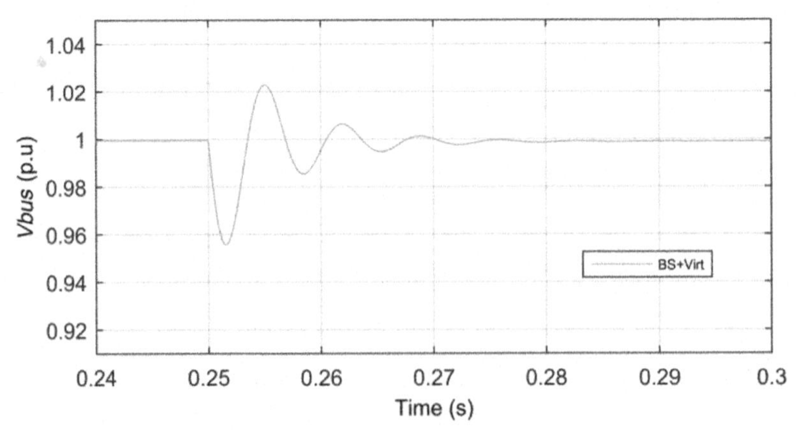

Figure 5.43 Voltage virtual distribution BS cascaded – Ideal CPL – Step 10.3 → 17.8 MW.

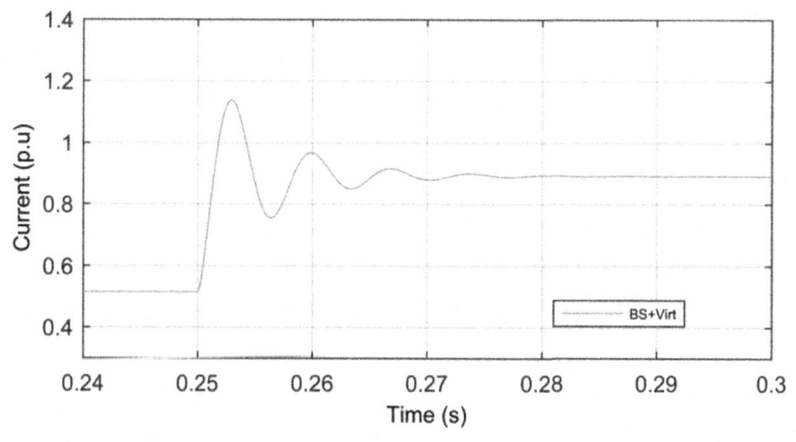

Figure 5.44 Current virtual distribution BS cascaded – Ideal CPL – Step 10.3 → 17.8 MW.

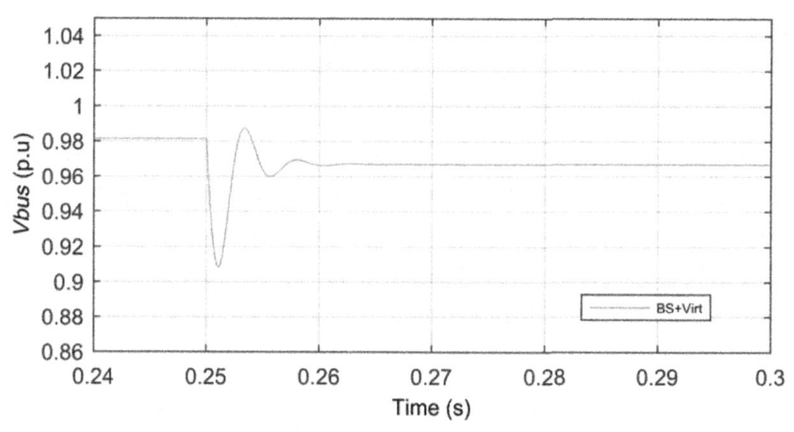

Figure 5.45 ISPS V_{bus} virtual distribution BS for ideal CPL – Step 30.9 → 53.4 MW.

5.6 ADAPTIVE BACKSTEPPING CONTROL

In the following subsections the reader is introduced into the theory behind Adaptive Backstepping [51]. This adaptive formulation of the Backstepping procedure allows a decoupling of the ISPS, and enables a decentralized control scheme which is no longer dependent on the virtual current source/virtual disturbance approach used in Sections 5.4 and 5.5. Instead of estimating a current source, now the power demand of the nonlocal network is estimated. Therefore, the estimation process is now

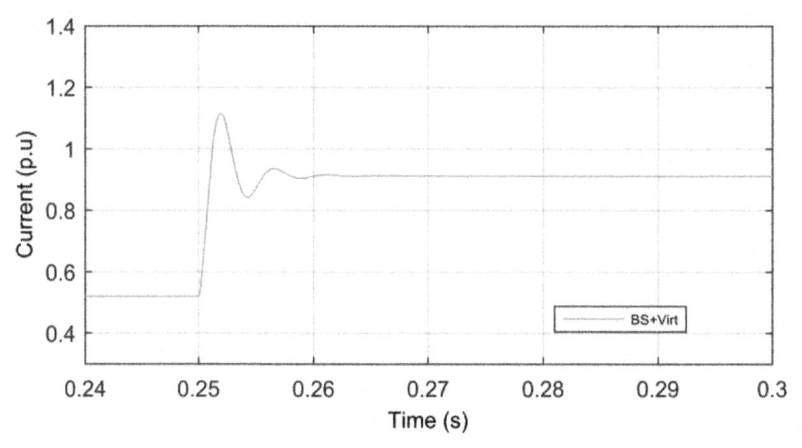

Figure 5.46 ISPS $I_{L1=L2=L3}$ virtual distribution BS for ideal CPL and 30.9 → 53.4 MW.

no longer linear as it was when using the steady-state Kalman filter but it is nonlinear through the CPL characteristic. In a nutshell the procedure is the application of the normal Backstepping algorithm and introducing an estimate with its corresponding estimation error, to formulate it, in an adaptive algorithm that is able to track the unknown time-varying load power changes. The theoretic foundation for a decentralized adaptive control algorithm with unmodeled dynamics was presented in [63].

In Section 5.6.1 the procedure of Adaptive Backstepping is explained. The reader who is familiar with those theoretical concepts will find in Section 5.6.2 the application of the Adaptive Backstepping theory on the ISPS circuit.

5.6.1 Procedure Adaptive Backstepping

At this point, a new dynamic controller design is introduced. The advantage of this approach is that the steady-state error resulting from uncertainties is avoided. This can be achieved by adaptively changing the uncertain parameters in the controller as described in [51]. The same example as in Section 5.5.2 is used which is based on [6,61,64].

In (5.147), it is assumed that the model was completely known. This is of course not always the case. This time, a model of the form (5.172) is examined, which is given in strict feedback form. For strict feedback form it is important that nonlinearities f_i and h_i of the derivative of state x_i are only dependent on the previous states which are fed back.

$$\begin{aligned}\dot{x}_1 &= \theta_{f_1}^T \boldsymbol{f}_1(x_1) + h_1(x_1)\cdot x_2 \\ \dot{x}_2 &= \theta_{f_2}^T \boldsymbol{f}_2(x_1, x_2) + h_2(x_1, x_2)\cdot x_3 \\ &\vdots \\ \dot{x}_k &= \theta_{f_k}^T \boldsymbol{f}_k(x_1, x_2, \ldots, x_3) + h_k(x_1, x_2, \ldots, x_k)\cdot u\end{aligned} \tag{5.172}$$

This time the known functions $\boldsymbol{f}_k$ return vectors. Also, the unknown parameter θ is considered to be a vector. The output is defined as $y = x_1$ which should track a reference signal $y_{ref}(t)$. This tracking control problem can be transformed to a regulation problem by introducing the tracking error variable. The error signal is defined as $z_i = x_i - \alpha_{i-1}$ (e.g., $z_1 = x_1 - y_{ref}$). Afterwards the first system equation is rewritten:

$$\dot{z}_1 = \dot{x}_1 - \dot{y}_{ref} = \theta_{f_1}^T \left[\boldsymbol{f}_1\left(z_1 + y_{ref}\right)\right] + \left[h_1\left(z_1 + y_{ref}\right)\right](z_2 + \alpha_1) - \dot{y}_{ref} \tag{5.173}$$

For simplicity $\boldsymbol{f}_1\left(z_1 + y_{ref}\right)$ will be referred to simply as $\boldsymbol{f}_1$, and similarly for h_1,$\boldsymbol{f}_2$,h_2.

At this stage the Backstepping algorithm is expanded, by introducing an estimate $\hat{\theta}_{f_1}$ of θ_{f_1} with the estimation error $\tilde{\theta}_{f_1} = \theta_{f_1} - \hat{\theta}_{f_1}$, (with $\dot{\tilde{\theta}}_{f_1} = -\dot{\hat{\theta}}_{f_1}$). By applying now the typical Lyapunov function candidate, that not only penalizes the tracking errors but also the estimation errors:

$$V_1\left(z_1, \hat{\theta}_{f_1}\right) = \frac{1}{2} z_1^2 + \frac{1}{2} \tilde{\theta}_{f_1}^T \Gamma_{f_1}^{-1} \tilde{\theta}_{f_1}^T \tag{5.174}$$

The newly introduced term $\Gamma = \Gamma^{\mathrm{T}} > 0$ is the adaptation gain matrix, and determines how fast the parameters of $\hat{\theta}$ are adapted. This leads to the following Lyapunov derivative:

$$\begin{aligned}\dot{V}_1 &= z_1 \dot{z}_1 + \tilde{\theta}^T \Gamma^{-1} \dot{\tilde{\theta}} \\ &= z_1 \left(\theta_{f_1}^T \boldsymbol{f}_1 + h_1(z_2 + \alpha_1) - \dot{y}_{ref}\right) - \tilde{\theta}_{f_1}^T \Gamma^{-1} \left(\dot{\hat{\theta}}_{f_1} - \Gamma_{f_1} \boldsymbol{f}_1 z_1\right)\end{aligned} \tag{5.175}$$

To cancel the indefinite terms, virtual control α_1, and the intermediate update laws $\tau_{f_{11}}$ are defined as:

$$\alpha_1 = \frac{1}{h_1}\left(-c_1 z_1 - \hat{\theta}_{f_1}^T \boldsymbol{f}_1 + \dot{y}_{ref}\right); c > 0 \tag{5.176}$$

$$\tau_{f_{11}} = \dot{\hat{\theta}}_{f_1} = \Gamma_{f_1} \boldsymbol{f}_1 z_1 \tag{5.177}$$

Eq. (5.177) is often referred to in the literature as a so-called tuning function [64]. Like the control law, these parameter update laws are built recursively. Substituting into the derivative of V_1leads to:

$$\dot{V}_1 = -c_1 z_1 + h_1 z_1 z_2 - \tilde{\theta}_{f_1}^T \Gamma_{f_1}^{-1} \left(\dot{\hat{\theta}}_{f_1} - \tau_{f_{11}} \right) \tag{5.178}$$

If the system would be of relative degree $=1$ this would be the final design step, the update laws would cancel the indefinite term while $z=0$ and the derivate would be reduced to:

$$\dot{V}_1 = -c_1 z_1^2 \tag{5.179}$$

which implies that the z_1 subsystem would be stabilized.

Thus, the next logical step lies on assuring that the z_2 subsystem converges; by using α_1 it is possible to step back to the second equation, which can be written after the introduction of the error variables $z_i = x_i - \alpha_{i-1}$:

$$\dot{z}_2 = \theta_{f_2}^T \boldsymbol{f}_2 + h_2(z_3 + \alpha_2) - \dot{\alpha}_1 \tag{5.180}$$

The second Lyapunov function can be therefore written as:

$$V_2 = V_1 + \frac{1}{2} z_2^2 + \frac{1}{2} \tilde{\theta}_{f_2}^T \Gamma_{f_2}^{-1} \tilde{\theta}_{f_2}^T \tag{5.181}$$

The derivative of (5.181) along the solutions of (5.173) and (5.180) corresponds to:

$$\begin{aligned} \dot{V}_2 = & -c_1 z_1^2 + h_1 z_2 \dot{z}_2 - \tilde{\theta}_{f_1}^T \Gamma^{-1} \left(\dot{\hat{\theta}}_{f_1} - \tau_{f_{11}} + \Gamma_{f_1} \boldsymbol{f}_1 \frac{\partial \alpha_1}{\partial x_1} z_2 \right) \\ & + z_2 \left[\hat{\theta}_{f_2}^T \boldsymbol{f}_2 + h_2(z_3 + \alpha_2) + \mu_1 \right] \\ & - \tilde{\theta}_{f_2}^T \Gamma^{-1} \left(\dot{\hat{\theta}}_{f_2}^T - \Gamma_{f_2} \boldsymbol{f}_2 z_2 \right) \end{aligned} \tag{5.182}$$

The term μ_1 represents the dynamics of $\dot{\alpha}_1$ and is calculated by the partial derivatives:

$$\mu_1 = \frac{\partial \alpha_1}{\partial t} = \frac{\partial \alpha_1}{\partial x_1} \left(\hat{\theta}_{f_1}^T \boldsymbol{f}_1 + h_1 x_2 \right) + \frac{\partial \alpha_1}{\partial \hat{\theta}_{f_1}} \dot{\hat{\theta}}_{f_1} + \frac{\partial \alpha_1}{\partial y_{ref}} \dot{y}_{ref} + \frac{\partial \alpha_1}{\partial \dot{y}_{ref}} \ddot{y}_{ref} \tag{5.183}$$

At this point it has to be mentioned that (5.175) is not necessarily negative definite, but the whole system with z_2 is asymptotically stable. Therefore now an update law for α_2 has to be defined. The virtual control and intermediate update laws are selected as:

$$\alpha_2 = \frac{1}{h_2}\left(-c_2 z_2 - h_1 z_1 - \hat{\theta}_{f_2}^T \boldsymbol{f}_2 + \mu_1\right); c > 0 \tag{5.184}$$

$$\tau_{f12} = \tau_{f11} - \Gamma_{f_1} \boldsymbol{f}_1 \frac{\partial \alpha_1}{\partial x_1} z_2 = \Gamma_{f_1} \boldsymbol{f}_1 \left(z_1 - \frac{\partial \alpha_1}{\partial x_1} z_2\right) \tag{5.185}$$

$$\tau_{f22} = \Gamma_{f_3} \boldsymbol{f}_2 z_2 \tag{5.186}$$

$$\dot{\hat{\theta}}_{f_2} = \tau_{f12} + \tau_{f22} \tag{5.187}$$

Taking the time derivative of (5.182) and inserting (5.197)–(5.199) results in:

$$\begin{aligned} \dot{V}_2 = & -cz_1^2 - c_2 z_2^2 + \boldsymbol{h}_2 \boldsymbol{z}_2 \boldsymbol{z}_3 - \tilde{\boldsymbol{\theta}}_{f_1}^T \boldsymbol{\Gamma}_{f_1}^{-1}\left(\dot{\hat{\boldsymbol{\theta}}}_{f_1} - \boldsymbol{\tau}_{f_{12}}\right) \\ & - \tilde{\boldsymbol{\theta}}_{f_2}^T \boldsymbol{\Gamma}_{f_2}^{-1}\left(\dot{\hat{\boldsymbol{\theta}}}_{f_2} - \boldsymbol{\tau}_{f_{22}}\right) \end{aligned} \tag{5.188}$$

Those steps are identically repeated until step n, where the real control input u enters the system, which for a second-order corresponds to the second step so that (5.197) changes to:

$$u = \alpha_2 = \frac{1}{h_2}\left(-c_2 z_2 - h_1 z_1 - \hat{\theta}_{f_2}^T \boldsymbol{f}_2 + \mu_1\right); c > 0 \tag{5.189}$$

This means that by setting $z_3 = 0$ and by substituting (5.198), (5.199) and (5.202) into (5.201), its blue part disappears and the following Lyapunov derivate is negative definite:

$$\dot{V}_2 = -c_1 z_1^2 - c_2 z_2^2 \tag{5.190}$$

At this point it should be mentioned explicitly that convergence of the parameter estimate $\hat{\theta}$ is guaranteed, but not necessarily convergence to the real value of θ.

One problem that arises when using Adaptive Backstepping is that more than one update law is obtained for each unknown constant. An example for this effect is given in the following section, where the unknown power is estimated by two update laws. It is possible to remove the overparameterization by choosing appropriate tuning functions, but it was shown that overparametrized controllers can reduce cost [65].

5.6.2 Application to MVDC System – Adaptive Backstepping With Power Estimation

In contrast to the Backstepping controller which was designed so far, the following control approach tries to incorporate the nonlinearity of the CPL and take advantage of it. Therefore, the constant power is estimated twice using one update law for each estimate. The update laws are retrieved from the Lyapunov condition for stability and are derived in the following. The overall structure of the controller which is designed in the end is depicted in Fig. 5.47.

The circuit model of the decentralized buck converter modeled in averaged mode with this corresponding DC link is depicted in Fig. 5.48. The model includes the constant power P which is estimated in the complete controller.

State-space representation (5.191) is derived according to the model in Fig. 5.48, where the current going through the CPL is $I_{CPL} = P/V$.

$$
\begin{aligned}
\dot{x}_1 &= \frac{x_2}{C_f} - \frac{x_1}{R_L C_f} - \frac{P}{x_1 C_f} \\
\dot{x}_2 &= \frac{dE}{L_f} - \frac{x_2 R_f}{L_f} - \frac{x_1}{L_f}
\end{aligned}
\tag{5.191}
$$

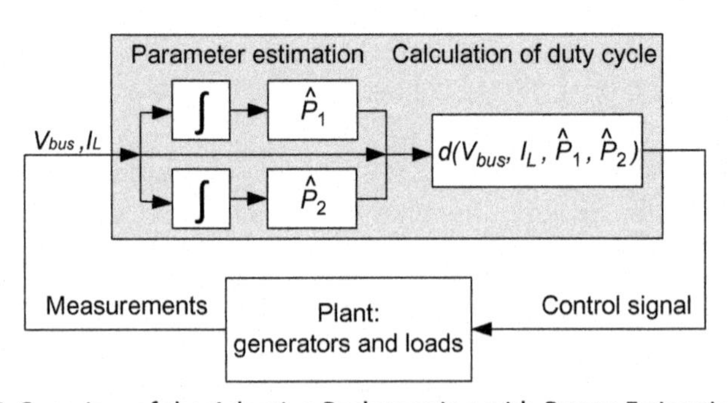

Figure 5.47 Overview of the Adaptive Backstepping with Power Estimation scheme.

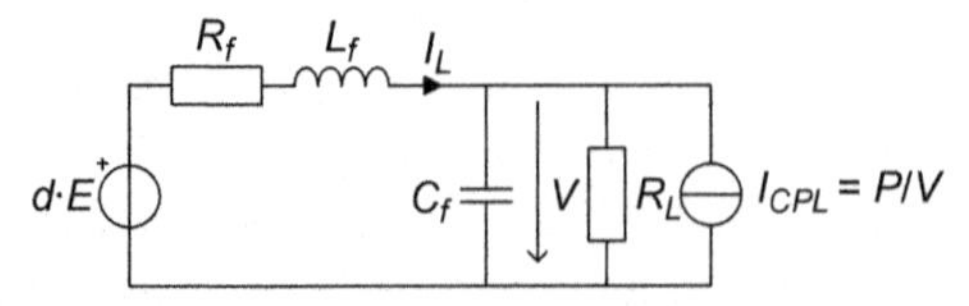

Figure 5.48 Averaged converter model with disturbance power.

Variable x_1 corresponds to V and x_2 to I_L. Next, the error variable $z_1$is introduced, for shifting the system's coordinates are shifted by V_{ref}. This is done in the same way as for the previous Backstepping approach.

$$z_1 = x_1 - V_{ref} \tag{5.192}$$

Since the reference voltage is supposed to be constant, $\dot{V}_{ref}$ is zero. Deriving (5.192) leads to

$$\dot{z}_1 = \frac{x_2}{C_f} - \frac{\boldsymbol{x_1}}{\boldsymbol{R_L C_f}} - \frac{\boldsymbol{P}}{\boldsymbol{x_1 C_f}}. \tag{5.193}$$

The goal is to set $\dot{z}_1 = -c_1 z_1$, where $c_1 > 0$ is a design parameter. In order to remove the unwanted terms in blue, the virtual control function was selected in the following way:

$$\alpha(x_1) = \frac{x_2}{C_f} = -c_1 z_1 + \frac{x_1}{R_L C_f} + \frac{\hat{P}_1}{x_1 C_f} \tag{5.194}$$

In this case it is not possible to insert the parameter P in the virtual control law to drive the error z_1 to zero as it is unknown as it was described in Section 5.6.1. For this reason, it is replaced with the estimate $\hat{P}_1$ instead.

The typical Lyapunov function is selected again:

$$V_1(x_1) = \frac{1}{2} z_1^2. \tag{5.195}$$

Subsequently, an estimation error is induced in the system. That is why z_1 cannot be rendered to $\dot{z}_1 = -c_1 z_1 + z_2$. Inserting (5.194) into (5.193) leads to the term which is used instead:

$$\dot{z}_1 = -c_1 z_1 + z_2 - \left(P - \hat{P}_1\right) \cdot \frac{1}{x_1 C_f} \tag{5.196}$$

To guarantee Lyapunov stability, the error between the estimate $\hat{P}_1$ and the real value of P has to be considered in the Lyapunov function. This allows the possibility to select an update law for the estimate which stabilizes the system. Considering the estimation error in the Lyapunov function

$$V_1 = \frac{1}{2} z_1^2 + \frac{1}{2\Gamma_1} \left(P - \hat{P}_1\right)^2 \tag{5.197}$$

Differentiation with respect to time leads to:

$$\begin{aligned}\dot{V}_1 &= z_1\dot{z}_1 + \left(P - \hat{P}_1\right)\frac{\dot{\hat{P}}_1}{\Gamma_1} \\ &= -c_1 z_1^2 + z_1 z_2 + \left(P - \hat{P}_1\right)\left(-\frac{z_1}{x_1 C_f} - \frac{\dot{\hat{P}}_1}{\Gamma_1}\right)\end{aligned} \tag{5.198}$$

In line with the method applied in the previous chapter the update law is chosen in such a way that the last term is set to zero, therefore the update function of the CPL power $\hat{P}_1$ which ensures stability is obtained:

$$\dot{\hat{P}}_1 = -\Gamma_1 \frac{z_1}{x_1 C_f} \tag{5.199}$$

The second step requires another error variable which incorporates a second estimate as well. The second error variable can defined as:

$$z_2 = \frac{x_2}{C_f} - \alpha = \frac{x_2}{C_f} + c_1 z_1 - \frac{x_1}{R_L C_f} - \frac{\hat{P}_1}{x_1 C_f} \tag{5.200}$$

with the derivative:

$$\begin{aligned}\dot{z}_2 = \frac{\dot{x}_2}{C_f} - c_1^2 z_1 + c_1 z_1 - \left(P - \hat{P}_1\right)\frac{c_1}{x_1 C_f} - \frac{x_2}{R_L C_f^2} + \\ \frac{x_1}{R_L^2 C_f^2} + \frac{P}{x_1 R_L C_f^2} - \left(\frac{\dot{\hat{P}}_1}{x_1 C_f} - \frac{\hat{P}_1}{x_1^2 C_f}\right)\end{aligned} \tag{5.201}$$

Again, the Lyapunov function is not just extended by the second error variable but by the second estimate as well:

$$V_2 = \dot{V}_1 + \frac{1}{2}z_2^2 + \frac{1}{2\Gamma_2}\left(P - \hat{P}_2\right)^2 \tag{5.202}$$

Now the derivative of the Lyapunov function for the whole system needs to be calculated while inserting (5.201) into (5.202):

$$\begin{aligned}\dot{V}_2 &= \dot{V}_1 + z_2\dot{z}_2 - \left(P - \hat{P}_2\right)\frac{\dot{\hat{P}}_2}{\Gamma_2} \\ &= -c_1 z_1^2 + z_1 z_2 + z_2\left(\frac{\dot{x}_2}{C_f} - c_1^2 z_1 + c_1 z_2 - \left(P - \hat{P}_1\right)\frac{c_1}{x_1 C_f}\right.\end{aligned}$$

$$-\frac{x_2}{R_L C_f^2}+\frac{x_1}{R_L^2 C_f^2}+\frac{P}{x_1 R_L C_f^2}-\frac{\dot{\hat{P}}_1}{x_1 C_f}+\frac{\hat{P}_1}{x_1^2 C_f})$$

$$-\left(P-\hat{P}_2\right)\frac{\dot{\hat{P}}_2}{\Gamma_2}=-c_1 z_1^2-c_2 z_2^2+z_2$$

$$\left(\frac{\dot{x}_2}{C_f}-\left(c_1^2-1\right)z_1+(c_1+c_2)z_2-\frac{c_1\hat{P}_2}{x_1 C_f}+\frac{c_1\hat{P}_1}{x_1 C_f}\right.$$

$$-\frac{x_2}{R_L C_f^2}+\frac{x_1}{R_L^2 C_f^2}+\frac{\hat{P}_2}{x_1 R_L C_f^2}-\frac{\dot{\hat{P}}_1}{x_1 C_f}+\frac{\hat{P}_1}{x_1^2 C_f})$$

$$+\left(P-\hat{P}_2\right)\left(-\frac{\dot{\hat{P}}_2}{\Gamma_2}-\frac{c_1 z_2}{x_1 C_f}+\frac{z_2}{x_1 R_L C_f^2}\right) \tag{5.203}$$

Analyzing (5.203), it can be concluded that stability is assured if the terms in brackets and marked in blue are set to zero. Then it is guaranteed that Eq. (5.203) ≤ 0 because the design parameters $c_1, c_2 > 0$. In order to render the terms in brackets to zero, the variables d and the update law for the second estimate $\hat{P}_2$ are set to cancel out the rest of the terms marked in blue:

$$d=\frac{L_f C_f}{E}\left(\left(c_1^2-1\right)z_1-(c_1+c_2)z_2+\left(\frac{1}{L_f C_f}-\frac{1}{R_L^2 C_f^2}\right)x_1\right.$$

$$+\left(\frac{R_f}{L_f C_f}+\frac{1}{R_L C_f^2}\right)x_2+\frac{c_1\hat{P}_2}{x_1 C_f}-\frac{c_1\hat{P}_1}{x_1 C_f}-\frac{\hat{P}_2}{x_1 R_L C_f^2}+\frac{\dot{\hat{P}}_1}{x_1 C_f}-\frac{\hat{P}_1}{x_1^2 C_f}) \tag{5.204}$$

$$\dot{\hat{P}}_2=\Gamma_2\left(-\frac{c_1 z_2}{x_1 C_f}+\frac{z_2}{x_1 R_L C_f^2}\right) \tag{5.205}$$

An interesting observation can be made by analyzing the complete system composed of the error variables (5.206) and the update laws for the estimates (5.199) and (5.205).

$$\dot{z}_1 = -c_1 z_1 + z_2 - \left(P - \hat{P}_1\right) \cdot \frac{1}{x_1 C_f}$$
$$\dot{z}_2 = -z_1 - c_2 z_2 + \left(P - \hat{P}_2\right)\left(-\frac{c_1 x_1}{C_f} + \frac{\hat{w}_1 x_1}{C_f^2}\right) \tag{5.206}$$

As can be seen from (5.207) and (5.208), the matrix, which is multiplied by the estimation error before adding it to the error variables $\boldsymbol{z}$, is the same as the update matrix for the estimates.

$$\begin{bmatrix} \dot{z}_1 \\ \dot{z}_2 \end{bmatrix} = \begin{bmatrix} -c_1 & 1 \\ -1 & -c_2 \end{bmatrix} \begin{bmatrix} z_1 \\ z_2 \end{bmatrix} + \begin{bmatrix} -\dfrac{x_1}{C_f} & 0 \\ 0 & -\dfrac{c_1 x_1}{C_f} + \dfrac{\hat{w}_1 x_1}{C_f^2} \end{bmatrix} \begin{bmatrix} P - \hat{P}_1 \\ P - \hat{P}_2 \end{bmatrix} \tag{5.207}$$

$$\begin{bmatrix} \dot{\hat{P}}_1 \\ \dot{\hat{P}}_2 \end{bmatrix} = \begin{bmatrix} -\Gamma_1 \dfrac{1}{x_1 C_f} & 0 \\ 0 & \Gamma_2 \left(-\dfrac{c_1}{x_1 C_f} + \dfrac{1}{x_1 R_L C_f^2}\right) \end{bmatrix} \begin{bmatrix} z_1 \\ z_2 \end{bmatrix} \tag{5.208}$$

5.6.3 Simulation Results

5.6.3.1 Cascaded System

A cascaded converter setup has been modeled in Matlab-SIMULINK with the parameters described in Table 6.1. In the simulation, a step variation of the CPL load has been performed. The variation of the bus voltage during the step change of the CPL load is described in Fig. 5.49. At $t = 0.25$ s the load step of 7.5 MW was performed, demonstrating that the control stabilizes the system towards the reference point $V_{ref} = 1$. The dynamic of the inductor current is described in Fig. 5.50.

5.6.3.2 Shipboard Power System

A Shipboard Power System has been modeled in Matlab\SIMULINK, based on the parameters in Table 6.2. The load steps of 22.5 MW have been performed at $t = 0.25$ s. The coefficients of the droop controllers were chosen to obtain an even current sharing among the converters. All converters equally share the total load power between them.

The effect of the increasing and decreasing of the CPL load on the load voltage is described in Fig. 5.51, demonstrating that the voltage

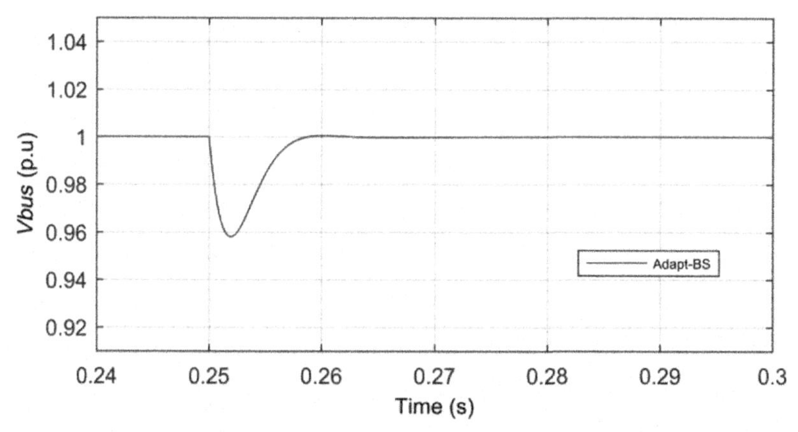

Figure 5.49 Voltage AdaptBS cascaded – Ideal CPL – Step 10.3 → 17.8 MW.

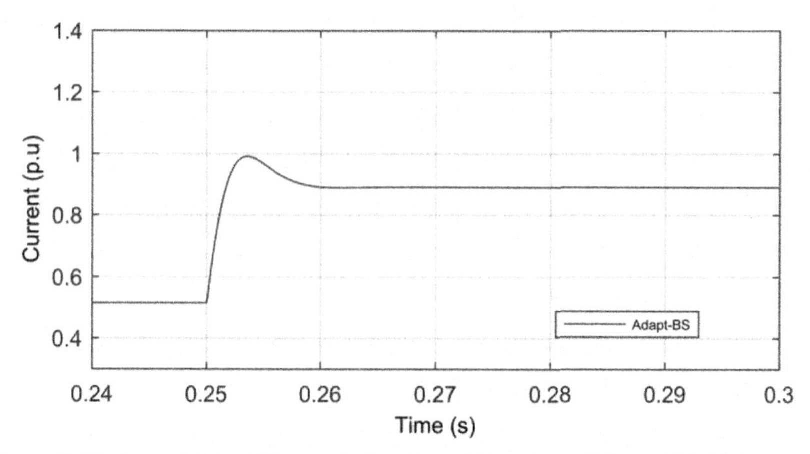

Figure 5.50 Current AdaptBS cascaded – Ideal CPL – Step 10.3 → 17.8 MW.

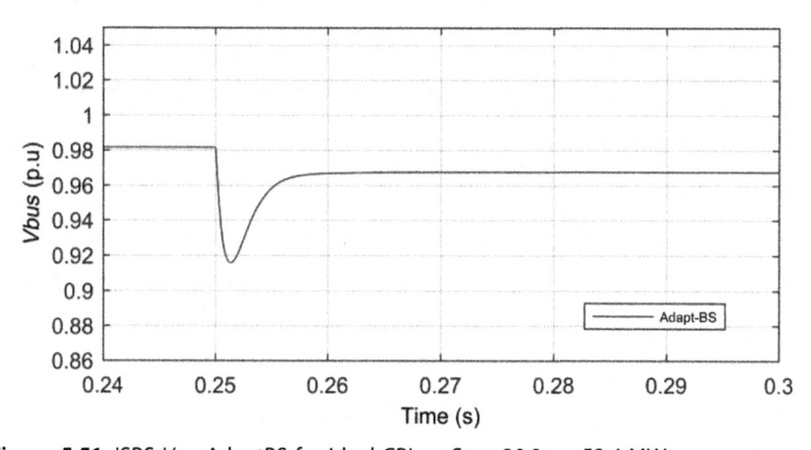

Figure 5.51 ISPS V_{bus} AdaptBS for ideal CPL – Step 30.9 → 53.4 MW.

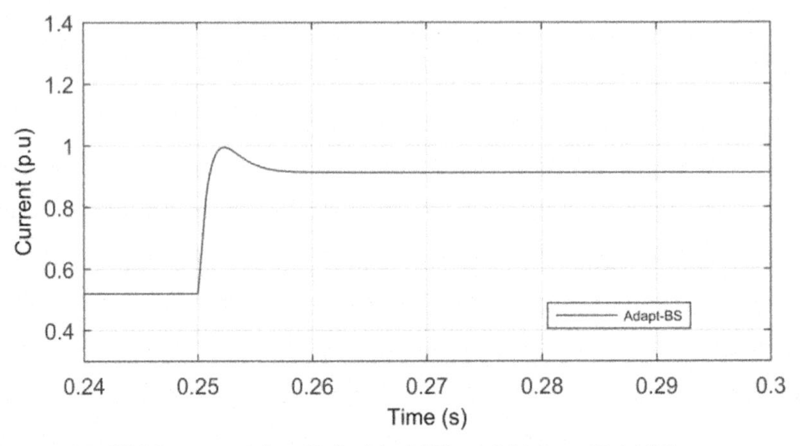

Figure 5.52 ISPS $I_{L1=L2=L3}$ AdaptBS for ideal CPL and 30.9 → 53.4 MW.

remains stable and reaches the voltage set-points that are gradually modified by the action of the droop control. In Fig. 5.52 one can observe the transient for the current.

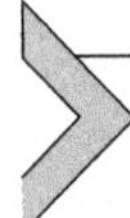

5.7 H$_\infty$ OPTIMAL CONTROL (CONTRIBUTOR SRIRAM KARTHIK GURUMURTHY)

In this section, the application of H$_\infty$ optimal control to power electronic converters is covered. An initial introduction to the history articulating how the control was developed, its first application to a control problem in general, and then to a power electronic converter and grid is covered. The fundamental concepts involved in H$_\infty$ optimal control are introduced and then a short example on how the method can be applied to a simple power electronic converter is shown. Finally, implementation of the technique for a MVDC grid is undertaken and the simulation test cases show the validity and usefulness of such a control technique from a power grid perspective.

5.7.1 Development of H$_\infty$ Control

The H$_\infty$ technique was first developed by Zames in the early 1980s [66–68]. Throughout the 1980s various computation methods and topics in linear matrix inequalities were undertaken and developed [69]. A major breakthrough in the field happened with the introduction of state-space solutions to H_2 and H_∞ problem. An elegant formulation of the theory through the works of Doyle [70] enabled the application of the technique in a more practical sense. The H_∞ controller found its first aviation application with the VSTOL (vertical short takeoff and landing) aircraft model through the work of Hyde in 1995 [70].

The H_∞ controller does not make assumptions on the statistics. The basic philosophy of the H_∞ approach is to perform minimization of a cost function for the worst-case disturbance input and hence the H_∞ approach is a mini-max optimization problem where the approach is pessimistic. These generalities that the H_∞ assumes about nature of the noise makes it more robust towards noise and uncertainties. This is of supreme importance for futuristic MVDC grids where one deals not only with fast dynamics but also with the presence of noise of unknown statistics. Apart from the influence of noise, the nonlinearities of power converters are also an important issue especially with the trend of increase in switching frequency of converters. The H_∞ approach allows modeling of parametric uncertainties and frequency-dependent uncertainties arising due to small signal approximations. The application of the H_∞ control problem to the stabilization of DC link in power electronic converters is investigated in [71].

5.7.2 Preliminaries

Firstly, we define a few mathematical preliminaries before going into norm optimization based control synthesis. A notation for state-space to transfer function transformation is defined as follows:

$$\mathrm{G(s)} := \begin{bmatrix} A & B \\ C & D \end{bmatrix} := C(sI-A)^{-1}B+D \tag{5.209}$$

A few definitions highlighting the important matrix norms and the notation of Riccati and Hamiltonian operators are described in a standardized way which will enable the reader to understand detailed theory from other literature.

5.7.2.1 H_2 Norm

The H_2 spaces are set of all matrix values functions G where the H_2 norm is bounded and analytic in open right-half plane. The H_2 norm is defined as follows:

$$\|G\|_2 = \sqrt{\frac{1}{2\pi}\int_{-\infty}^{\infty} Tr\{G^H(j\omega)G(j\omega)\}d\omega} \tag{5.210}$$

where G^H refers to Hermitian of G (complex conjugate transpose).

5.7.2.2 H_∞ Norm

The H_∞ spaces are set of all matrix values functions G where the H_∞ norm is bounded and analytic in open right-half plane. It is the set of all stable and proper rational transfer functions referred to as RH_∞. The H_∞ norm is defined as follows:

$$\|G\|_\infty := \sup_\omega \overline{\sigma} G(j\omega) \tag{5.211}$$

The upper singular value represented by $\overline{\sigma}$ is determined by singular value decomposition (SVD) of a matrix. The supremum operator denoted by *sup* denotes the peak $\overline{\sigma}$ as the frequency ω is varied over the entire range. Bisection algorithm is used to compute the infinity norm in an efficient manner [69]. In both H_2 and H_∞ theory, state-space formulation is done by solving Algebraic Riccati Equations (ARE). Hence, we define a Hamiltonian matrix (5.212) to denote the ARE of the form (5.213).

$$H := \begin{bmatrix} A & R \\ -Q & -A^H \end{bmatrix} \tag{5.212}$$

$$A^H X + XA + XRX + Q = 0 \tag{5.213}$$

The stabilizing solution X of the Riccati equation is defined by the Riccati operator (Ric) on the Hamiltonian H, which basically solves the eigenvalue problem in Eq. (5.212).

$$X := Ric(H) \tag{5.214}$$

5.7.3 Linear Fractional Transformation

Consider the generalized plant P and the controller K as shown in Fig. 5.53. The key idea behind norm based optimal control is to

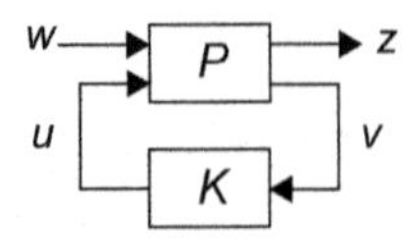

Figure 5.53 Lower LFT structure.

synthesize class of controller K which must satisfy conditions of internal stability and secondly minimize a specific matrix norm of interest. The controller input is the error signal v and its output is the control signal u, whilst w and z are exogenous inputs and outputs respectively. The input–output relation could be mapped through the P matrix from Eq. (5.215).

$$\begin{bmatrix} z \\ v \end{bmatrix} = P(s)\begin{bmatrix} w \\ u \end{bmatrix} = \begin{bmatrix} P_{11}(s) & P_{12}(s) \\ P_{21}(s) & P_{22}(s) \end{bmatrix}\begin{bmatrix} w \\ u \end{bmatrix} \tag{5.215}$$

In order to establish a direct mapping between w and z, one needs to eliminate the second row of the Eq. (5.215). The steps involved to eliminate u are as follows:

$$\begin{aligned} u = Kv \Rightarrow v = K^{-1}u \\ v = P_{21}w + P_{22}u \end{aligned} \tag{5.216}$$

$$\begin{aligned} K^{-1}u = P_{21}w + P_{22}u \\ (K^{-1} - P_{22})u = P_{21}w \\ \Rightarrow u = (K^{-1} - P_{22})^{-1}P_{21}w \\ u = K(I - P_{22}K)^{-1}P_{21}w \end{aligned} \tag{5.217}$$

Substituting the equation for u in the equation of z (first row), we get the following result:

$$\begin{aligned} z = P_{11}w + P_{12}u = P_{11}w + P_{12}K(I - P_{22}K)^{-1}P_{21}w \\ \Rightarrow z = (P_{11} + P_{12}K(I - P_{22}K)^{-1}P_{21})w \end{aligned} \tag{5.218}$$

The above rule is called as lower Linear Fractional Transformation (LFT) of K with P. The lower LFT basically describes the transfer function T_{zw} from exogenous inputs w to exogenous outputs z. By using the MIMO rule, one can express the exogenous outputs in terms of exogenous inputs to derive the generalized plant P and this is explained in the next section.

$$z = F_l(P, K)w \tag{5.219}$$

$$F_l(P, K) = (P_{11} + P_{12}K(I - P_{22}K)^{-1}P_{21}) \tag{5.220}$$

5.7.3.1 H_2 Optimal Control Problem

The H_2 optimal control statement is then to find a controller K which minimizes the 2-Norm of the transfer function T_{zw}.

$$\underset{K}{\mathrm{Min}} \left\| F_l(P,K) \right\|_2 \tag{5.221}$$

Both H_2 and LQG paradigm do not have guaranteed stability margins as far as robustness is concerned [72,73]. The LQG control is only a special case of H_2 control, where the noise is uncorrelated white noise with zero mean.

5.7.3.2 H_∞ Optimal Control Problem

Minimizing the infinity norm the transfer function T_{zw} parametrized by the controller K, leads to the H_∞ optimal controller.

$$\underset{K}{\mathrm{Min}} \left\| F_l(P,K) \right\|_\infty \tag{5.222}$$

The H_∞ is designed with pessimistic approach as it tries to minimize the worst-case disturbance on the system. The H_∞ control guarantees robust stability margins. The infinity norm satisfies multiplicative property and hence can be used to represent unstructured uncertainty. This is not possible with H_2 norm.

$$\left\| A(s)B(s) \right\|_\infty \leq \left\| A(s) \right\|_\infty \left\| B(s) \right\|_\infty \tag{5.223}$$

When the plant has low margins, bandwidth due to presence of RHP zero, [74] proposes design of pre- and postcompensator to condition the plant and then design of a weighting function to apply a mixed sensitivity approach for control design.

5.7.4 Weighted Sensitivity H_∞ Control

In this section, the H_∞ weighted sensitivity control is explained. In this approach, dynamic weighting functions, which are basically filters, are used to modify the loop shape of the plant in the desired manner. We are considering the output feedback structure, where only the output quantity is measured and used for negative feedback, the full information of the plant, which is the state measurements, are not required. The sensitivity transfer function $S(j\omega)$ is the transfer function from the disturbance d to the output y, similarly the complementarity sensitivity function $T(j\omega)$ represents the command tracking transfer function from the reference w to output y. The sum of S and T add up to unity for all frequencies as given in Eq. (5.224)

$$S(j\omega) + T(j\omega) = I \tag{5.224}$$

T should be high for good command tracking at low frequencies, but be low for high frequencies since we are not interested in tracking noise from the reference command. Ideally, we would prefer low values for S to eliminate the impact of disturbance. There is a conflict at higher frequencies that both S and T are required to be low. This is not possible due to the complementarity nature of S and T as evident from Eq. (5.224). To resolve this conflict, it is ideal to have S with high pass characteristics and T with low pass characteristics. The transfer function from noise source to the control input is known as the noise sensitivity transfer function $R(j\omega)$. R is required to have low pass characteristics, since the noise generally belongs to the higher side of the spectrum. Hence a low pass R would imply that the control input would not be sensitive to measurement noise. The expressions of the three key transfer functions are shown in (5.225).

$$\begin{aligned} S(j\omega) &= [I+GK]^{-1} \\ T(j\omega) &= GK[I+GK]^{-1} = I - S \\ R(j\omega) &= K[I+GK]^{-1} = KS \end{aligned} \tag{5.225}$$

5.7.4.1 Generalized Plant Modeling

In the weighted sensitivity approach, some crucial control signals of interest are augmented with dynamic weighting functions. In the previous section, the three sensitivity transfer functions were introduced. These sensitivity functions can be modified by choosing appropriate weighting functions. The explanation of how modifying each weighting function would modify its respective characteristic function is explained in the next section. In this section, the modeling of generalized plant or in other words the augmented plant is briefed. The weighting functions W_e, W_T, and W_u are frequency-dependent weightings of error signal v, plant output y, and the control input u. The augmented or generalized plant P consists of the actual plant G and the three weighting functions. It can be seen from Fig. 5.54, that the input for P are the reference command w and u and the outputs are z_1, z_2, z_3, and v. By using the MIMO rule, one can write the transfer functions from each output to input. The transfer function of the augmented plant $P(s)$ represents the mapping from the exogenous inputs (w,u) to the exogenous outputs (z_1, z_2, z_3, v) and it is given by Eq. (5.226).

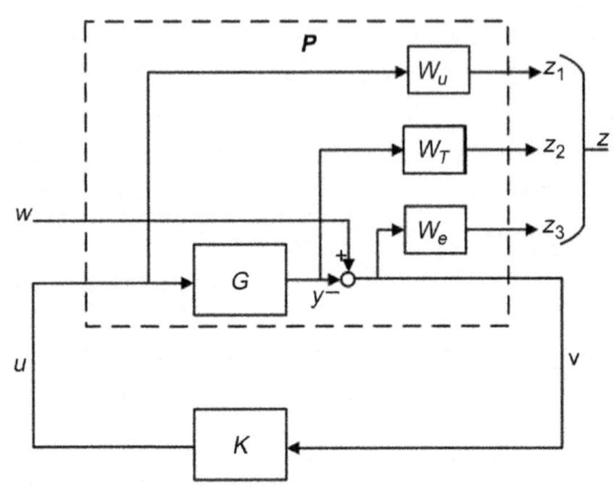

Figure 5.54 Generalized plant structure.

$$\begin{bmatrix} z_1 \\ z_2 \\ z_3 \\ v \end{bmatrix} = P(s)\begin{bmatrix} w \\ u \end{bmatrix} = \begin{bmatrix} 0 & W_u I \\ 0 & W_T G \\ W_e I & W_e G \\ -I & -G \end{bmatrix}\begin{bmatrix} w \\ u \end{bmatrix} \tag{5.226}$$

5.7.4.2 Description of Weighting Functions

The impact of having these weighting functions on the closed-loop transfer function can be understood by considering LFT of P and K. Thus, in the minimization of T_{zu}, the weightings can be used as punishments to achieve desired loop shape.

$$T_{zw} = \begin{bmatrix} W_u R \\ W_T T \\ W_e S \end{bmatrix}$$
$$z = \begin{bmatrix} z_1 \\ z_2 \\ z_3 \end{bmatrix} = T_{zw} w \tag{5.227}$$

It is desired to have low values of S at lower frequencies and low values of T at higher frequencies. To achieve the desired loop shape of S, W_e should have the characteristics of a low pass filter. Analogously, W_T should have high pass characteristics to get the desired loop shape of T. W_u must have high pass behavior to limit the control effort to a lower bandwidth, that is to achieve the desired low pass behavior for R. For the design of the weighting functions, we adopt first-order filter design. Higher-order filters are also possible, but it will result in a higher-order controller.

$$
\begin{aligned}
W_e(s) &= \frac{s/M_e + \omega_e}{s + \omega_e \epsilon} \\
W_T(s) &= \frac{s + \omega_T/M_T}{\epsilon s + \omega_T} \\
W_u(s) &= \frac{s + \omega_u/M_u}{\epsilon s + \omega_u}
\end{aligned}
\tag{5.228}
$$

The parameters ω_e, ω_u, and ω_T represent the bandwidth required and the parameter ϵ is a small positive number used to introduce a left half plane pole in the high pass filter (HPF) transfer function W_e. The parameters M_e, M_u, and M_T represent the peak sensitivity values. The design of these weighting functions will be explained in the next section with a practical example of an MVDC grid.

5.7.5 Application to MVDC System

In this section, we show the application of the H_∞ weighted sensitivity control strategy for the voltage stability of MVDC grid. The same system as in previous chapters is considered and the control design is done. The primary control consists of the H_∞ control and the secondary control consists of the virtual resistance droop controller for power sharing. Firstly, we explain how the weighting functions are designed to compute the H_∞ controller and then time domain simulations of the MVDC system show the validity and effectiveness of the control design. The overview of the H_∞ control technique is represented in Fig. 5.55.

5.7.5.1 Design of Weighting Functions

The design of weighting function is explained in this section. As explained earlier, it is desired to have S with high pass characteristics and T with low pass characteristics. Hence the bandwidth of S must be smaller than that of T. The bandwidth of S and T are chosen as 45 Hz and 100 Hz respectively. The bandwidth of S is more crucial than T, since it gives the actual sense of aggressiveness of the system. The higher the bandwidth of S, the greater is the sensitivity to disturbance.

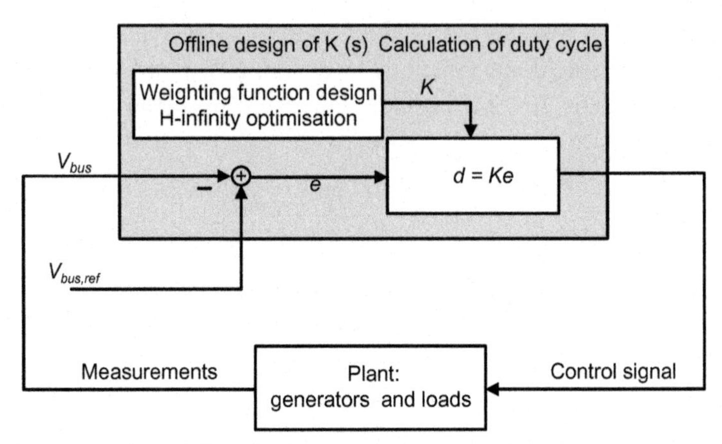

Figure 5.55 Overview of the mixed sensitivity H_∞ control design.

Similarly, higher the value of M_s, the greater is the aggressiveness. Typically, for achieving good performance the peak sensitivities are chosen at less than 6 dB and the value of M_T is chosen to be smaller than M_s. The values of M_s and M_T are 4.5 and 3 dB respectively. The choice of parameter ε must be an arbitrary positive number preferably close to zero, since it makes the ideal HPF transfer function of S proper by introducing a pole into the LHP. Similarly, in the low pass filter (LPF) transfer functions, the integrator introduces a pole at zero. For internal stability, pole at zero is not allowed and hence the parameter ε provides a time constant for the integrator by moving the pole into the LHP. For this scenario, we choose ε as 0.003. The noise sensitivity R is designed with a low pass behavior. Typically, our assumption is that the measurement noise exists significantly at higher frequencies than at lower frequencies. Hence the design of Wu is LPF. The cut off frequency is chosen as 1000 Hz (Figs. 5.56 and 5.57).

The H_∞ controller is designed by first forming the augmented plant or generalized plant P, which requires the information of weighting functions and the actual plant/converter model G. The MATLAB routine augw performs this function. The H_∞ controller can be designed using the *hinfsyn* command. The resulting controller K is a fifth-order controller with five poles and four zeroes. As expected, K is internally stabilizing since it satisfies the conditions of internal stability. The gain margin and phase margin of the controller are 19.8 dB and 68.8 degrees respectively (Figs. 5.58 and 5.59).

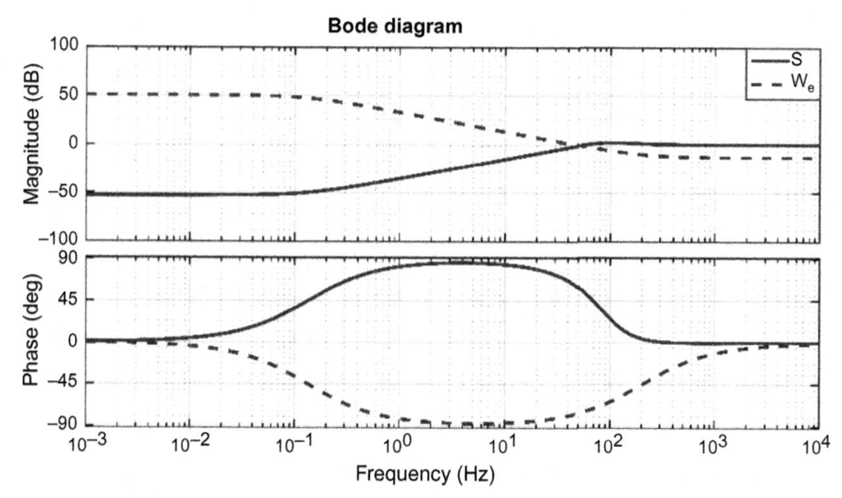

Figure 5.56 Sensitivity function and the corresponding weighting function W_e.

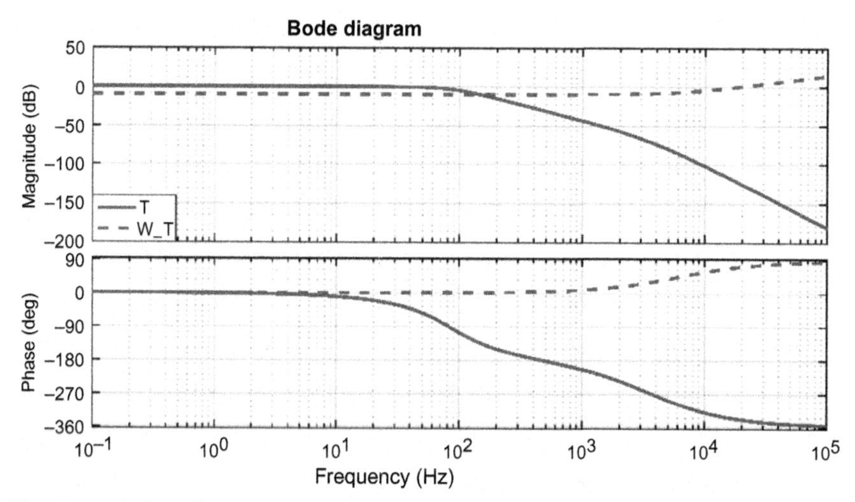

Figure 5.57 Complementarity sensitivity function and the corresponding weighting function W_T.

5.7.6 Simulation Results

5.7.6.1 Cascaded System

A cascaded converter setup has been modeled in Matlab-SIMULINK with the parameters described in Table 6.1. In the simulation, a step variation of the CPL load has been performed. The variation of the bus voltage during the step change of the CPL load is described in Fig. 5.60. At

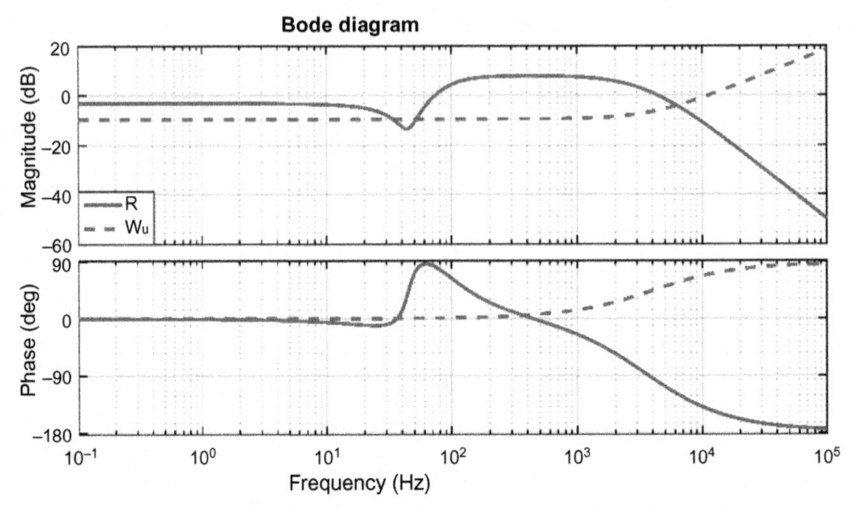

Figure 5.58 Noise sensitivity function and the corresponding weighting function W_u.

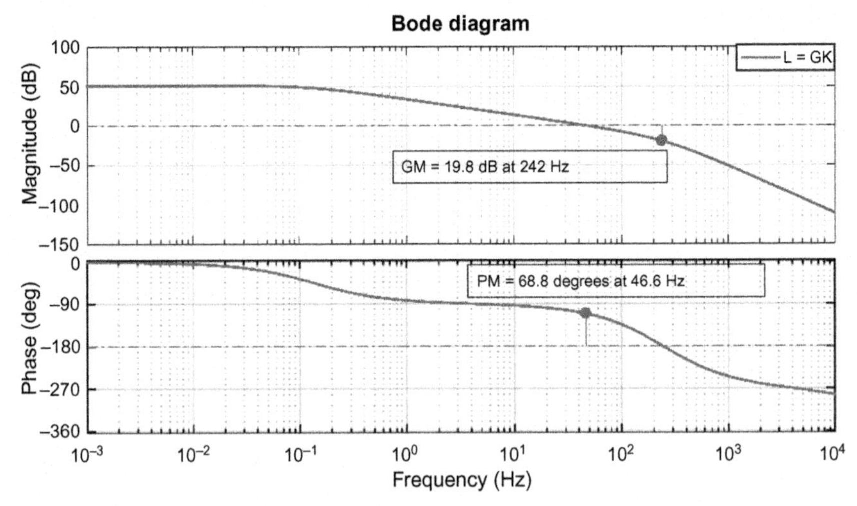

Figure 5.59 Loop transfer function.

$t = 0.25$ s the load step of 7.5 MW was performed, demonstrating that the H_∞ control stabilizes the system towards the reference point $V_{ref} = 1$. The dynamic of the inductor current is described in Fig. 5.61.

5.7.6.2 Shipboard Power System

For this simulation scenarios, we consider the same single bus Shipboard Power System, with three generators and all loads represented in a lump

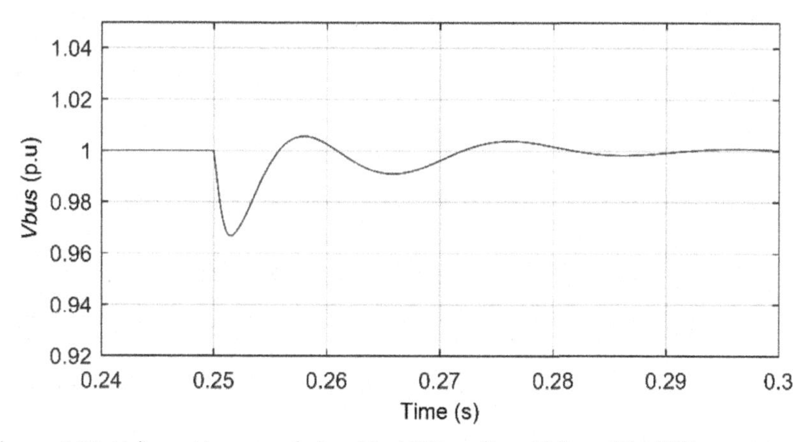

Figure 5.60 Voltage H_∞ cascaded – Ideal CPL – Step 10.3 → 17.8 MW.

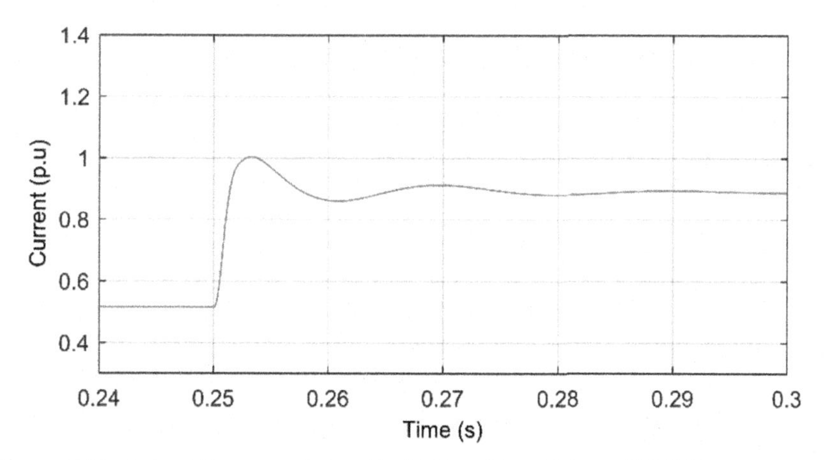

Figure 5.61 Current H_∞ cascaded – Ideal CPL – Step 10.3 → 17.8 MW.

manner. We are interested in observing how the H_∞ controller behaves when a nonlinear load such an ideal CPL is introduced into the system. Additionally, the impact of measurement uncertainties is tested to observe robustness properties. The droop resistances are set equal to provide equal voltage references to the primary controller. A voltage drop of 0.3 p.u at full load was considered while designing the droop controller. This value provides sufficient resolution in primary voltage references at low loads to prevent improper current sharing and at the same time, furthermore the drop does not create instability at peak loads. A load step increase of

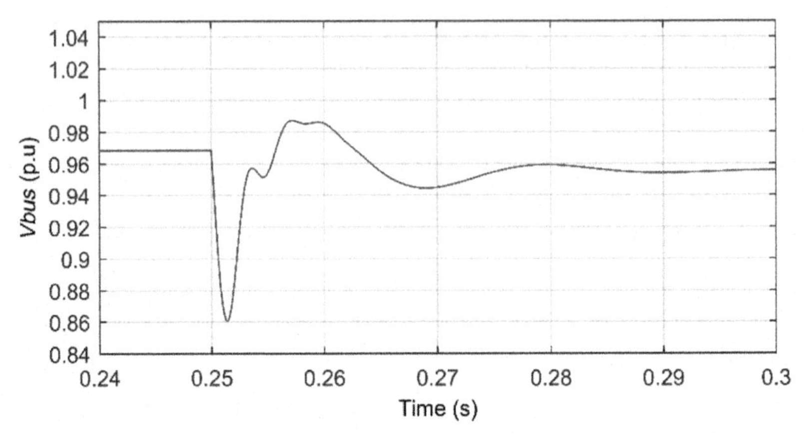

Figure 5.62 ISPS V_{bus} H_∞ for ideal CPL – Step 30.9 → 53.4 MW.

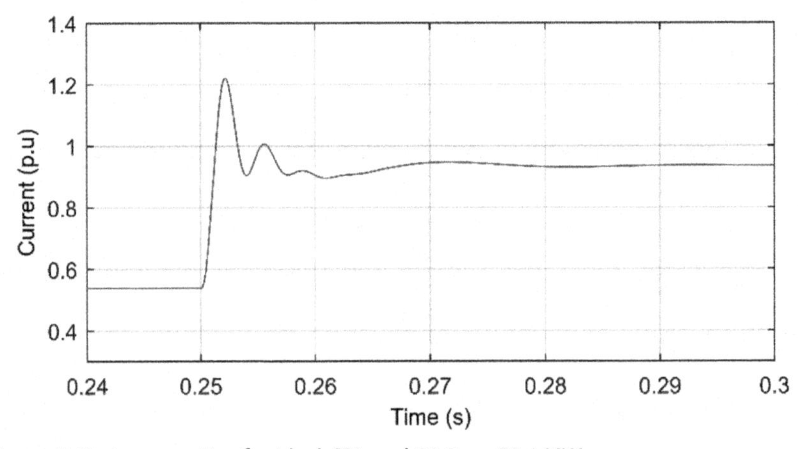

Figure 5.63 $I_{L1=L2=L3}$ H_∞ for ideal CPL and 30.9 → 53.4 MW.

22.5 MW is applied at the load side at $t = 0.25$ s. The effect of the increasing and decreasing of the CPL load on the load voltage is described in Fig. 5.62, demonstrating that the voltage remains stable and reaches the voltage set-points that are gradually modified by the action of the droop control. In Fig. 5.62 one can observe the transient for the current (Fig. 5.63).

5.7.6.3 Impact of Measurement Noise

In this scenario, the robustness properties of the controller are presented. Pink noise corrupts the measurement of each primary controller. A band

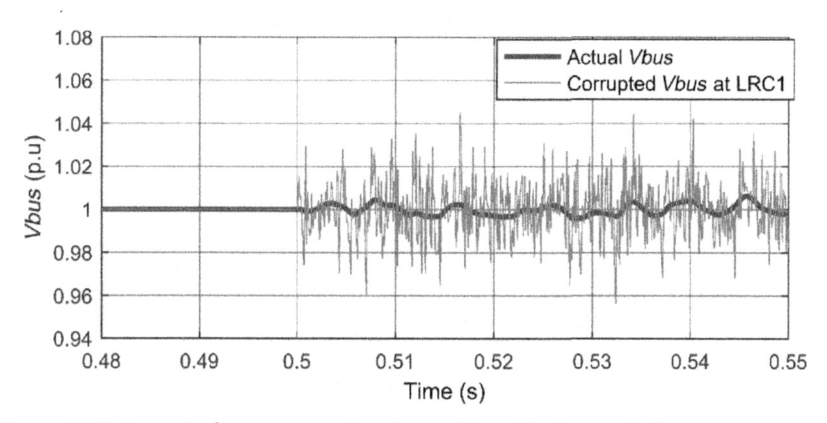

Figure 5.64 Impact of measurement noise on V_{bus} with H_{∞} control.

limited white noise source is low pass filtered to create the said pink noise. Different noise sources are used for each primary controller and then the bus voltage is observed. Remember that our control design has good noise suppression abilities as seen from R. The noise sources of all primary controllers are turned on simultaneously at time $t = 0.5$ s. The individual primary controllers are able to filter out the noisy inputs as observed in Fig. 5.64. The blue waveform shows the corrupted bus voltage measurement of the first converter.

5.8 SLIDING MODE CONTROL

Sliding Mode (SM) controllers switch back and forth between two control laws as a function of the state vector. Due to this switching sliding states occur which have as an advantage the robustness of the control in relation to parameter fluctuations of the plant. A disadvantage is the high-frequency switching of the control element, which on mechanical switches leads to additional wear and tear.

SM control provides an approach that complies with the nonlinear nature of the switching-mode source power converters. In practice, the converter switches are driven as a function of the instantaneous values of the state variables in a way that forces the system trajectory to stay on a suitable selected surface in the state-space known as the sliding surface or manifold. SM was originally designed for linear plants where afterwards extension for nonlinear plants were developed. We will only consider linear control loops, as these systems allow us to determine the essential basics of sliding state control. In certain cases, the design of sliding state controllers for nonlinear control loops is not more difficult than that for linear control loops. This applies, e.g., to input linear controlled systems according to the methodology described in Section 5.1.

5.8.1 Design Approach

The SM control can be divided into three stages:

1. The arrival phase, in which the trajectories converge to the switching surface or switching plane and reach them in finite time.
2. The sliding state, in which the trajectory slides on the manifold into the resting position.
3. The equilibrium point $x_e = 0$, in which the system remains stable.

For a globally stable equilibrium point $x_e = 0$, it must be ensured that all trajectories of the system go to the switching plane in finite time and then to the equilibrium point $x_e = 0$, in the sliding state. We consider linear control loops

$$\dot{\boldsymbol{x}} = \boldsymbol{A}\boldsymbol{x} + \boldsymbol{b}\boldsymbol{u} \tag{5.229}$$

with the switching surface

$$s(\boldsymbol{x}) = 0 \tag{5.230}$$

and the control law

$$\boldsymbol{u}(\boldsymbol{x}) = \begin{cases} u_{+}(\boldsymbol{x}); & for \;\; s(\boldsymbol{x}) > 0 \\ u_{-}(\boldsymbol{x}); & for \;\; s(\boldsymbol{x}) > 0 \end{cases} \tag{5.231}$$

For a nonlinear control loop the linearization methodology presented in Section 5.1.1 can be used, and afterwards the SM control is applied on this linearized system. The manipulated variable u is not defined here for $s(x) = 0$. In most cases, the following switching function is used:

$$s(\boldsymbol{x}) = r^T \boldsymbol{x} \tag{5.232}$$

Which corresponds to a switching surface of $\boldsymbol{r}^T \boldsymbol{x} = \boldsymbol{0}$. First of all, the reachability of the switching surface for all trajectories of the state-space has to be ensured. A necessary condition for this is that the trajectories of the control loop run from both sides towards the switching surface, as it must be

$$\begin{aligned} \dot{s} &< 0, \quad for \quad s(\boldsymbol{x}) > 0 \\ \dot{s} &> 0, \quad for \quad s(\boldsymbol{x}) < 0 \end{aligned} \tag{5.233}$$

If you summarize both conditions, you get

$$s\dot{s} < 0, \tag{5.234}$$

as a condition for a sliding state. The following applies

$$\dot{s} = grad^T s(\boldsymbol{x}) \cdot \dot{\boldsymbol{x}} \tag{5.235}$$

and for the case of $s(\boldsymbol{x}) = r^T \boldsymbol{x}$; it follows that $\dot{s} = r^T \dot{\boldsymbol{x}}$. Unfortunately, the condition $s\dot{s} < 0$, does not secure for every conceivable case that the trajectories reach the switching surface in finite time. The condition is therefore necessary, but not sufficient for a sliding state control.

There are different approaches to ensure the reachability of the sliding surface for all trajectories in finite time. A very common approach is that mentioned in [75]. Here, the decrease of the switching function along the trajectories $\boldsymbol{x}(t)$ is specified. It applies

$$\dot{s}(\boldsymbol{x}) = -q \ sgn(s(\boldsymbol{x})) - ks(\boldsymbol{x}) \tag{5.236}$$

with positive constants q and k. Obviously the sliding surface s in combination with Eq. (5.236) satisfies the necessary condition.

$$s\dot{s} = -q|s| \cdot -ks^2 < 0 \tag{5.237}$$

Since the Eq. (5.236) has a decrease rate of $\dot{s} < -q$ or an increase rate of $\dot{s} > q$ even for very small values of $|s|$, the trajectories $\boldsymbol{x}(t)$ reach the sliding surface also in finite time. If you consider that:

$$\dot{s}(\boldsymbol{x}) = grad^T s(\boldsymbol{x}) \cdot \dot{\boldsymbol{x}} = grad^T s(\boldsymbol{x}) \cdot (\boldsymbol{A}\boldsymbol{x} + \boldsymbol{b}u) \tag{5.238}$$

then one obtains from (5.236) the control law:

$$u(\boldsymbol{x}) = -\frac{grad^T s(\boldsymbol{x}) \cdot \boldsymbol{A}\boldsymbol{x} + q \ sgn(s(\boldsymbol{x})) + ks(\boldsymbol{x})}{grad^T s(\boldsymbol{x}) \cdot \boldsymbol{b}} \tag{5.239}$$

In the case of a switching hyperplane as switching surface

$$u(\boldsymbol{x}) = -\frac{\boldsymbol{r}^T \boldsymbol{A}\boldsymbol{x} + q \ sgn\left(\boldsymbol{r}^T \boldsymbol{x}\right) + k\boldsymbol{r}^T \boldsymbol{x}}{\boldsymbol{r}^T \cdot \boldsymbol{b}} \tag{5.240}$$

The freely selectable positive parameters q and k can be used to influence the dynamics of the control.

5.8.2 Dynamics in the Sliding State

When the trajectories $\boldsymbol{x}(t)$ reach the switching line and the sliding state begins, the following question arises: Which dynamics does the control loop have during the sliding state? The problem with this question is the discontinuity of the differential equation

$$\dot{\boldsymbol{x}} = \boldsymbol{A}\boldsymbol{x} + \boldsymbol{b}\boldsymbol{u}; \quad \boldsymbol{u}(\boldsymbol{x}) = \begin{cases} u_+(\boldsymbol{x}); \; for \;\; s(\boldsymbol{x}) > 0 \\ u_-(\boldsymbol{x}); \; for \;\; s(\boldsymbol{x}) > 0 \end{cases} \tag{5.241}$$

of the closed loop on the switching surface.

$$s(\boldsymbol{x}) = \boldsymbol{r}^T\boldsymbol{x} = 0 \tag{5.242}$$

The differential equation is obviously not defined on the sliding surface, i.e., the existence and uniqueness of its solution is not guaranteed there. There are several methods to solve this problem [75], such as Filippov's method [76].

We determine the dynamics of the system in the sliding state according to the following frequently used method. The controlled system is transformed into the control standard form

$$\begin{aligned} \dot{x}_1 &= x_2 \\ &\vdots \\ \dot{x}_{n-1} &= x_n \\ \dot{x}_n &= -a_0x_1 - a_1x_2 - \cdots - r_{n-1}x_{n-1} \end{aligned} \tag{5.243}$$

Furthermore, if solving Eq. (5.244) for x_n one obtains Eq. (5.245)

$$\boldsymbol{r}^T\boldsymbol{x} = r_1x_1 + \cdots + r_{n-1}x_{n-1} + r_nx_n = 0 \tag{5.244}$$

$$x_n = -r_1x_1 - \cdots - r_{n-1}x_{n-1} \tag{5.245}$$

Without further restriction to the generality the coefficient r_n was selected such that $r_n = 1$. Now we substitute in the controllable canonical form of the plant (5.243) the state variable x_n with Eq. (5.245). We obtain now the differential equation system (5.246).

$$\begin{aligned} \dot{x}_1 &= x_2 \\ &\vdots \\ \dot{x}_{n-2} &= x_{n-1} \\ \dot{x}_{n-1} &= r_1x_1 - \cdots - r_{n-1}x_{n-1} \end{aligned} \tag{5.246}$$

Eq. (5.246) describes the dynamics of the control loop during the sliding motion. It has to be highlighted that the state variable in this case is given by the algebraic Eq. (5.245) and the equation is therefore omitted. The differential Eq. (5.246) are no longer dependent on the parameters of the controlled system. This means that the control loop (5.246) is robust against parameter fluctuations of the controlled system. It is also noteworthy that the system order has decreased by one degree to order $n-1$ and the coefficients r_t of the switching surface (5.244) form the coefficients of the characteristic polynomial of the linear dynamics (5.246) in the case of the sliding state.

5.8.3 Proof of Robustness

The main advantage of SMCs is their robustness against variation of the control loop parameters $\boldsymbol{\Delta A}$ or external disturbances $\boldsymbol{d}(t)$, if these appear in the system description:

$$\dot{\boldsymbol{x}} = (\boldsymbol{A} + \Delta\boldsymbol{A})\boldsymbol{x} + \boldsymbol{b}u + \boldsymbol{d} \tag{5.247}$$

This means that the dynamics of the closed loop during the sliding motion depend as in Eq. (5.246), only from the parameters of the switching surface (5.242). The dynamics are independent of $\boldsymbol{\Delta A}$ and $\boldsymbol{d}(t)$. The robustness therefore applies if both of the following conditions are met [23]:

1. There exists a vector $\boldsymbol{p}$, such that $\boldsymbol{\Delta A} = \boldsymbol{b}\boldsymbol{p}^T$
2. There exists an $\alpha(t)$, so that $\boldsymbol{d}(t) = \boldsymbol{b}\alpha(t)$.

For example, if the system is *cab* written in the controllable canonical form, the condition of $\boldsymbol{\Delta A} = \boldsymbol{b}\boldsymbol{p}^T$ can be met if only the coefficients a_i vary.

$$\dot{\boldsymbol{x}} = \begin{bmatrix} 0 & 1 & 0 & \cdots & 0 \\ 0 & 0 & 1 & \cdots & 0 \\ \vdots & \vdots & \vdots & \ddots & \vdots \\ 0 & 0 & 0 & \cdots & 1 \\ -a_0 & -a_1 & -a_2 & \cdots & -a_{n-1} \end{bmatrix} \boldsymbol{x} + \begin{bmatrix} 0 \\ 0 \\ \vdots \\ 0 \\ 1 \end{bmatrix} u \tag{5.248}$$

5.8.4 Application to DC–DC Converters

In this section we apply the SMC to the buck converter supplying a resistive load (Fig. 5.65).

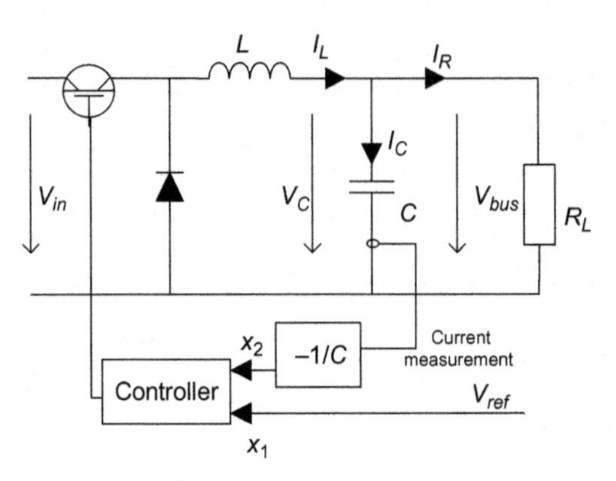

Figure 5.65 Buck converter with resistive load.

The purpose of the circuit is to set the output voltage V_{bus} based on a constant reference voltage V_{ref} such that $V_{bus} = V_{ref}$ applies. For the derivative of the voltage over the capacitor the following equation is valid:

$$u(\boldsymbol{x}) = \frac{1}{C} I_C = \frac{1}{C}(I_L - I_{CPL}) = \frac{1}{C}\left(\int \frac{zV_{in} - V_{bus}}{L} dt - \frac{V_{bus}}{R_{Ld}}\right) \tag{5.249}$$

One obtains from Eq. (5.249)

$$\ddot{V}_{bus} = -\frac{1}{LC} V_{bus} - \frac{1}{R_L C} \dot{V}_{out} + \frac{zV_{in}}{LC} \tag{5.250}$$

We define as state variables

$$\begin{aligned} x_1 &= V_{bus} - V_{ref} \\ x_2 &= \dot{x}_1 = \dot{V}_{bus} \end{aligned} \tag{5.251}$$

With the Eqs. (5.250) and (5.251) one obtains the state-space model

$$\begin{bmatrix} \dot{x}_1 \\ \dot{x}_2 \end{bmatrix} = \begin{bmatrix} 0 & 1 \\ -\frac{1}{LC} & -\frac{1}{R_L C} \end{bmatrix} \begin{bmatrix} x_1 \\ x_2 \end{bmatrix} + \begin{bmatrix} 0 \\ \frac{1}{LC} \end{bmatrix} u \tag{5.252}$$

where the actuating variable u is composed of

$$u = zV_{in} - V_{ref} \tag{5.253}$$

is the same.

The system described by Eqs. (5.252) and (5.253) is used for the application of a sliding state control, as the digital actuated variable d specifies a switching action. The goal of the control is to reach the operating point

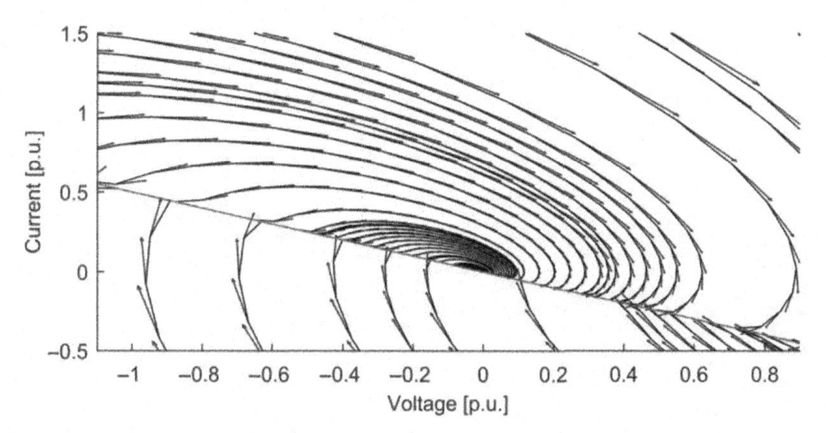

Figure 5.66 The trajectories of the two areas converges to the switching line.

$$x_1 = V_{bus} - V_{ref} = 0 \tag{5.254}$$

The system (5.252) has a stable equilibrium point x_e for $z = 1$ where the following applies:

$$x_{e1} = \begin{bmatrix} V_{bus} - V_{ref} \\ 0 \end{bmatrix} \tag{5.255}$$

In the case of $z = 0$ the following applies:

$$x_{e2} = \begin{bmatrix} -V_{ref} \\ 0 \end{bmatrix} \tag{5.256}$$

The switching line, with $r_1 > 1$, separates the state-space in two areas (Fig. 5.66).

$$\begin{gathered} s(\boldsymbol{x}) = r_1 x_1 + x_2 = 0 \\ z = \begin{cases} 0; & for \;\; s(\boldsymbol{x}) > 0 \\ 1; & for \;\; s(\boldsymbol{x}) > 0 \end{cases} \end{gathered} \tag{5.257}$$

Now it is necessary to determine the parameter r_1 of the switching line (5.257) from the necessary condition for the existence of a sliding state, $\dot{s}s < 0$. It follows from Eq. (5.258) in combination with the system Eqs. (5.252) and (5.253):

$$\dot{s}s = s(\boldsymbol{x})\left[\left(r_1 - \frac{1}{R_L C}\right)x_2 - \frac{1}{LC}x_1 + \frac{zV_{in} - V_{ref}}{LC}\right] < 0 \tag{5.258}$$

By choosing $r_1 = \frac{1}{R_L C}$ it is possible to decouple the condition of existence (5.258) from the state $x_2 = \dot{V}_{bus}$. This selection in combination with Eq. (5.251) leads to the following condition that needs to be satisfied:

$$s\dot{s} = s(\boldsymbol{x})\frac{1}{LC}(zV_{in} - V_{bus}) < 0 \tag{5.259}$$

From this condition we obtain for the case 1: $s(\boldsymbol{x}) < 0$ and $z = 1$.

$$V_{in} > V_{bus} \tag{5.260}$$

For the case 2 $s(\boldsymbol{x}) > 0$ and $z = 0$, the condition (5.259) is used to obtain

$$V_{bus} > 0 \tag{5.261}$$

For the existence of a sliding necessary condition (5.259) next to the Eqs. (5.260) and (5.261) has to be fulfilled. Furthermore from (5.262) it can be seen that for case 1 has a decrease rate of $-\frac{V_{bus}}{LC} < 0$. For case 2 an increase rate of $\frac{V_{IN} - V_{bus}}{LC} > 0$ is present. The trajectories of the control converge in finite time towards the sliding state.

$$\dot{s} = \frac{1}{LC}(zV_{in} - V_{bus}) \tag{5.262}$$

Ideally, a converter will switch at infinite frequency with its phase trajectory moving on the sliding surface when it enters SM operation However, in the presence of finite switching time and time delay, this ideal behavior is not possible. The discontinuity in the feedback control will produce a particular dynamic behavior in the vicinity of the surface trajectory known as chattering.

If the chattering is uncontrolled, the converter system will become self-oscillating at a very high switching frequency corresponding to the chattering dynamics. This undesirable high switching frequency will result in excessive switching losses, inductor and transformer core losses, and EMI noise issues. To solve these problems, the control law in (5.257) is redefined as:

$$u = \begin{cases} 0 \ for \ s(x) < -\beta \\ 1 \ for \ s(x) > \beta \end{cases} \tag{5.263}$$

The purpose of the arbitrary small value β in (5.263) is to introduce a hysteresis band with the boundary conditions. By doing so, it provides a form of control to the switching frequency of the converter. This method alleviates the chattering effect of SM control. With this choice of the control law, the operation is so that if the parameters of the state variables are such that $\sigma(x) > \beta$, the switch of the buck converter will turn on. Conversely, it will turn off when $\sigma(x) < -\beta$. In the region $-\beta < \sigma(x) < \beta$, the switch remains in its previous state. Therefore, the introduction of a no-switching region $-\beta < \sigma(x) < \beta$ allows the maximum switching frequency of the SM controller to be controlled, which alleviates the effect of chattering. Furthermore, it is possible to control the frequency of the operation by varying the magnitude of β.

In a practical converter control, only the inductor current I and the output voltage V are sensed. Generation of current reference and voltage reference is needed to implement the following manifold:

$$s = \mathrm{V} - V_{ref} + k \cdot \left(\mathrm{I} - I_{ref}\right) \tag{5.264}$$

In practice, generation of reference inductor current signal is challenging, since it depends on the converter operating point (output power and input voltage). Classically, the reference signal is usually derived directly from the inductor current by using a low pass [77].

5.8.5 Simulation Results

In our simulation examples we use a slightly different approach than in the previous example. For the Cascaded System and ISPS we use the state I_d which is the so-called disturbance current. This disturbance current is estimated by the Kalman filter in the framework of the virtual disturbance. The resulting manifold is described by:

$$s = V - V_{ref} + k \cdot I_d \tag{5.265}$$

5.8.5.1 Cascaded System

A cascaded converter setup has been modeled in Matlab-SIMULINK with the parameters described in Table 6.1. In the simulation, a step variation of the CPL load has been performed. The variation of the bus voltage during the step change of the CPL load is described in Fig. 5.67 At

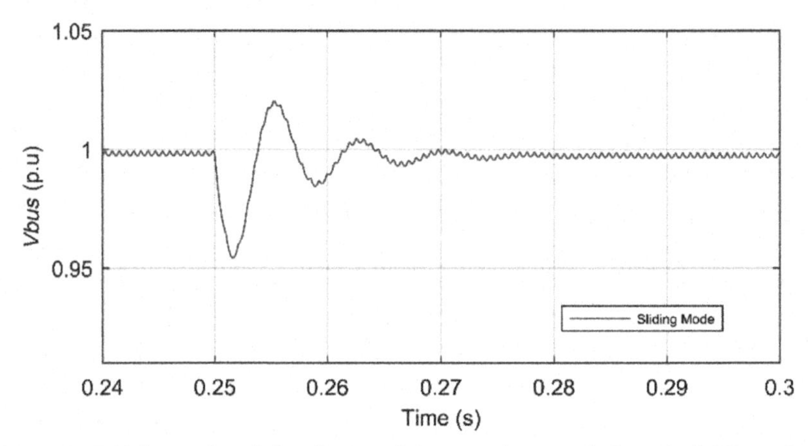

Figure 5.67 Voltage virtual distribution sliding mode cascaded – Ideal CPL – Step 10.3 → 17.8 MW.

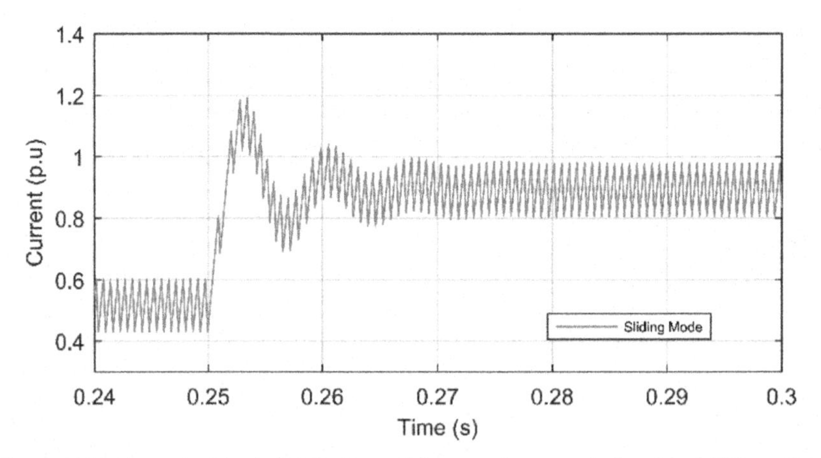

Figure 5.68 Current virtual distribution sliding mode cascaded – Ideal CPL – Step 10.3 → 17.8 MW.

$t = 0.25$ s the load step of 7.5 MW was performed, demonstrating that the SM control is able to stabilize the system towards the voltage reference $V_{ref} = 1$. The dynamic of the inductor current is described in Fig. 5.68.

In order to better clarify the effect of the parameter β previously defined, the phase plane plot of the cascaded simulation is described in Fig. 5.69.

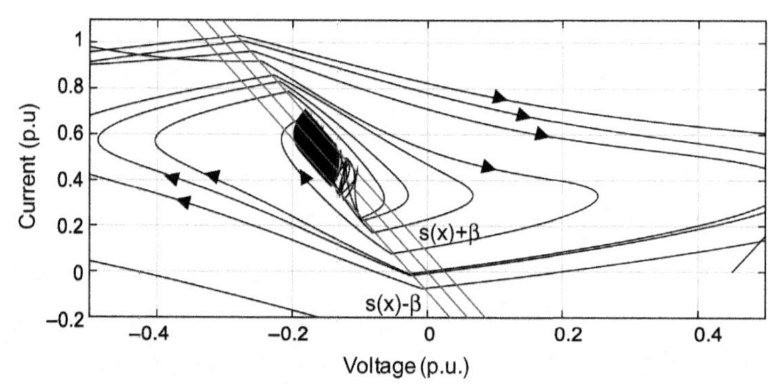

Figure 5.69 Phase plot virtual distribution sliding mode cascaded – Ideal CPL – Step 10.3 → 17.8 MW.

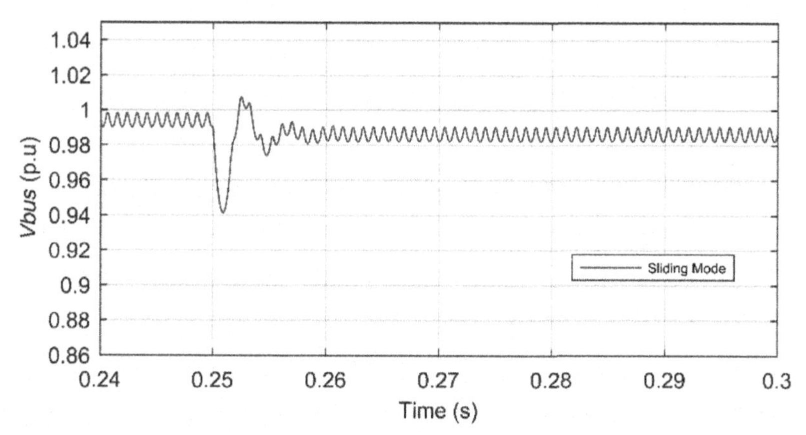

Figure 5.70 ISPS V_{bus} virtual distribution sliding mode for ideal CPL – Step 30.9 → 53.4 MW.

5.8.5.2 Shipboard Power System

A Shipboard Power System has been modeled in Matlab\SIMULINK, based on the parameters in Table 6.2. The load steps of 22.5 MW have been performed at $t = 0.25$ s. The coefficients of the droop controllers were chosen to obtain an even current sharing among the converters. All converters equally share the total load power between them.

The effect of the increasing and decreasing of the CPL load on the load voltage is described in Fig. 5.70, demonstrating that the voltage remains stable and reaches the voltage set-points that are gradually modified by the action of the droop control. In Fig. 5.71 one can observe the transient for the current.

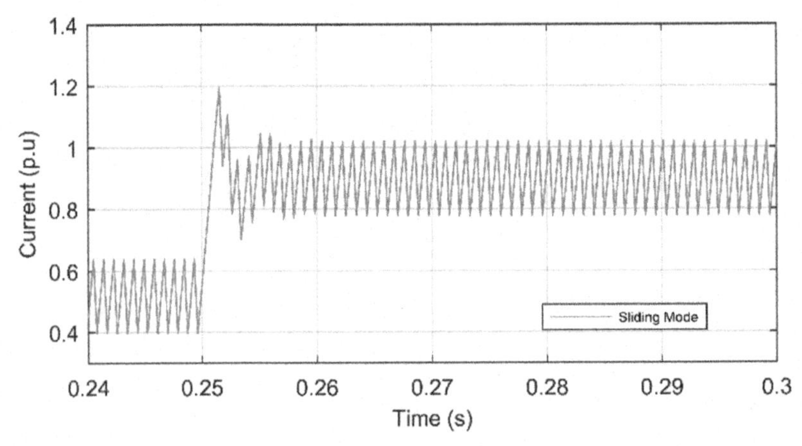

Figure 5.71 ISPS $I_{L1=L2=L3}$ virtual distribution sliding mode for ideal CPL and 30.9→53.4 MW.

5.9 SUMMARY

Throughout this chapter several control approaches were presented: First their theoretical foundation and then their application on MVDC ISPS.

This chapter started with the LSF which follows the idea of compensating the nonlinearity so that a linear system is created on which linear control theory can be applied. This control relies on the measurement of all states and an exact load model. The synthesis procedure is straightforward. A decentralized application does not seem to be feasible as perfect knowledge of the system is needed for appropriate nonlinearity cancelation.

The second approach was the synergetic control which is a nonlinear control design method that includes the switching characteristic of a system into the design. The synergetic control law seems from the centralized design procedure very complex, especially if several converters in parallel have to be coordinated. While the parameter tuning for the LSF is mainly a matter of pole placement the selection for the appropriate synergetic control parameters is coupled to multiple algebraic restrictions, as not only disturbance rejection is influenced through them but also current sharing modes. A decentralized variation of the synergetic control is in the context of the virtual disturbance also possible. The third approach was nonlinear Immersion and Invariance controller, which has the advantage of ensuring, by definition, the global asymptotic stability of the system. However, it has a complicated control law whose implementation might be challenging in a digital controller.

The fourth approach was an observer-based control which relies on the concept of 2DOF controller. Here, only the bus voltage is measured while the local inductor current is estimated. This is because in systems with switching converters the current has a strong ripple and a direct measurement has to be filtered; thus, it could be replaced by a software measurement in form of the Kalman filter. Its design compared to the LSF is more complex as steady-state Kalman filter, covariance matrices, and correlation parameters have to be selected. The usage of the Kalman filter is not limited to software measurement; it also performs in its augmented version the additional estimation of the nonlocal part of the power system (i.e., the estimation of the other converter currents), and generates, in a 2DOF structure, also the set-point trajectory. It should be noted that such set-point trajectory generation relies on phase-minimal system behavior. Furthermore, due to the application of steady-state Kalman estimation, the estimated system is always linearized in the operating point. Both centralized and decentralized solutions are possible; a centralized control has knowledge about all inductor current estimates while the decentralized control has only knowledge of the local inductor current estimate. Both variations of this control approach rely fully on linear control theory.

The fifth approach combines the virtual disturbance concept with a nonlinear control law, namely, Backstepping. Backstepping is a procedure which is based on Lyapunov Control Function; its synthesis procedure is less complex than the previous decentralized approach, especially as it does not need a set-point trajectory while it also uses estimated inductor currents.

The sixth approach is Adaptive Backstepping, where the nonlocal part is no longer assumed as a current source and estimated by a Kalman filter, but corresponds to Power Estimation. This leads to a complete nonlinear decentralized controller while also having a nonlinear plant. For both Backstepping approaches a centralized realization does not seem possible as those control laws require the system matrix to be written in a lower triangular form.

The seventh approach was the H_∞ approach where the derived controllers are internally stable and provide guaranteed stability margins. One of the drawbacks of the LQG control approach is that it seeks to optimize a performance index over a span of time while assuming that the model is perfect. In this context the H_∞ approach assumes that the model is imperfect and it tries to optimize the stability and quality given uncertainty in the plant model, actuators and feedback sensors.

The eight approach is the SMC which is theoretically robust because of less dependence on the system parameters. However, the standard formulation of SM control demonstrates some difficulties in the practical applications primarily because of the chattering phenomena that can degrade the quality of the control. In order to solve this problem, many different adjustments to the first formulation of the SM control have been derived which comprehend the use of a high-order SM control, methods for controlling the duty cycle. These methods are model dependent which reduces the high robustness of the SM control. Furthermore the implementation of the SM control in an ISPS is only possible by the usage of the virtual disturbance.

REFERENCES

[1] H.K. Khalil, Nonlinear Systems, Prentice Hall, Upper Saddle River, NJ, 2002.
[2] A.M. Rahimi, G.A. Williamson, A. Emadi, Loop-cancellation technique: a novel nonlinear feedback to overcome the destabilizing effect of constant-power loads, IEEE Trans. Vehic. Technol. 59 (2) (2010) 650–661.
[3] G. Sulligoi, D. Bosich, Zhu, L., M. Cupelli, A. Monti, Linearizing control of shipboard multi-machine MVDC power systems feeding Constant Power Loads, in: 2012 IEEE Energy Conversion Congress and Exposition (ECCE), pp. 691, 697, 15–20 Sept. 2012.
[4] A. Emadi, A. Khaligh, C.H. Rivetta, G.A. Williamson, Constant power loads and negative impedance instability in automotive systems: definition, modeling, stability, and control of power electronic converters and motor drives, IEEE Trans. Vehic. Technol. 55 (4) (2006) 1112–1125.
[5] J.G. Ciezki, R.W. Ashton, The application of feedback linearization techniques to the stabilization of DC-to-DC converters with constant power loads, in: Proceedings of the 1998 IEEE International Symposium on Circuits and Systems, 1998. ISCAS '98., vol. 3, no., pp. 526, 529 vol. 3, 31 May-3 Jun 1998.
[6] J. Adamy, Nichtlineare Regelungen, Springer, Berlin; Heidelberg, 2009.
[7] A. Isidori, Nonlinear Control Systems, third ed., Springer, Berlin; New York, 1995.
[8] T. Wey, Nichtlineare Regelungssysteme: Ein differentialalgebraischer Ansatz, Teubner, 2002.
[9] H. Schwarz, Nichtlineare Regelungssysteme, R. Oldenbourg, 1991.
[10] G. Sulligoi, D. Bosich, V. Arcidiacono, G. Giadrossi, Considerations on the design of voltage control for multi-machine MVDC power systems on large ships, 2013 IEEE Electric Ship Technologies Symposium (ESTS), pp. 314, 319, 22–24 April 2013.
[11] G. Sulligoi, D. Bosich, G. Giadrossi, L. Zhu, M. Cupelli, A. Monti, Multiconverter medium voltage DC power systems on ships: constant-power loads instability solution using linearization via state feedbackcontrol, IEEE Trans. Smart Grid 5 (5) (2014) 2543–2552.

[12] A. Kolesnikov, et al., Modern Applied Control Theory: Synergetic Approach in Control Theory, vol. 2, TSURE Press, 2000.
[13] I. Kondratiev, R. Dougal, Synergetic Control Strategies for Shipboard DC Power Distribution Systems, in: American Control Conference, 2007. ACC '07, pp. 4744, 4749, 9–13 July 2007.
[14] I. Kondratiev, E. Santi, R. Dougal, G. Veselov, Synergetic control for DC-DC buck converters with constant power load, in: 2004 IEEE 35th Annual Power Electronics Specialists Conference, 2004. PESC 04., vol. 5, no., pp. 3758, 3764 Vol. 5, 20–25 June 2004.
[15] I. Kondratiev, A Synergetic Control for Parallel-connected DC-DC Buck Converters, PhD Dissertation, University of South Carolina, College of Engineering and Information Technology, 2005.
[16] M. Cupelli, M. Moghimi, A. Riccobono, A. Monti, A comparison between synergetic control and feedback linearization for stabilizing MVDC microgrids with constant power load, in: 2014 IEEE PES Innovative Smart Grid Technologies Conference Europe (ISGT-Europe), pp. 1, 6, 12–15 Oct. 2014.
[17] E. Santi, A. Monti, D. Li, K. Proddutur, R.A. Dougal, Synergetic control for power electronics applications: a comparison with the sliding mode approach, J. Circuits Systems Computers 13 (04) (Aug. 2004) 737–760.
[18] K.J. Åström, B. Wittenmark, Computer-controlled Systems: Theory and Design, third ed., Dover Publications, Mineola, NY, 2011.
[19] A. Nusawardhana, S.H. Zak, W.A. Crossley, Nonlinear synergetic optimal controllers, J. Guidance Control Dynamics 30 (4) (2007) 1134–1147.
[20] E. Santi, D. Li, A. Monti, A.M. Stankovic, A Geometric Approach to Large-signal Stability of Switching Converters under Sliding Mode Control and Synergetic Control, IEEE 36th Power Electronics Specialists Conference, 2005. PESC '05, pp. 1389, 1395, 16-16 June 2005.
[21] E. Santi, A. Monti, D. Li, K. Proddutur, R.A. Dougal, Synergetic control for DC-DC boost converter: implementation options, IEEE Trans. Ind. Appl. 39 (6) (Nov.-Dec. 2003) 1803–1813.
[22] C. Edwards, S. Spurgeon, Sliding Mode Control: Theory and Applications, Taylor & Francis, London, 1998.
[23] J. Slotine, W. Li, Applied Nonlinear Control, Prentice-Hall, Englewood Cliffs, NJ, 1991.
[24] C. Tse, Complex Behavior of Switching Power Converters, CRC Press, London, 2000.
[25] I. Kondratiev, R. Dougal, General synergetic control strategies for arbitrary number of paralleled buck converters feeding constant power load: implementation of dynamic current sharing, in: 2006 IEEE International Symposium on Industrial Electronics, vol. 1, no., pp. 257, 261, 9–13 July 2006.
[26] A. Astolfi, D. Karagiannis, R. Ortega, Nonlinear and Adaptive Control With Applications, Springer Science & Business Media, 2007.
[27] K.D. Young, V.I. Utkin, U. Ozguner, A control engineer's guide to sliding mode control, in: IEEE Transactions on Control Systems Technology, vol. 7, no. 3, pp. 328–342, May 1999.
[28] A. Astolfi, R. Ortega, Immersion and invariance: a new tool for stabilization and adaptive control of nonlinear systems, in: IEEE Transactions on Automatic Control, vol. 48, no. 4, pp. 590–606, April 2003.
[29] A. Kwasinski, C.N. Onwuchekwa, Dynamic behavior and stabilization of DC microgrids with instantaneous constant-power loads, in: IEEE Transactions on Power Electronics, vol. 26, no. 3, pp. 822–834, 2011.
[30] J.M. Guerrero, et al., Hierarchical control of droop-controlled AC and DC microgrids—A general approach toward standardization, IEEE Trans. Ind. Electr. 58.1 (2011) 158–172.

[31] J. Liu, D. Obradovic, A. Monti, Decentralized LQG control with online set-point adaptation for parallel power converter systems, in: 2010 IEEE Energy Conversion Congress and Exposition (ECCE), pp. 3174, 3179, 12–16 Sept. 2010.
[32] J. Liu, Measurement System and Technique for Future Active Distribution Grids, PhD. Thesis, RWTH Aachen University, 2013.
[33] J.B. Burl, Linear Optimal Control: H_2 *and H[infinity] Methods*, Addison Wesley Longman, Menlo Park, CA, 1999.
[34] B. Tomescu, H.F. VanLandinham, Disturbance rejection and robustness considerations in DC/DC converters, in: 30th Annual IEEE Power Electronics Specialists Conference, 1999. PESC 99., vol. 2, no., pp. 1204, 1209 vol. 2, 1999.
[35] J. Liu, A. Benigni, D. Obradovic, S. Hirche, A. Monti, State estimation and branch current learning using independent local kalman filter with virtual disturbance model, IEEE Trans. Instrument. Measur. 60 (9) (2011) 3026–3034.
[36] J. Liu, A. Benigni, D. Obradovic, S. Hirche, A. Monti, State estimation and learning of unknown branch current flows using decentralized Kalman filter with virtual disturbance model, in: 2010 IEEE International Workshop on Applied Measurements For Power Systems (AMPS), pp. 31, 36, 22–24 2010.
[37] L. Zhu, J. Liu, M. Cupelli, A. Monti, Decentralized linear quadratic Gaussian control of multi-generator MVDC shipboard power system with Constant Power Loads, in: IEEE Electric Ship Technologies Symposium (ESTS), 2013, pp. 308, 313, 22–24 April 2013.
[38] M. Cupelli, M. de Paz Carro, A. Monti, Hardware in the loop implementation of a disturbance based control in MVDC grids, in: 2015 IEEE Power and Energy Society General Meeting (PES), pp. 1, 5, 26–30 July 2015.
[39] J.V. Breakwell, J.L. Speyer, A.E. Bryson, Optimization and control of nonlinear systems using the second variation, SIAM J. Control 1 (2) (1963) 193–223.
[40] H.J. Kelley, Guidance theory and extremal fields, IRE Trans. Automatic Control 7 (5) (Oct 1962) 75–82.
[41] B.D.O. Anderson, Linear Optimal Control, Prentice-Hall, Englewood Cliffs, NJ, 1971.
[42] G. Rigatos, P. Siano, Distributed state estimation for condition monitoring of nonlinear electric power systems, in: 2011 IEEE International Symposium on Industrial Electronics (ISIE), pp. 1703, 1708, 27–30 June 2011.
[43] G. Rigatos, P. Siano, N. Zervos, Sensorless control of distributed power generators with the derivative-free nonlinear kalman filter, IEEE Trans. Ind. Electr. 61 (11) (2014) 6369–6382.
[44] S. Sarkka, On unscented kalman filtering for state estimation of continuous-time nonlinear systems, IEEE Trans. Automatic Control 52 (9) (2007) 1631–1641.
[45] Y. Bar-Shalom, X.-R. Li, T. Kirubarajan, Estimation With Applications to Tracking and Navigation, Wiley, New York, 2001.
[46] R.E. Kalman, A new approach to linear filtering and prediction problems, Trans. ASME, J. Basic Eng. 82 (1960) 35–45.
[47] J.R. Leigh, Control Theory: A Guided Tour, third ed., The Institution of Engineering and Technology, London, UK, 2012.
[48] K. Ogata, Modern Control Engineering, fifth ed., Prentice-Hall, Boston, 2010.
[49] K.J. Aström, R.M. Murray, Feedback Systems, Princeton University Press, 2009.
[50] C.L. Phillips, H.T. Nagle, A. Chakrabortty, Digital Control System Analysis & Design, fourth ed., Pearson Prentice Hall, Boston, 2015.
[51] M. Krstić, I. Kanellakopoulos, P.V. Kokotović, Nonlinear and Adaptive Control Design, Wiley, New York, 1995.
[52] H. El Fadil, F. Giri, M. Haloua, H. Ouadi, Nonlinear and adaptive control of buck power converters, in: Proceedings. 42nd IEEE Conference on Decision and Control, 2003, vol. 5, no., pp. 4475, 4480 Vol. 5, 9–12 Dec. 2003.

[53] L. Fall, O. Gehan, E. Pigeon, M. Pouliquen, M. M'Saad, Nonlinear state-feedback with disturbances estimation for DC-DC Buck converter, in: 2012 2nd International Symposium on Environment Friendly Energies and Applications (EFEA), vol., no., pp. 25, 30, 25–27 June 2012.
[54] M.L. McIntyre, M. Schoen, J. Latham, Simplified adaptive backstepping control of buck DC:DC converter with unknown load, in: 2013 IEEE 14th Workshop on Control and Modeling for Power Electronics (COMPEL), pp. 1, 7, 23–26 June 2013.
[55] M. Cupelli, M.M. Mirz, A. Monti, Application of Backsteppping to MVDC Ship Power Systems with Constant Power Loads, in: Electrical Systems for Aircraft, Railway and Ship Propulsion (ESARS), 2015, pp. 1, 5, 3–5 Mar. 2015.
[56] M. Cupelli, M.M. Mirz, A. Monti, A comparison of backstepping and LQG control for stabilizing MVDC microgrids with constant power loads, in: 2015 IEEE Eindhoven PowerTech (POWERTECH), pp. 1, 6, 29 June -2 July 2015.
[57] R.M. Murray, Z. Li, S. Sastry, A Mathematical Introduction to Robotic Manipulation, CRC Press, Boca Raton, FL, 1994.
[58] Z. Artstein, Stabilization with relaxed controls, Nonlin. Anal. Theory Methods Appl. 7 (11) (1983) 1163–1173.
[59] E.D. Sontag, A universal construction of Artstein's theorem on nonlinear stabilization, Systems Control Letters 13 (1989) 117–123.
[60] R.A. Freeman, J.A. Primbs, Control Lyapunov functions: New ideas from an old source, in: Proc. 35th Conference on Decision and Control, pp. 3926–3931, Dec. 1996.
[61] L. Sonneveldt, Adaptive Backstepping Flight Control for Modern Fighter Aircraft, *PhD Dissertation*, Technical University Delft, 2010.
[62] O. Härkegard, Flight Control Design Using Backstepping, *Master Thesis*, Linköping University, 2001.
[63] J. Zhou, C. Wen, Decentralized backstepping adaptive output tracking of interconnected nonlinear systems, IEEE Trans. Autom. Control 53 (10) (2008) 2378–2384.
[64] M. Krstić, I. Kanellakopoulos, P.V. Kokotović, Adaptive nonlinear control without overparametrization, Systems Control Letters 19 (1992) 177–185.
[65] F. Beleznay, M. French, Overparameterised adaptive controllers can reduce nonsingular costs, Systems Control Letters 48 (1) (2003) 12–25.
[66] G. Zames, Feedback and optimal sensitivity: Model reference transformations, multiplicative seminorms, and approximate inverses, in: IEEE Transactions on Automatic Control, vol. 26, no. 2, pp. 301–320, April 1981.
[67] B. Francis, G. Zames, On H^{∞}-optimal sensitivity theory for SISO feedback systems, in: IEEE Transactions on Automatic Control, vol. 29, no. 1, pp. 9–16, January 1984.
[68] S. Boyd, V. Balakrishnan, P. Kabamba, A bisection method for computing the H∞ norm of a transfer matrix and related problems, in: Mathematics of Control, Signals and Systems, 1989.
[69] J.C. Doyle, K. Glover, P.P. Khargonekar, B.A. Francis, State-space solutions to standard H2 and H∞ control problems, in: IEEE Transactions on Automatic Control, vol. 34, no. 8, pp. 831–847, Aug 1989.
[70] R.A. Hyde, K. Glover, G.T. Shanks, VSTOL first flight on an H_{∞} control law, Comput. Control Eng. J. 6 (1) (1995) 11–16.
[71] H. Mosskull, Optimal DC-link stabilization design, in: IEEE Transactions on Industrial Electronics, vol. 62, no. 8, pp. 5031–5044, 2015.
[72] S. Skogestad, I. Postlethwaite, Multivariable Feedback Control – Analysis and Design, second edition, Ed. by Wiley, 2005.
[73] J. Doyle, Guaranteed margins for LQG regulators, in: IEEE Transactions on Automatic Control, vol. 23, no. 4, pp. 756–757, Aug 1978.

[74] D. McFarlane, K. Glover, A loop-shaping design procedure using H∞ synthesis, in: IEEE Transactions on Automatic Control, vol. 37, no. 6, pp. 759–769, Jun 1992.
[75] J.Y. Hung, W. Gao, J.C. Hung, Variable structure control: a survey, in: IEEE Transactions on Industrial Electronics, vol. 40, no. 1, pp. 2–22, Feb 1993.
[76] A.F. Filippov, Differential Equations with Discontinuous Right-hand Sides, Kluwer Academic Publishers, 1988.
[77] P. Mattavelli, L. Rossetto, G. Spiazzi, Small-signal analysis of DC-DC converters with sliding mode control, in: IEEE Transactions on Power Electronics, vol. 12, no. 1, pp. 96–102, 1997.

FURTHER READING

N. Doerry, Naval power systems: integrated power systems for the continuity of the electrical power supply, IEEE Electrif. Mag. 3 (2) (2015) 12–21.
IEEERecommended Practice for 1 kV to 35 kV Medium-Voltage DC Power Systems on Ships, in: IEEE Std. 1709–2010, pp. 1, 54, Nov. 2 2010.
S.D. Sudhoff, K.A. Corzine, S.F. Glover, H.J. Hegner, H.N. Robey Jr., DC link stabilized field oriented control of electric propulsion systems, IEEE Trans. Energy Conversion 13 (1) (1998).
I. Batarseh, K. Siri, H. Lee, Investigation of the output droop characteristics of parallel-connnected DC-DC converters, in: 25th Annual IEEE Power Electronics Specialists Conference, PESC '94 Record., pp. 1342, 1351 vol. 2, 20–25 Jun 1994.
V. Arcidiacono, A. Monti, G. Sulligoi, Generation control system for improving design and stability of medium-voltage DC power systems on ships, IET Electr. Systems Transport. 2 (3) (2012) 158–167.
C.H. Rivetta, A. Emadi, G.A. Williamson, R. Jayabalan, B. Fahimi, Analysis and control of a buck DC-DC converter operating with constant power load in sea and undersea vehicles, IEEE Trans. Ind. Applicat. 42 (2) (2006) 559–572.
R.W. Erickson, D. Maksimović, Fundamentals of Power Electronics, second ed., Kluwer Academic, Norwell, MA, 2001.
M. Steurer, F. Bogdan, M. Bosworth, O. Faruque, J. Hauer, K. Schoder, et al., Multifunctional megawatt scale medium voltage DC test bed based on modular multilevel converter (MMC) technology, in: 2015 International Conference on Electrical Systems for Aircraft, Railway, Ship Propulsion and Road Vehicles (ESARS), pp. 1, 6, 3–5 March 2015.
V. Staudt, M.K. Jager, A. Rothstein, A. Steimel, D. Meyer, R. Bartelt, C. Heising, Short-circuit protection in DC ship grids based on MMC with full-bridge modules, in: 2015 International Conference on Electrical Systems for Aircraft, Railway, Ship Propulsion and Road Vehicles (ESARS), pp. 1, 5, 3–5 March 2015.
Y. Tang, A. Khaligh, On the feasibility of hybrid Battery/Ultracapacitor Energy Storage Systems for next generation shipboard power systems, in: 2010 IEEE Vehicle Power and Propulsion Conference (VPPC), pp. 1, 6, 1–3 Sept. 2010.
C.H. van der Broeck, R.W. De Doncker, S.A. Richter, J. von Bloh, Unified control of a buck converter for wide load range applications, in: IEEE Transactions on Industry Applications, - in press.
K. Zhou, Essentials of Robust Control, Ed. by Prentice-Hall, May 25, 1999.
M. Kishnani, S. Pareek, R. Gupta, Optimal tuning of DC motor via simulated annealing, in: International Conference on Advances in Engineering and Technology Research (ICAETR), 2014, pp. 1, 5, 1–2 Aug. 2014.
F. Barati, D. Li, R.A. Dougal, Voltage regulation in medium voltage DC systems, in: 2013 IEEE Electric Ship Technologies Symposium (ESTS), pp. 372, 378, 22–24 April 2013.

D. Bosich, M. Gibescu, N. Remijn, I. Fazlagic, J. de Regt, Modeling and simulation of an LVDC shipboard power system: voltage transients comparison with a standard LVAC solution, in: Electrical Systems for Aircraft, Railway and Ship Propulsion (ESARS), 2015, pp. 1, 5, 3–5 Mar. 2015.

M. Cupelli, L. Zhu, A. Monti, Why ideal constant power loads are not the worst case condition from a control standpoint, in: IEEE Transactions on Smart Grid, in press.

H. Xin, Modeling and Control of the Multi-terminal MVDC System, *Master Thesis*, Institute for Automation of Complex Power Systems - RWTH Aachen University, 2012.

H.J. Marquez, Nonlinear Control Systems: Analysis and Design, John Wiley, 2003.

Mathworks, 2017, Convert Model From Continuous to Discrete Time, accessed 17 October 2017 available at https://de.mathworks.com/help/control/ref/c2d.html.

N. Kazantzis, C. Kravaris, Time-discretization of nonlinear control systems via Taylor methods, Comput. Chem. Eng. 23 (6) (1999) 763–784.

R.F. Stengel, Optimal Control and Estimation, Courier Corporation, 2012.

Mathworks, 2017, Design Discrete Linear-quadratic (LQ) Regulator for Continuous Plant, accessed 17 October 2017 available https://de.mathworks.com/help/control/ref/lqrd.html.

Mathworks, 2017, Design Discrete Kalman Estimator for Continuous Plant, accessed 17 October 2017 available at: https://de.mathworks.com/help/control/ref/kalmd.html.

Mathworks, 2017, *Linear-Quadratic-Gaussian (LQG) Design*, accessed 17 October 2017 available at https://de.mathworks.com/help/control/ref/lqg.html.

CHAPTER SIX

Simulation

In this chapter the controllers presented in Chapter 1, Overview—Voltage Stabilization of Constant Power Loads (CPLs), are evaluated. The chapter is structured in such way that in Section 6.1 the results for a cascaded system are presented. In Section 6.2 and its subsections the shipboard power system is simulated for averaged and switched converter representations. Section 6.3 summarizes this chapter.

The scope of these simulations lies solely in assessing the performance of the stabilizing control action. Therefore, each generator is assumed as an ideal voltage source that via a buck converter supplies the DC bus, which operates under voltage control and in continuous conduction mode (CCM). The electromechanical generation system is considered too sluggish to react on the fast dynamics by the DC bus [9], thus it appears during the fast DC bus transient as a constant. This assumption is confirmed by the analysis performed in [10]; there, an examination with a detailed generation model including speed governor and auto voltage regulator versus a stiff voltage source was done. The conclusion was that the maximum transient voltage error was 2% while applying a DC load perturbation from 10% to 40%. Hereafter, the ideal voltage source was used to reduce the study complexity and to focus purely on the control for voltage stabilization. Hence, to evaluate the performance of the controller, no limit on di/dt was imposed nor saturation on the current was put in place, as those limits come from the physical realization, and as technology evolves, this would skew the analysis from a pure performance point.

For guaranteeing CCM the linear resistance was chosen in such way that discontinuous conduction mode (DCM) is avoided [3]. This is because in shipboard power systems the components are sized in a way that they function at their highest efficiency in normal operating conditions, and DCM would therefore show an irregular operation where low load is requested and high losses occur. Due to the restrictions of the Sliding Mode Control described in Section 5.8 we decided to omit it from the simulation in this chapter.

Modern Control of DC-Based Power Systems.
DOI: https://doi.org/10.1016/B978-0-12-813220-3.00006-5

6.1 CASCADED SYSTEM EVALUATION

The cascaded system evaluation is especially useful for decentralized control approaches, since it is possible to observe directly what the best potential estimate is, as the estimate directly reflects the load.

The cascaded system consists of one POL converter and one LRC in which the stabilizing control is evaluated. It offers valuable insight on how direct changes in the POL control will affect a decentralized control strategy. The single-generator models are simulated using averaged converters models. The simulations were run with system parameters mentioned in Table 6.1.

For evaluating the control performance so as to guarantee DC bus stability, the load demand of the POL converter was increased in two steps. Starting from a base load of 2.8 MW, the load demand is increased by a step of 7.5 MW at the time instant of $t = 0.2$ s resulting in a load factor of 0.515. The second increase in load demand occurs at $t = 0.25$ s with the same step size and results in a load factor of 0.89 which equals to a load demand of 17.8 MW.

All subsequent figures show the transient response on the second load step, as this moves the system closer to the stability limit and therefore the over- and undershoots are bigger than for a load step with the same magnitude at a lower base load. Fig. 6.1 shows the measurements of the bus voltages and Fig. 6.2 for the inductor currents.

It should be highlighted that the synergetic control achieves the best bus voltage transient performance. Its undershoot stops at 0.975 p.u. which is better than the Adaptive Backstepping. The rise time is the

Table 6.1 Cascaded System Parameters

	LRC (Buck)	**POL (Buck)**
P_n (MW)	20	20
V_{in} (kV)	8.91	6
V_{out} (kV)	6	3
f_s (kHz)	1	1
R_f (mΩ)	99.7	33.2
L_f (mH)	2.1	1.1
C_f (μF)	659.72	2000
C_{if} (μF)	–	3608
Base load (MW)	2.8	–
Load increase (MW)	–	7.5

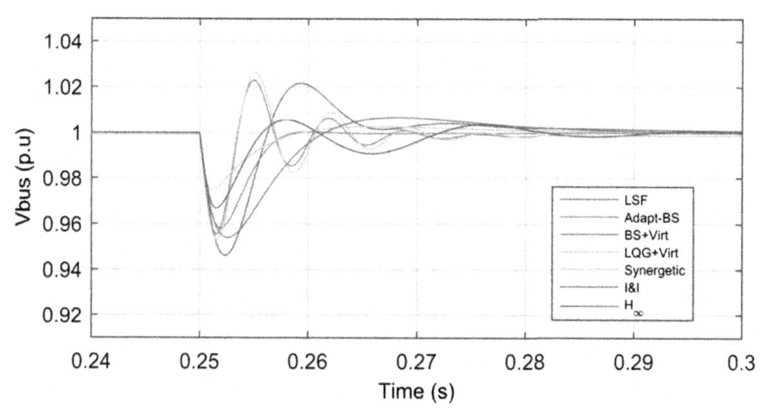

Figure 6.1 Voltage—ideal CPL—Step 10.3 → 17.8 MW.

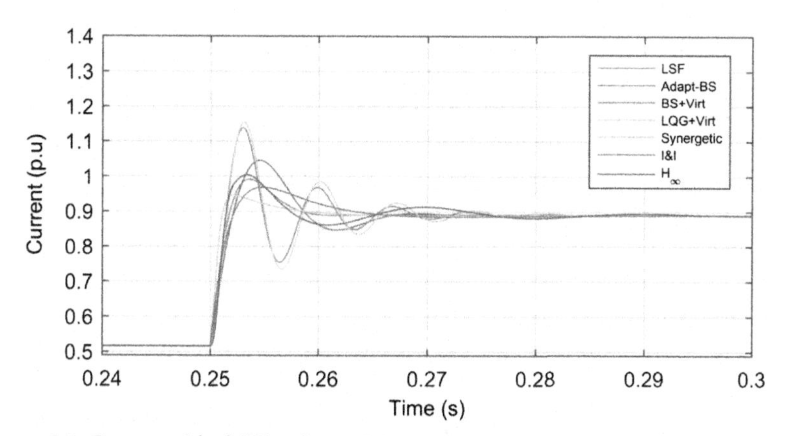

Figure 6.2 Current—ideal CPL—Step 10.3 → 17.8 MW.

fastest of all the controllers while achieving a negligible overshoot of 0.0025 p.u. that is one decade better than the second best. The rise time is faster than all the controllers which exhibit a bigger overshoot. The Adaptive Backstepping controller features the lowest rise time, explained by its error rate having an exponential convergence rate which is reflected by no overshoot, but at a cost of a longer rise time. The perceived low performance of the LSF lies solely on the chosen poles which link to a desired damping. Hence, a trade-off between rise time and over-/undershoot is feasible.

A special attention should be also given to the estimates of the control methods which use estimation techniques. In Fig. 6.3 the estimates of the LQG virtual disturbance controller are depicted. In incidence of average

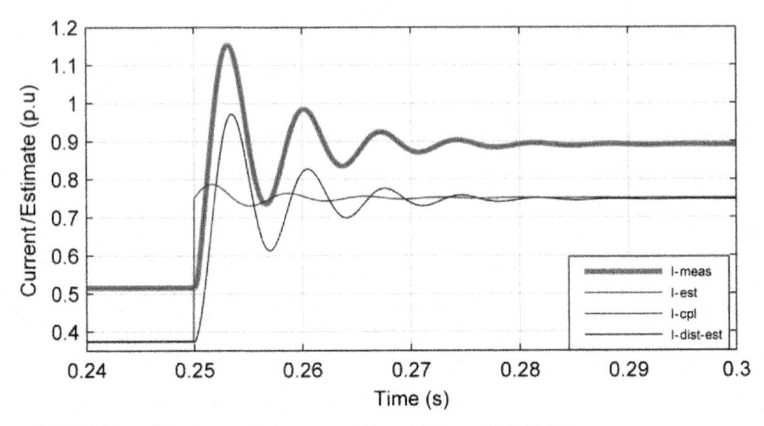

Figure 6.3 LQG estimate and current—Step 10.3 → 17.8 MW.

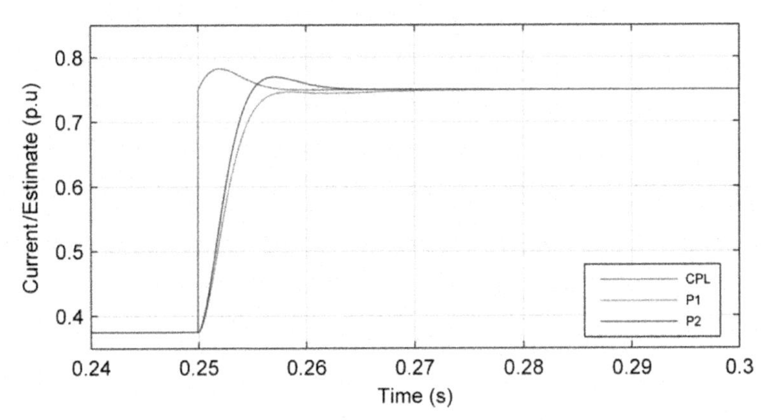

Figure 6.4 Adaptive backstepping estimates and current—Step 10.3 → 17.8 MW.

converter models and an ideal CPL, the measurement and the estimate of the inductor current are identical, denoted by the blue and red curve. In Fig. 6.3 it can be observed how the initial disturbance estimation error, after the step increase in CPL current, converges after 0.02 s to the true value of the CPL current. It can also be observed that the offset between the red and green curve is exactly the linear current part. The shape of estimates for the Backstepping virtual disturbance controller does not differ from the LQG estimates as both rely on an identical Kalman filter.

Fig. 6.4 depicts the estimates of the Adaptive Backstepping and the CPL current, where there exists a slight difference between the estimates $P1$ and $P2$ and here the exponential convergence rate of the estimated values can be observed. An interesting observation is that both estimates converge to the same value which is equal to the actual CPL current. As mentioned in Subsection 5.6.1 the estimates do not necessarily converge to the true value.

In Figs. 6.5 and 6.6 the ideal CPL was replaced with a PI controlled buck converter supplying a resistive load. It can be observed that for the LSF control a bandwidth interaction takes place, since the undershoot is bigger than in the ideal CPL case.

The Synergetic control shows still the lowest undershoot. Furthermore, the virtual disturbance control exhibits lower undershoot than the Adaptive Backstepping; the reason is that this control uses an ideal CPL model from which the PI-controlled POL already deviates. With the introduction of a controlled load its bandwidth has now an

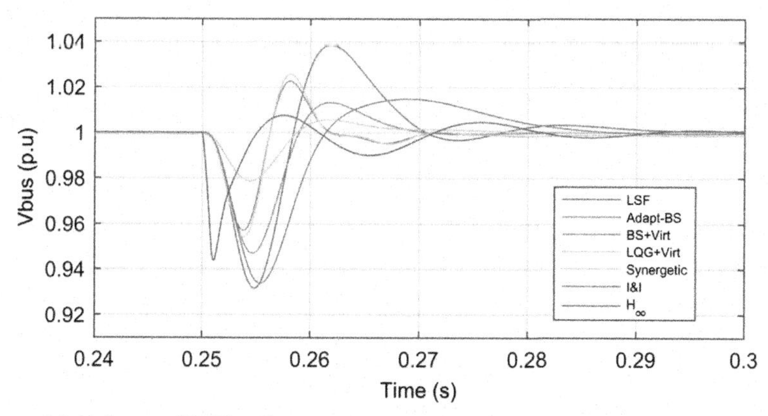

Figure 6.5 Voltage—PI CPL—Step 10.3 → 17.8 MW.

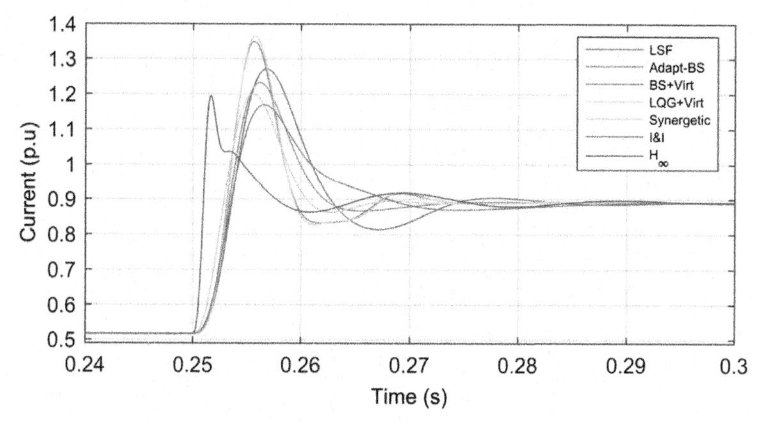

Figure 6.6 Voltage and current—PI CPL—Step 10.3 → 17.8 MW.

influence on the control loop of the LRC as explained in Chapter 2, Small-Signal Analysis of Cascaded Systems. There is an interaction in bandwidths of the two control systems. The virtual disturbance-based controllers exhibit over-/undershoots similar in magnitude to in the ideal CPL case although the transient time is slightly longer. An interesting observation is that due to the PI control, the estimation error now converges with fewer oscillations to its true value as presented in Fig. 6.6, where the current exhibits for both virtual disturbance controllers, after $t = 0.26$, fewer oscillations until the final value is reached.

The last load model considered is represented by a PID-controlled buck converter. The results for this simulation are depicted in Fig. 6.7 for the voltages and in Fig. 6.8 for the currents. Due to the higher bandwidth, the observed time evolution of the bus voltage transient resembles the shape observed for the ideal CPL, although with the introduction of the differentiating term the short time dynamics are worse than with the ideal CPL assumption.

The transient response parameters for the load step are mentioned in Tables 6.2–6.4. Additionally, the values for three commonly used performance indexes, Integral Squared Error (ISE), Integral Absolute Error (IAE), and Integral Time-weighted Absolute Error (ITAE), are given for voltages and currents to assess the disturbance impact on the control [11]. The already small magnitudes for the performance indexes of the disturbance rejection in the bus voltage suggest that a further optimization would not yield at extremely different results, except perhaps for the LSF control, which performs around one decade worse on the ISE index.

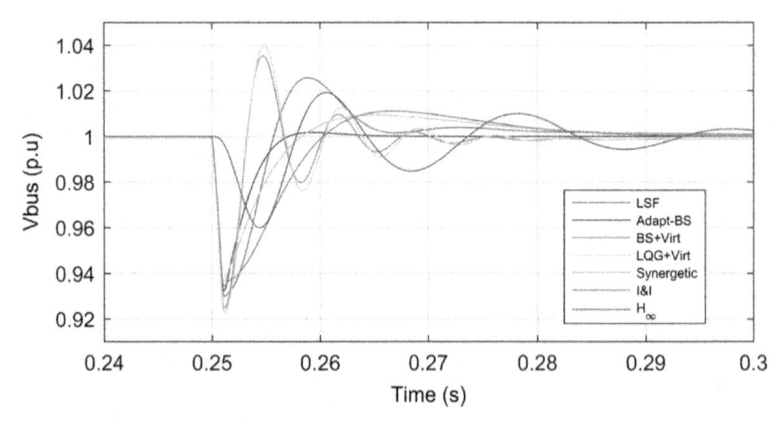

Figure 6.7 Voltage—PID CPL—Step 10.3 → 17.8MW.

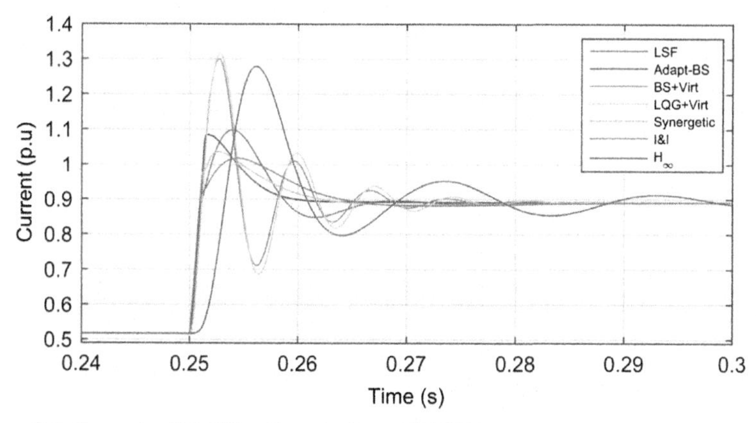

Figure 6.8 Current—PID CPL—Step 10.3 → 17.8MW.

It is perceived that the voltage overshoot is kept small by all the control strategies, and the settling time, defined as the time until the step response stays within a range of 2.5% of the final, is rather fast.

The transients of current and voltage for the virtual disturbance approaches display nearly overlapping transients, the differences in magnitude of voltage and current are negligible in the averaged model. The time evolution of both virtual disturbance control strategies is identical, this resides in the fact that both control strategies use a steady-state Kalman Filter. Consequently, the input in the control system is a brutally-linearized software measurement. The Kalman Filter reflects itself in a remaining steady-state error which is caused by an estimation deviation of 0.5%.

In contrast to the virtual disturbance-based controls the Adaptive Backstepping is able to achieve a zero steady-state error under all CPL implementations; furthermore, one can observe a different behavior, which lies on the fact that the Adaptive Backstepping uses a nonlinear model for estimation combined with a nonlinear control law, while the Backstepping with virtual disturbance estimation uses a linear plant model. This strengthens the assumption that the difference in control performance may originate from the disturbance modeling and its estimation.

The comparison between different CPL implementations (ideal, PI, PID) reveals the interdependency between control bandwidths and their interactions. Selecting a PID control with a more realistic model of the CPL causes higher disturbances compared to the ideal CPL. On the other

Table 6.2 Load Step Response Characteristics Under Second Load Step—CPL

Ideal CPL	Bus Voltage (1 p.u.)						
	Over-shoot (%)	**Under-shoot (%)**	**Set. Time (s) (2.5%)**	**Rise Time (s) (Step → 100%)**	**ISE (e-06)**	**IAE (e-04)**	**ITAE (e-06)**
LSF	102.16	94.62	0.0048	0.0064	10.831	4.1557	7.0735
Adapt-BS	100.04	95.82	0.0041	0.0088	5.2235	1.7461	1.4286
BS + Virt	102.29	95.55	0.0027	0.0035	4.6357	2.1571	2.3437
LQG + Virt	102.65	95.29	0.0055	0.0035	5.7716	2.5846	3.0646
Synergetic	100.27	97.48	0.0014	0.0094	1.8207	1.5357	2.2760
I&I	100.66	95.39	0.0050	0.0117	8.7435	5.4351	4.1490
Hinf	100.60	95.67	0.003	0.0055	3.266	2.21	57.72
	Inductor Current (0.5 → 0.9 p.u.)						
LSF	104.44	84.77	0.0154	0.0024	872.51	32	20.254
Adapt-BS	99.14	–	0.0071	0.0020	788.35	25	10.265
BS + Virt	113.76	75.90	0.0153	0.0016	927.13	34	21.376
LQG + Virt	115.34	73.56	0.0217	0.0016	970.33	36	26.017
Synergetic	94.52	–	0.0051	0.0012	755.62	24	9.6628
I&I	98.21	–	0.0085	0.0024	787.05	28	15.883
Hinf	100.4	3.85	0.0135	0.0076	596	43	116.11

Table 6.3 Load Step Response Characteristics Under Second Load Step—PI CPL

PI	CPL	Bus Voltage (1 p.u.)						
		Over-shoot (%)	**Under-shoot (%)**	**Set. Time(s) (2.5%)**	**Rise Time (s) (Step → 100%)**	**ISE (e-06)**	**IAE (e-04)**	**ITAE (e-06)**
	LSF	102.63	93.08	0.0153	0.0048	15.165	4.7695	7.7207
	Adapt-BS	100.18	93.23	0.0071	0.0059	7.6906	1.8799	1.4396
	BS + Virt	103.54	92.48	0.0052	0.0063	10.828	3.1241	3.3119
	LQG + Virt	103.98	92.26	0.0078	0.0063	12.740	3.6412	4.1838
	Synergetic	100.56	97.87	-	0.0088	1.7478	1.4863	2.0544
	I&I	101.49	93.39	0.022	0.012	20.607	5.847	9.6373
	Hinf	100.7	94.37	0.0025	0.0049	4.93	2.38	61.67
		Inductor Current (0.5 → 0.9 p.u.)						
	LSF	127.10	81.41	0.0231	0.0036	1600	55	54.080
	Adapt-BS	123.20	86.66	0.0134	0.0034	1400	45	33.822
	BS + Virt	135.04	83.16	0.0188	0.0032	1600	48	37.266
	LQG + Virt	136.50	82.90	0.0188	0.0032	1200	43	33.510
	Synergetic	119.94	86.35	0.015	0.0029	1200	41	27.318
	I&I	116.88	—	0.018	0.0035	1168	48	43.693
	Hinf	135.79	—	0.0236	0.0085	454	31.8	842

Table 6.4 Load Step Response Characteristics Under Second Load Step—PID CPL

PID	CPL	Bus Voltage (1 p.u.)						
		Over-shoot (%)	**Under-shoot (%)**	**Set. Time(s) (2.5%)**	**Rise Time (s) (Step → 100%)**	**ISE (e-06)**	**IAE (e-04)**	**ITAE (e-06)**
	LSF	103.91	93.25	0.0098	0.0088	22.725	6.3231	10.959
	Adapt-BS	101.31	94.69	0.0035	0.0071	10.365	3.2390	3.9068
	BS + Virt	102.26	95.70	0.0066	0.0033	5.5479	2.2888	2.6339
	LQG + Virt	102.58	95.41	0.0099	0.0033	6.5261	2.4771	2.8707
	Synergetic	100.93	93.61	0.0045	0.0084	10.290	3.9704	6.4381
	I&I	101.11	93.45	0.0140	0.0085	17.396	5.154	7.531
	Hinf	101.90	96.00	0.0062	0.0079	7.63	3.79	99.3
		Inductor Current (0.5 → 0.9 p.u.)						
	LSF	110.40	84.86	0.0154	0.0012	1600	55	54.268
	Adapt-BS	108.47	–	0.0073	0.0010	1400	45	33.835
	BS + Virt	130.00	25.63	0.0209	0.0011	1600	48	37.263
	LQG + Virt	136.15	68.75	0.0219	0.0012	1600	49	38.486
	I&I	101.76	–	0.0095	0.0012	824.5	31	21.13
	Hinf	142.00	88.50	0.018	0.0035	1041	45.4	1187

hand, when selecting the ideal CPL model, the interaction between the generation side and load side converter is neglected, and the dynamics of the generation side are projected on the load. Thus, the ideal CPL model does not necessarily represent the worst case in terms of system stability as explained in [12].

A very interesting case shows the PI-controlled POL, which due to the missing derivative part is slower than the PID control; therefore, the load is not as tightly controlled as with the PID; in this case the Adaptive Backstepping controller performs very differently, this is because it uses a load model which differs too much from the imposed behavior.

6.2 SHIPBOARD POWER SYSTEM

A shipboard power system consists typically of many LRC in parallel which supply the common MVDC bus as depicted previously in Figs. 2.1 and 6.3. Prior to the simulation, it is necessary to mention which load sharing strategy was chosen in shipboard power systems.

6.2.1 Load Sharing

The strategies used in parallel converter systems for load current sharing can be categorized into droop control and active current sharing. Active current sharing can be either master–slave or democratic current sharing. The main difference between the master–slave and democratic methods is providing a variable or fixed ratio of current during transient and static operation [5]; and the difference between droop and active current sharing is that droop uses r_d as a virtual resistance for current sharing while active current sharing uses the voltage of the source, hence, for realizing droop the converters are controlled independently [4].

In this study, democratic current sharing is applied in synergetic control. While droop was used for all other stabilizing control methods.

The effect of the droop control is to create a virtual voltage drop on the converters output proportional to the converters output current [13], so the controller increases or decreases the setpoint by $r_{dk} \cdot I_k$, to maintain

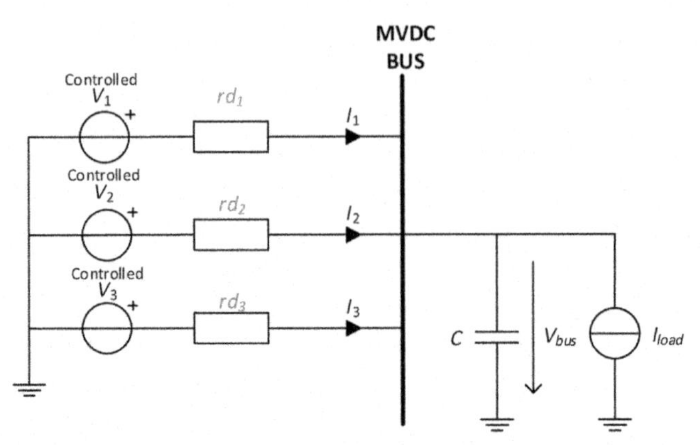

Figure 6.9 Equivalent circuit of the shipboard power system under droop control.

the desired power sharing ratio. It has to be kept in mind that the maximum value of r_d to ensure stability of the loop is defined by:

$$r_d < \frac{E_k - V_{bus}}{I_{max}} = 0.878 \tag{6.1}$$

Subsequently, the tuning of r_d results always in a between very tight load sharing (high r_d) or a low steady-state voltage drop (low r_d). It is possible to compensate the steady-state voltage drop due to additionally introduced resistance r_d with a secondary control action that consists of an additional PI controller on the droop control loop.

As the selected control architectures are two-layered controllers which feature a fast stabilizing control as a primary action and a slower secondary loop which is responsible for the proper sharing between the generators, an analysis with respect to the load sharing has to be undertaken. The analyzed case consists of three LRCs which are connected to the same MVDC bus, feeding both CPL and resistive loads. The sharing of the power among the converters is achieved by a droop control that is integrated in each controller; the representation of the model under droop control is shown in Fig. 6.9, where *Controlled* $V_{1,2,3}$ represent the voltage output of the corresponding converters after the stabilizing control action is applied.

6.2.2 Averaged Model

Initially, to evaluate the different control strategies the grid model is implemented in Matlab. The grid model represents the integrated power system of an all-electrical ship as it is proposed in [2]. This case study

Table 6.5 Shipboard Power System Parameters (Matlab)

	LRC$_{1,2,3}$ (Buck)	**POL (Buck)**
P_n (MW)	20	60
V_{in} (kV)	8.91	6
V_{out} (kV)	6	3
f_s (kHz)	1	1
R_f (mΩ)	99.7	33.2
L_f (mH)	2.1	1.1
C_f (μF)	659.72	2000
C_{if} (μF)	–	3608
Base load (MW)	2.8	–
Load increase (MW)	–	22.5

features the control implementations of Chapter 1, Overview—Voltage Stabilization of Constant Power Loads (CPLs). The converter models are simulated using averaged converters models. The simulations were run with the system parameters stated in Table 6.5 which are based on [12]. The test cases are reflecting large changes in the operating conditions; on one hand the sudden connection of large amount of load, and on the other hand the unexpected disconnection/failure of available generation capacity.

At this stage it should be clarified, that each control could exhibit varying performances in terms of overshoot, undershoot, settling time, rise time, and oscillations, depending on the selection of its control coefficients. The scope of this overview is to highlight that decentralized controllers do not necessarily perform worse than centralized controllers and that under simplified conditions stabilization with good results can be achieved by decentralized control architectures, while the system undergoes large signal perturbations.

The three LRCs have a load sharing of 33%/33%/33% set via droop. For evaluating the achieved control performance with respect to guaranteeing DC bus stability the load demand of the POL converter was increased in two steps. Starting from a base load of 2.8 MW per LRC, which is has be selected to guarantee CCM, the load demand is increased by a step of 22.5 MW at the time instant of $t = 0.1$ s resulting in a load factor of *0.515*.

The second increase in load demand occurs at $t = 0.25$ s with same step size, this results in a load factor of 0.89 which is equal to a load

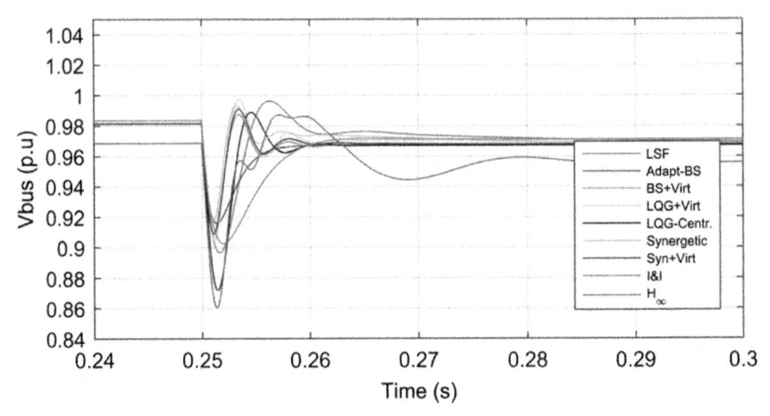

Figure 6.10 ISPS V_{bus} for ideal CPL—Step 30.9 → 53.4 MW.

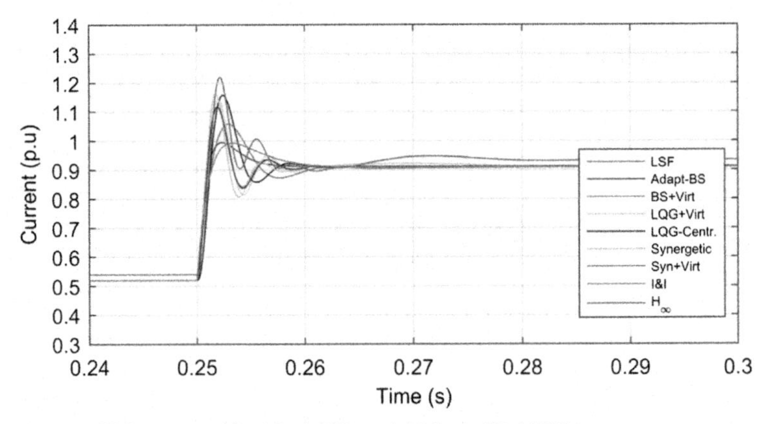

Figure 6.11 ISPS $I_{L1=L2=L3}$ for Ideal CPL and 30.9 → 53.4 MW.

demand of 53.4 MW. Figs. 6.10 and 6.11 display the measurements of bus voltage and inductor current for the second load step.

In the ISPS test cases two additional controllers are considered in the performance evaluation: the LQG-Centralized Controller, and the Synergetic Virtual Disturbance Controller. The first uses a centralized formulation of the virtual disturbance concept while the second replaces the LQR + Setpoint Trajectory or the Backstepping part with a Synergetic Control law.

The steady-state error on the voltage in Fig. 6.10 after the load step increase appears throughout all implemented controllers and is a result of the inserted droop resistances. It can be observed that the Adaptive Backstepping achieves the best results in terms of overshoot, undershoot, and settling time while the LQG-Centralized control exhibits the highest voltage drop. The LSF exhibits the slowest dynamics due to the chosen dynamics through pole placement and the nonlinearity compensation. All three decentralized virtual disturbance-based controls display a behavior comparable to that already observed in the cascaded system setup and show in all aspects better performance compared to the centralized LQG formulation. Characteristic values, like those calculated in the cascaded system, are presented here in Table 6.6. It should be clarified that due to the steady-state error the numerical values are calculated for the end value of voltage and current, taken at $t = 0.39$ s.

To avoid the steady-state error after the load increase two options are possible. One could be to design a prefilter, which actually can only be designed for a specific load level and would mean that at other load levels an over- or undervoltage appears. The alternative is the implementation of a secondary control loop consisting of a PI controller with the feedback of the bus voltage as mentioned in Subsection 6.2.1. The latter option was chosen.

It's to be pointed that the introduction of a Droop—PI outer control loop for steady-state error suppression will lead to a change in transient behavior as another control loop interaction is introduced. The corresponding results can be analyzed in Figs. 6.12 and 6.13 where the reader can observe that due to the additional PI-Loop the Centralized LQG improves in performance; this resides in the fact that the centralized formulation profits more from an adjusted voltage setpoint. For the other controllers the decrease in voltage drop corresponds directly with the reduction in steady-state error. The rise times for the decentralized virtual disturbance and Synergetic control are 0.0022 s, while for the Centralized LQG it is 0.0032 s, and for the LSF it corresponds to 0.0037 s.

After presenting the possibility of mitigating the steady-state error in the bus voltage, all further test cases are performed without the Droop—PI as its inclusion would deviate the analysis from a pure stabilizing action point-of-view towards a system level analysis, and in certain combinations retuning control coefficients would be necessary.

In an ISPS consisting of multiple LRCs also the case of a loss of available generation capability needs to be considered. At the time instant $t = 0.3$ s LRC_3 is disconnected and $LRC_{1,2}$ are accountable to supply the 30.9 MW load. Thus, the load factor is 0.515 of the installed power before and 0.775 after the reduction of installed power. The results for the control strategies are shown in Figs. 6.14 and 6.15. Those results

Table 6.6 Numerical Results—Load Increase ($t = 0.25$ s)

Load Increase (30,9 → 53,4MW)		Over-shoot (%)	Under-shoot (%)	Settling time (2.5%)	Rise time (step → 100%)	ISE (e-06)	IAE (e-04)	ITAE (e-06)
	Bus Voltage (End Value = 0.9674 p.u.)							
	LSF	102.99	92.66	0.0072	0.0042	14.245	5.5532	9.6130
	Adapt-BS	101.47	94.67	0.0034	-	6.9140	2.4301	1.5341
	BS + Virt	102.06	93.89	0.0020	0.0024	5.4146	2.0254	1.1521
	LQG + Virt	103.09	94.54	0.0039	0.0024	6.7732	2.2767	1.4360
	LQG-Centr	102.20	90.12	0.0031	0.0035	15.521	3.1477	2.0329
	Synergetic	102.68	94.86	0.0020	0.0023	4.8723	3.0720	4.4029
	Syn + Virt	102.42	93.97	0.0035	0.0023	5.4757	2.1855	1.5453
	I&I	101.23	94.51	0.0078	-	14.801	4.0407	3.1594
	Hinf	101.20	89.43	0.0047	0.0061	30.801	5.6000	144
	Inductor Current (End Value = 0.9113 p.u.)							
	LSF	115.99	95.75	0.0095	0.0014	873.92	29	16.544
	Adapt-BS	109.10	—	0.0052	0.0011	829.91	24	8.2260
	BS + Virt	122.32	92.35	0.0068	0.0010	882.38	26	9.7587
	LQG + Virt	125.41	89.43	0.0073	0.0010	902.87	28	11.078
	LQG-Centr	127.05	94.06	0.0068	0.0013	957.11	29	11.725
	Synergetic	127.62	88.54	0.0051	0.0009	888.15	29	18.740
	Syn + Virt	122.57	91.86	0.0068	0.0010	882.93	26	10.178
	I&I	109.12	—	0.0075	0.0014	834.46	26	10.316
	Hinf	117.55	—	0.0211	0.0078	361.00	27.5	727.7

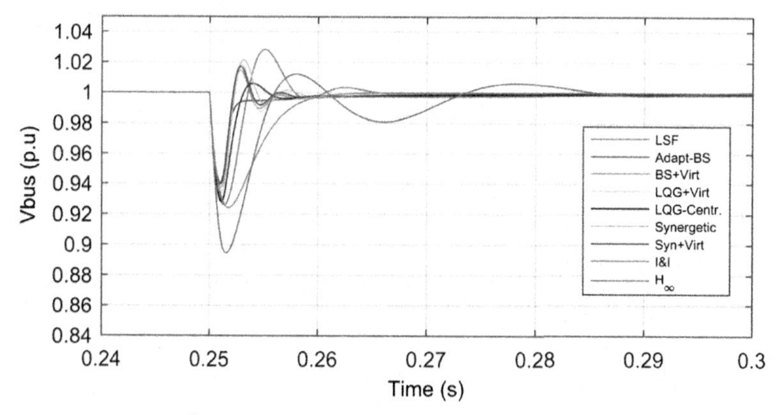

Figure 6.12 ISPS V_{bus} for ideal CPL—Step 30.9 → 53.4 MW (Droop–PI).

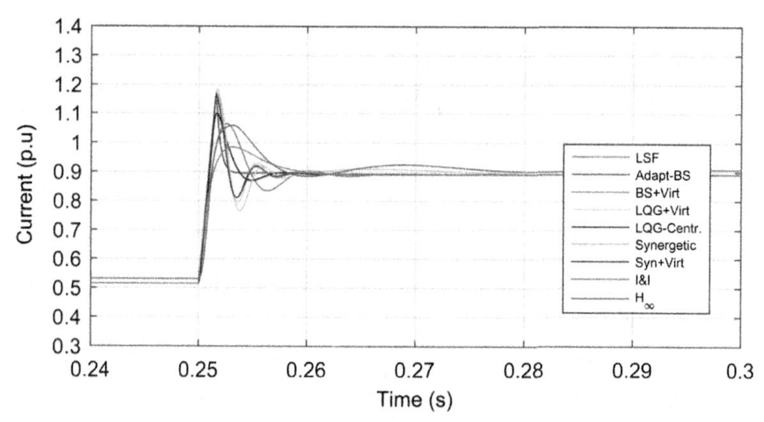

Figure 6.13 ISPS $I_{L1=L2=L3}$ for ideal CPL and 30.9 → 53.4 MW (Droop–PI).

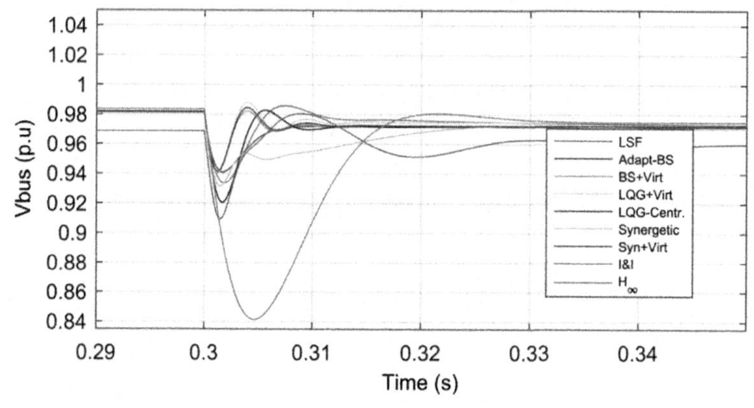

Figure 6.14 ISPS V_{bus} for ideal CPL—generator loss (60 → 40 MW) 30.9 MW base load.

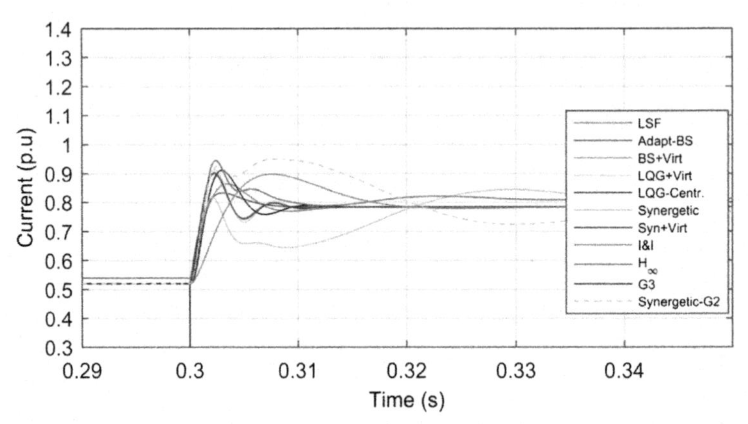

Figure 6.15 ISPS $I_{L1=L2}$ for ideal CPL—generator loss (60 → 40 MW) 30.9 MW base load.

corresponding to the test case with generator loss are similar to the test cases with load step-up, since the loss of one generator can be seen as a load step from the controller's point of view. Detailed response data can be found in Table 6.7. However, an exception appears in case of the Synergetic controller which will be explained in detail successively. In Fig. 6.14 the bus voltage in case of the Synergetic control shows a "dent" which occurrence could be understood looking at the currents in Fig. 6.15. There, an unexpected behavior of the currents for the Synergetic control appears; the remaining control strategies, where $I_{L1} = I_{L2}$ display a valid current behavior. This is because the Synergetic control is completely model-based as explained in Section 5.2; thus, with an unplanned disconnection of one converter, the model does not match to its initial definition anymore.

6.2.3 Switched Converter Model

The fundamental difference between the simulations in this subsection and those from the previous subsection is the move from averaged converter model towards switched converter model representations, and later, from the ideal CPL representation towards a PID-controlled buck converter. As a consequence, the voltage and currents signals have a switching ripple superimposed; hence, their signals have harmonic components. Such signals cannot be directly used in closed-loop control, they should be first filtered.

The decentralized controllers compute the estimates based on the bus voltage measurement; this puts a strong emphasis on the quality of the

Table 6.7 Numerical Results—Generation Loss ($t = 0.3$ s)

Loss of Generation (60MW -> 40MW)	Bus Voltage (End Value = 0.972 p.u.)	Over-shoot (%)	Under-shoot (%)	Set. Time (s) (2.5%)	Rise Time (s) (step → 100%)	ISE (e-06)	IAE (e-04)	ITAE(e-06)
	LSF	101.40	96.03	0.0533	0.055	4.7914	3.2106	5.0737
	Adapt-BS	100.99	96.78	0.0529	0.0601	2.9978	1.6531	1.0716
	BS + Virt	101.05	96.79	0.0518	0.0529	1.8838	1.2763	0.73535
	LQG + Virt	101.66	96.74	0.0519	0.0527	2.2189	1.3965	0.89069
	LQG-Centr	101.08	94.69	0.0532	0.0543	5.5763	2.0674	1.4028
	Synergetic	101.00	95.80	0.0529	0.0765	7.5156	4.2895	6.6365
	Syn + Virt	101.28	96.82	0.0518	0.0528	1.9199	1.3687	0.95508
	I&I	101.05	87.71	0.0541	0.0865	10.203	17.010	18.8962
	Hinf	101.9	93.79	0.0028	0.0052	1.14	1.85	49
	Inductor Current (End Value = 0.7837 p.u.)							
	LSF	110.07	98.05	0.0562	0.0517	396.49	19	9.3836
	Adapt-BS	106.18	–	0.0556	0.0515	381.23	17	5.8210
	BS + Virt	114.74	95.09	0.0559	0.0512	400.61	18	6.7721
	LQG + Virt	116.52	93.42	0.0562	0.0512	408.46	19	7.5912
	LQG-Centr	116.21	96.64	0.0579	0.0516	430.33	20	8.2614
	Synergetic	104.16	82.10	0.0894	0.0515	608.57	44	8.2541
	Syn + Virt	115.11	94.65	0.0562	0.0512	401.57	18	7.0432
	I&I	109.12	–	0.0562	0.0518	834.46	19	6.922
	Hinf	112.6	–	0.0133	0.0014	278.5	16	487.67

bus voltage measurement which is mainly perturbed by a noise signal influenced by the switching frequency. Therefore, a 10th-order notch filter (Butterworth) was used for the voltage. Additionally, for controllers which use measured currents these were second-order low pass filtered. Due to this filtering a degradation in performance will occur.

In Fig. 6.16 a comparison of the measured and filtered bus voltage in the case of an increase in load for the decentralized LQG Controller is depicted, where the measured voltage exhibits a ripple which is suppressed in the filtered signal. The filtered bus voltage is used inside the control loop.

The centralized controllers, which do not perform state estimation, use instead the measurement of inductor current. Likewise, the Adaptive Backstepping, as discussed in Subsection 7.6.1, relies on the measurement of the inductor current to calculate the power estimates. The inductor current does not merely contain the switching frequency components but also harmonics that are introduced due to the switching behavior. In addition, the frequency spectrum of the current is also dependent on the duty cycle, which is continuously changing. These characteristics lead to the selection of a second-order low pass filter to remove the high frequency components of the inductor current.

The figures presented in this subsection are the evaluation of controllers, using a switched converter model. They contain the measured quantities without smoothing.

The transient response and the time evolution of the virtual disturbance-based controllers is consistent, which solidifies the assumption

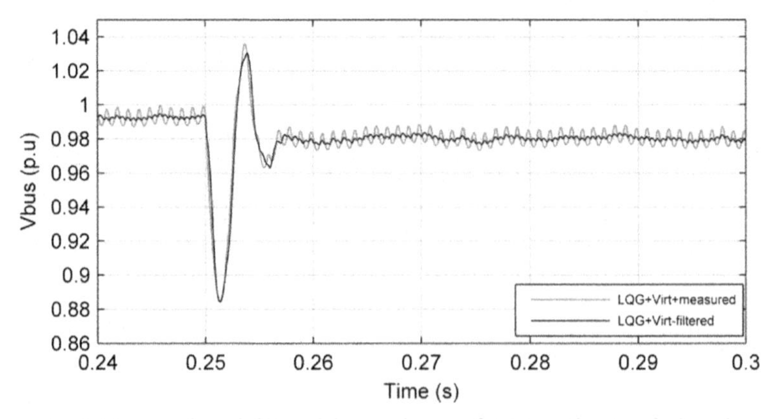

Figure 6.16 Measured and filtered bus voltage of LQG with virtual disturbance—ideal CPL.

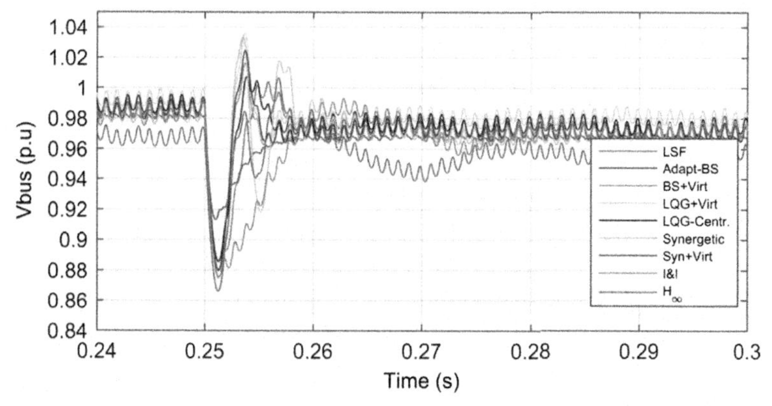

Figure 6.17 ISPS V_{bus} for switched LRC, ideal CPL—Step 30.9 → 53.4 MW load.

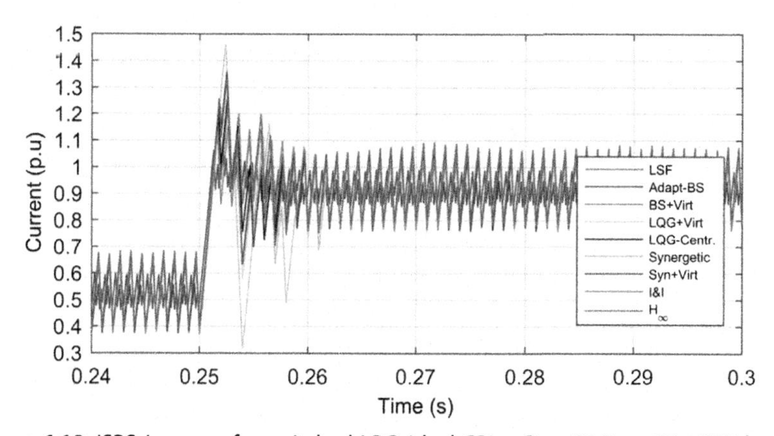

Figure 6.18 ISPS $I_{L1=L2=L3}$ for switched LRC, ideal CPL—Step 30.9 → 53.4 MW load.

that the limiting factor for achieving a better control performance is the form of Kalman filter used. In Fig. 6.17 it can be observed that the synergetic control exhibits larger oscillation before reaching its settling time.

When analyzing the currents in Fig. 6.18 the higher current ripple becomes obvious as do the oscillations after the load step for the centralized synergetic controller. Those changes from the averaged model are in line with the observations done in the cascaded system; accordingly, any deviation from the model which was used for control design has a bigger impact on the performance of the synergetic control. Thus, the decentralized formulation offers better robustness respect to model–reality deviations. The Adaptive Backstepping control seems to offer at this point the most desirable performance if the overshoot in the current is a major

Table 6.8 Load Step 30.9 → 53.4 MW—Ideal CPL Load Representation ($t = 0.25$ s)
Bus Voltage (End Value = 0.9751 p.u.), Ripple ~0.485%

	Over-shoot (%)	**Under-shoot (%)**	**Set. Time (s) (2.5%)**	**Rise Time (s) (step → 100%)**
LSF	102.96	89.75	0.005	0.0038
Adapt-BS	—	93.64	0.0041	0.0078
BS + Virt	104.19	90.27	0.0041	0.0027
LQG + Virt	106.25	90.77	0.0043	0.0026
LQG-Centr	103.27	90.81	0.0050	0.0032
Synergetic	106.04	89.68	0.0080	0.0025
Syn + Virt	105.01	90.27	0.0042	0.0026
I&I	102.51	89.74	0.0085	0.070
Hinf	100.6	86.66	0.0080	0.0028

Inductor Current (End Value = 0.9111 p.u.), Ripple ~ 15%

	Over-shoot (%)	**Under-shoot (%)**	**Settling time (15%)**	**Rise time (step → 100%)**
LSF	136.10	79.51	0.0018	0.0012
Adapt-BS	112.83	—	0.0022	0.001
BS + Virt	146.97	69.42	0.0080	0.0012
LQG + Virt	150.59	63.22	0.0080	0.0012
LQG-Centr	140.05	83.27	0.0100	0.0012
Synergetic	160.79	34.68	0.0112	0.0012
Syn + Virt	148.94	69.42	0.0080	0.0012
I&I	114.15	—	0.0022	0.0012
Hinf	130.15	—	0.0020	0.0018

concern for the system designer. The numerical results in Table 6.8 highlight that even under a switched converter evaluation the controllers' performance with exception of the synergetic control is adequate.

The Adaptive Backstepping Controller response displays a promising performance, but when analyzing the form of the duty cycle a strong oscillation between 0 and 1 can be observed as depicted in Fig. 6.19, where the duty cycles of all controllers are shown. The reasoning is that Power Estimation is dependent on the measured inductor current and the three decentralized virtual disturbance controllers are not affected by current ripple as they use the Kalman Filter, which brings an inherent way of dealing with noise.

A single switching converter with an ideal CPL was investigated for further analysis of the Adaptive Backstepping behavior. This analysis was conducted to exclude effects of other LRCs and the harmonics

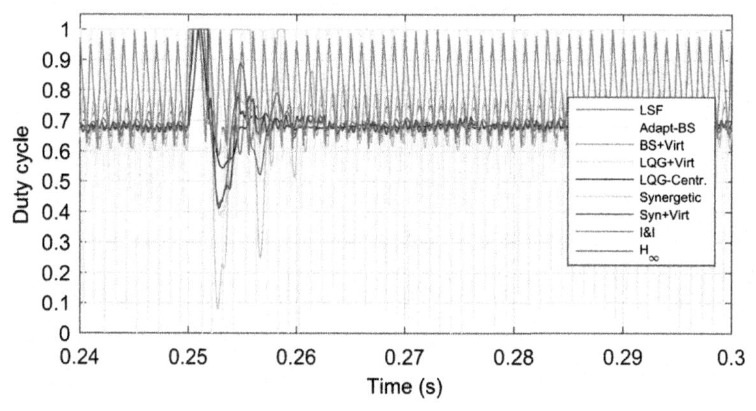

Figure 6.19 Duty cycle of all control methods.

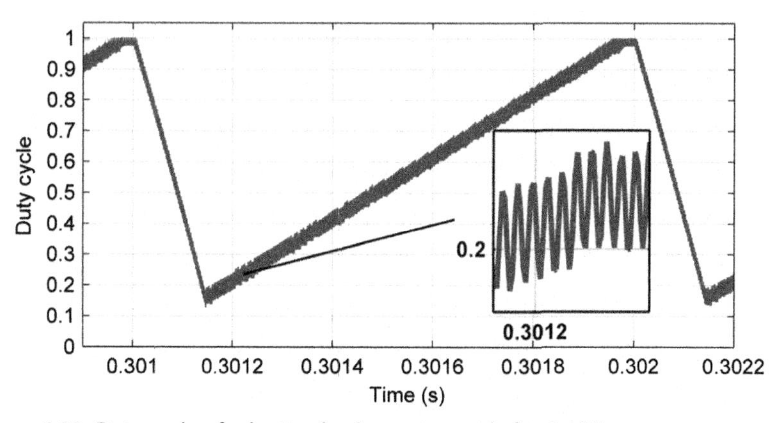

Figure 6.20 Duty cycle of adaptive backstepping with $f_s = 1$ MHz.

introduced by switching CPL representation. For this case a simulation was performed with a switching frequency of 1 MHz. The results depicted in Fig. 6.20 revealed that the hard 0.1−1 duty cycle oscillation of Fig. 6.19 is replaced by a saw tooth curve. It has a rising slope of $m_r = 0.001$ and a falling slope of $m_f = -5518$ in the time interval $t = 0.301 - 0.302$ representing steady-state operation. This diverges from the averaged simulations where the duty cycle has a constant value in steady state.

The decentralized virtual disturbances-based controllers seem to exhibit in the test case a similar response; then, a more detailed look on the quality of their estimates is given in Fig. 6.21 and in Fig. 6.22 during the closed-loop control. In both figures the estimate is not perfectly

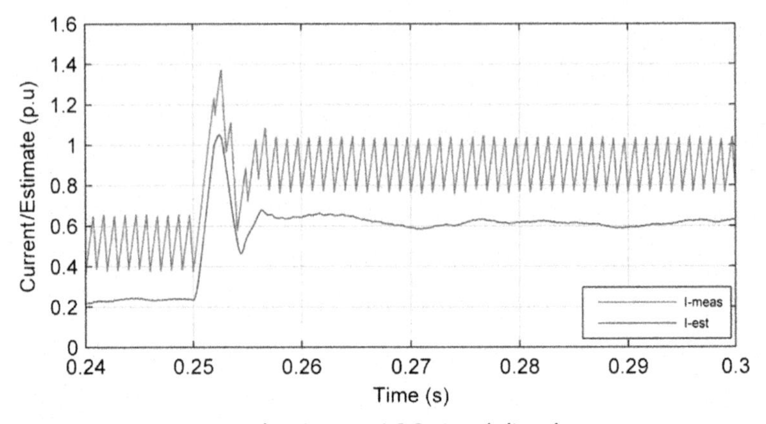

Figure 6.21 Measurement and estimate—LQG virtual disturbances.

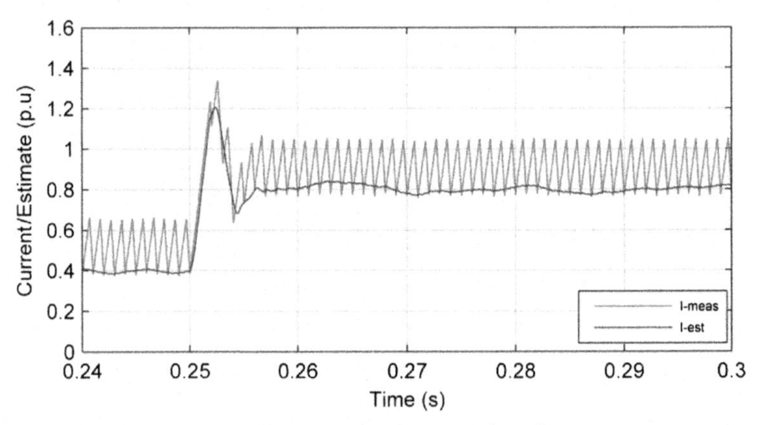

Figure 6.22 Measurement and estimate back virtual disturbances.

aligned with the measurement in its magnitude, even a steady-state error of nearly 0.3 p.u. in Fig. 6.21 is observable. The virtual Backstepping implementation offers a lower steady-state error of 0.0639 compared to the 0.3068 of the LQG controller.

So far the tests were undertaken with switched LRC models while the CPL was still implemented in its ideal representation. Substituting the ideal CPL with a buck converter and its switched model representation, the ripple in the bus voltage is further amplified due to an additional source of switching converter, as depicted in Fig. 6.23. The introduction of switching buck converter as load introduces a higher ripple on the bus voltage compared to the case when moving from an averaged source converter to its switched representation, because the input of the buck

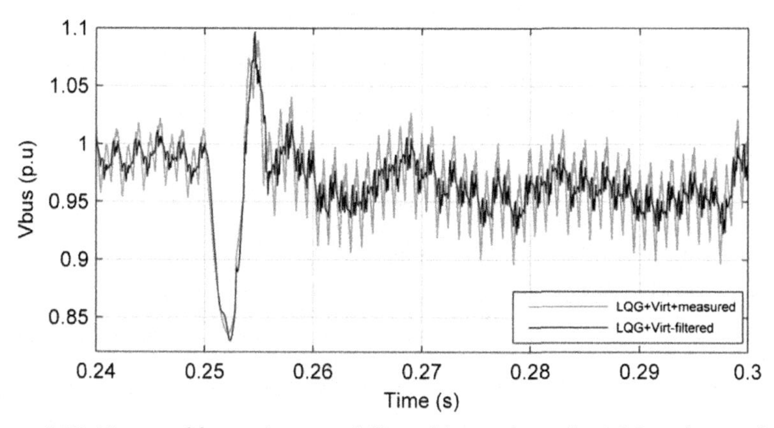

Figure 6.23 Measured bus voltage and filtered bus voltage for LQG with virtual disturbance—switched CPL representation.

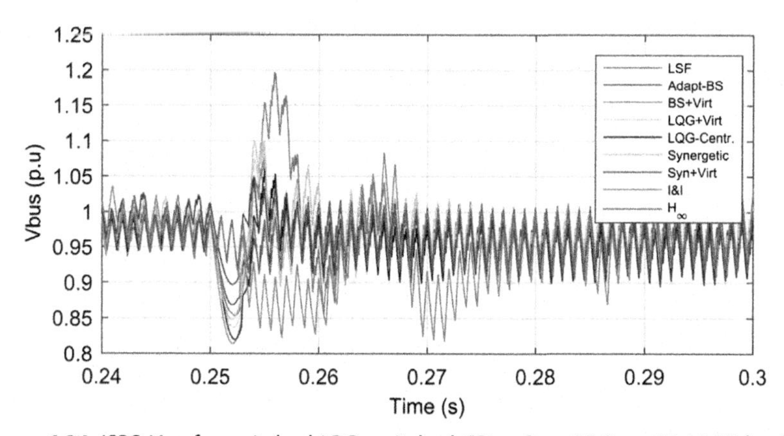

Figure 6.24 ISPS V_{bus} for switched LRC, switched CPL—Step 30.9 → 53.4 MW load.

converter is not smooth and consequently the ripple in the bus voltage increases by 4% and the current ripple faces an increment of 6% (Fig. 6.24).

This additional switching impacts the transient response especially on the LSF in Fig. 6.25 which now undergoes higher oscillations and longer settling time. Compared to the results depicted in Figs. 6.17 and 6.18 the oscillations of the Synergetic controller disappeared.

Once more, the virtual disturbance controllers offer a very similar behavior with respect to their transient behavior and time evolution, which strengthens the statement of [6] stating that the disturbance model has to be improved for the case of switching converters. The results of the transient response for a load connection are given in Table 6.9.

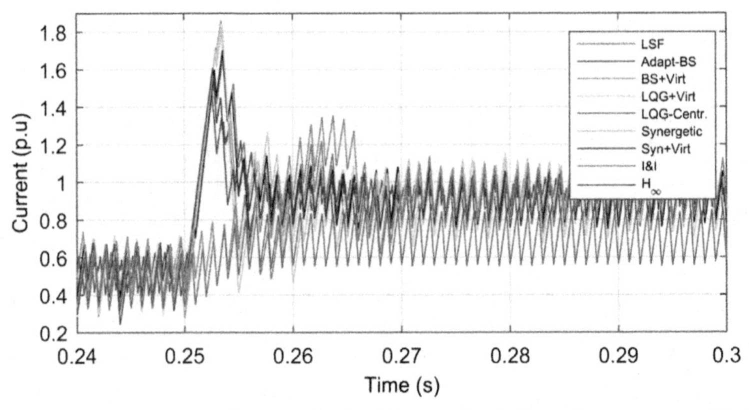

Figure 6.25 ISPS $I_{L1=L2=L3}$ for switched LRC, switched CPL—Step 30.9 → 53.4 MW load.

Table 6.9 Load Step 30.9 → 53.4 MW—Switched CPL Representation ($t = 0.25$ s)
Bus Voltage (End Value = 0.9607 p.u.) Ripple ~ 4,5%

	Overshoot (%)	Undershoot (%)	Settling time (7.5%)	Rise time (step → 100%)
LSF	124.60	88.81	0.0161	0.0036
Adapt-BS	–	90.38	0.0028	0.0041
BS + Virt	111.79	84.69	0.0052	0.0037
LQG + Virt	113.56	87.05	0.0052	0.0036
LQG-Centr	110.44	85.28	0.0061	0.0039
Synergetic	114.60	88.28	0.0101	0.0034
Syn + Virt	109.50	93.36	0.0060	0.0028
I&I	114.15	87.15	0.025	0.012
Hinf	102.15	88.15	0.022	0.015

Inductor Current (End Value = 0.9228 p.u.) Ripple ~ 21%

	Overshoot (%)	Undershoot (%)	Settling time (23.5%)	Rise time (step → 100%)
LSF	181.73	52.32	0.0136	0.0012
Adapt-BS	165.69	–	0.0062	0.0012
BS + Virt	201.67	73.16	0.0051	0.0014
LQG + Virt	199.72	62.71	0.0044	0.0014
LQG-Centr	184.33	74.09	0.0061	0.0015
Synergetic	197.44	44.44	0.0111	0.0013
Syn + Virt	174.47	83.20	0.0037	0.0011
I&I	158.25	–	0.008	0.0018
Hinf	115.25	–	0.008	-

6.3 SUMMARY

Throughout this chapter the application of the generation control strategies presented in Chapter 1, Overview—Voltage Stabilization of Constant Power Loads (CPLs), were shown on two test systems. The first consisted of the cascaded converter configuration presented in Section 6.1. In this system, the influence of different constant power load representations (ideal, PI, PID) was presented. The simulation results are in line with the conclusion of the theoretical analysis presented in throughout Chapter 2 and assuming that the ideal CPL representation would represent the worst case condition is not correct especially for control strategies that are model based, like the Synergetic Control. Under the simulation case of load increase from 10.3 MW → 17.8 MW all control methods were able to stabilize the bus voltage with a minor voltage drop during load increase, although none could be identified as "The best" since depending on system requirements each column in Table 6.2 could be weighted differently; expressly, certain aspects like straightforwardness of control design and synthesis are not to be underestimated.

The second test system was a parallel source multiconverter system representing the circumstances in ISPSs. Throughout the course of this book a great number of simulations have been performed (even/uneven power sharing, averaged/switched LRCs, filtered/non filtered measurements, PI−Droop/no-PI−Droop, ideal CPL/switched CPL, load increase/decrease, LRC disconnection); from these combinations a few scenarios were presented, to be exact:

- Even power sharing;
- Averaged LRC—with ideal CPL—no-PI−Droop—Load increase;
- Averaged LRC—with ideal CPL—with-PI−Droop—Load decrease;
- Averaged LRC—with ideal CPL—LRC disconnection;
- Switched LRC—with ideal CPL under filtered measurements—no-PI−Droop;
- Switched LRC with switched CPL under filtered measurements—no-PI−Droop.

All presented controllers achieve the goal of voltage stabilization under even and uneven power sharing conditions. Hence, generation side control asserts success in addressing the key technical issue of bus voltage regulation [1] and can enable the deployment of MVDC on ISPSs. The Synergetic control exhibited oscillations in the transient response for the case of ideal CPL–switched LRCs, such oscillations do not appear for the switched CPL–switched LRCs case.

All decentralized virtual disturbance-based controllers (LQG, Backstepping, Synergetic) exhibited a certain robustness towards changing scenario conditions; they exhibited also a robustness towards modeling error, like the disconnections of a LRC which made the control reaction of the Synergetic control not usable, as it is a model-based controller.

In the switched LRC simulations, the usage of steady-state Kalman filter in a closed-loop control did not exhibit any obvious restrictions even though the system is assumed by definition linear.

The switched LRC simulations also revealed that the Adaptive Backstepping controller cannot be used in its actual implementation for switching frequency of 1 kHz, as the duty cycle oscillates rapidly between 0 and 1 and such a behavior would severely limit the lifetime of the equipment. This explains why in all cited Adaptive Backstepping applications only low power converters in a cascaded configuration are used [7,8]. They operate with switching frequencies of 40 kHz–1 MHz. There the simple discretization procedure seems good enough; however, detailed plots of the duty cycle are not shown.

To adapt the Adaptive Backstepping controller a time-discrete Lyapunov design procedure based on sampled data would need to be undertaken without guaranteeing a 1 kHz switching frequency; moreover, it seems that with Adaptive Backstepping the duty cycle will always oscillate. The observed duty cycle in Fig. 6.20 is a sign that in the nonlinear process of power estimation in combination with variable structure systems, hidden dynamics are present which influence the results and do not let the power estimation process converge to a constant steady-state value.

REFERENCES

[1] N. Doerry, Naval Power Systems: integrated power systems for the continuity of the electrical power supply, IEEE Electr. Mag. 3 (2) (2015) 12. 21, June.
[2] IEEE Recommended Practice for 1 kV to 35 kV Medium-Voltage DC Power Systems on Ships, IEEE Std. 1709–2010, pp. 1, 54, Nov. 2 2010.
[3] R.W. Erickson, D. Maksimović, Fundamentals of Power Electronics, second ed., Kluwer Academic, Norwell, Mass, 2001.
[4] I. Kondratiev, A Synergetic Control for Parallel-connected DC-DC Buck Converters, *PhD Dissertation*, University of South Carolina, College of Engineering and Information Technology, 2005.
[5] I. Kondratiev, R. Dougal, General synergetic control strategies for arbitrary number of paralleled buck converters feeding constant power load: implementation of dynamic current sharing, 2006 IEEE Int. Symp. Ind. Electr. vol. 1 (2006). no., pp. 257, 261, 9-13 July.

[6] J. Liu, D. Obradovic, A. Monti, Decentralized LQG control with online set-point adaptation for parallel power converter systems, in 2010 IEEE Energy Conversion Congress and Exposition (ECCE), pp. 3174, 3179, 12–16 Sept. 2010.

[7] H. El Fadil, F. Giri, M. Haloua, H. Ouadi, Nonlinear and adaptive control of buck power converters, in Proceedings. 42nd IEEE Conference on Decision and Control, 2003, vol. 5, no., pp. 4475, 4480 Vol. 5, 9–12 Dec. 2003.

[8] M.L. McIntyre, M. Schoen, J. Latham, Simplified adaptive backstepping control of buck DC:DC converter with unknown load, 2013, in IEEE 14th Workshop on Control and Modeling for Power Electronics (COMPEL), pp. 1,7, 23–26 June 2013.

[9] F. Barati, L. Dan, R.A. Dougal, Voltage regulation in medium voltage DC systems, in 2013 IEEE Electric Ship Technologies Symposium (ESTS), pp. 372, 378, 22–24 April 2013.

[10] D. Bosich, M. Gibescu, N. Remijn, I. Fazlagic, J. de Regt, Modeling and Simulation of an LVDC Shipboard Power System: Voltage Transients Comparison with a Standard LVAC Solution, in Electrical Systems for Aircraft, Railway and Ship Propulsion (ESARS), 2015, pp. 1,5, 3–5 Mar. 2015.

[11] M. Kishnani, S. Pareek, R. Gupta, Optimal tuning of DC motor via simulated annealing, in International Conference on Advances in Engineering and Technology Research (ICAETR), 2014, pp. 1,5, 1–2 Aug. 2014.

[12] M. Cupelli, L. Zhu, A. Monti, Why ideal constant power loads are not the worst case condition from a control standpoint, IEEE Trans. Smart Grid 6 (6) (2015) 2596–2606.

[13] H. Xin, Modeling and Control of the Multi-terminal MVDC System, *Master Thesis*, Institute for Automation of Complex Power Systems - RWTH Aachen University, 2012.

CHAPTER SEVEN

Hardware In the Loop Implementation and Challenges

7.1 INTRODUCTION

This chapter introduces the implementation of the control strategies on digital controllers and the interaction with real-time digital simulators. The chapter describes some techniques for defining the discretized control algorithms to be performed in digital devices, such as FPGAs or microcontrollers.

This section is divided as follows: In the first part the Hardware in the Loop technique is presented, giving a general description. Thereafter, the discretization technique of the controllers is presented.

7.2 THE HARDWARE IN THE LOOP

The Hardware in the Loop (HiL) simulation is a technique that consists of replacing the physical system with its mathematical representation. The mathematical model is simulated in a real-time simulator where the measurements of the system are converted into physical signals that are sent towards the hardware device to be tested.

This technique presents the benefit of running tests that could be dangerous for the plant and gives the possibility of studying systems before developing a real one, in order to evaluate possible errors and find some changes to improve the quality of the overall system.

In the Electrical Power System the HiL can be used, as an example, for testing digital controllers, the integration of Distributed Generators on the grid and monitoring platforms.

Modern Control of DC-Based Power Systems.
DOI: https://doi.org/10.1016/B978-0-12-813220-3.00007-7

Ultimately, the advantages of the Hardware in the Loop can be summarized by these points:

- Increased Safety
- Enhanced Quality
- Reduced Time.

In the general configuration, the HiL is composed by a digital simulator, which performs in real time the simulation of the electrical power system, the digital controllers, and the interfaces between the digital simulator and the hardware devices.

In the simulation test, the HiL testing platform is based on the following components:

- A the real-time computer target which is a PC running an operating system provided by National Instruments executing graphical user interface with the possibility of changing the control parameters during runtime.
- The FPGA target, a Xilinx Virtex5 LX-30 FPGA board, where the control strategies are implemented.
- The Real-Time Digital Simulator (RTDS) and the the analog input/output block which connects the FPGA with RTDS. Simulation data and control signals are exchanged through the analog signal interfaces of RTDS (GTAI and GTAO) and the FPGAs. The RTDS performs the real-time simulation of the whole MVDC Ship Power System.

The HiL setup previously described is shown in Fig. 7.1.

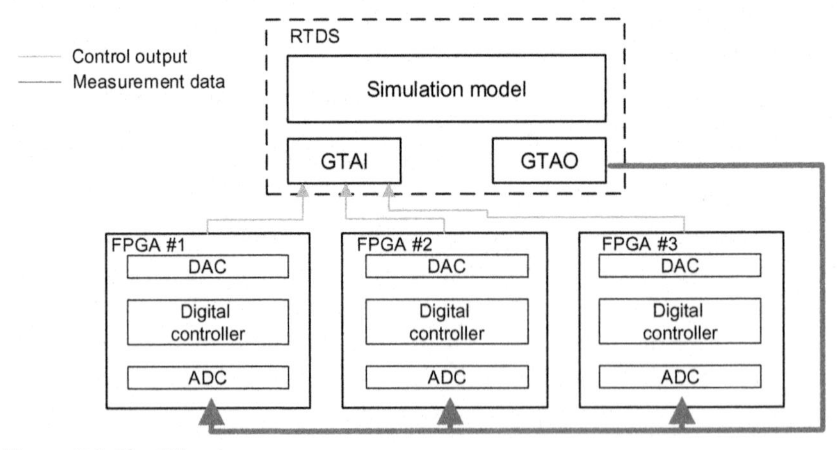

Figure 7.1 The HiL setup.

7.3 DISCRETIZATION

The result of each time-step of the real time simulator can be used by the digital controller to calculate the control output. These measurements can be, generally, analog or digital signals. In both cases, the digital controller samples the measurement signals and calculates the control output that is sent to the digital simulator, forming the HiL setup.

Considering a continuous-time system in the form:

$$\dot{x} = f(x(t), u(t)) \tag{7.1}$$

where the input u is mapped into the state x by means of $\Sigma: u \rightarrow x$, which represents the plant.

With the definition of the time sample T, the continuous-time system can originate the equivalent discete-time system, by using the samples obtained every T seconds. The resulting system has the following form [1]:

$$x(k+1) = f(x(k), u(k)) \tag{7.2}$$

The discretization process is described in Fig. 7.2, consisting of the blocks H, Σ, S, C.

- The block S, representing the *sampler*, produces the discrete output sequence of samples $x(kT) = x(k)$, obtained from the input x. The sampling is generally performed with regular intervals, determined by the *sampling period* T. As it can be observed from Fig. 7.2, the block S

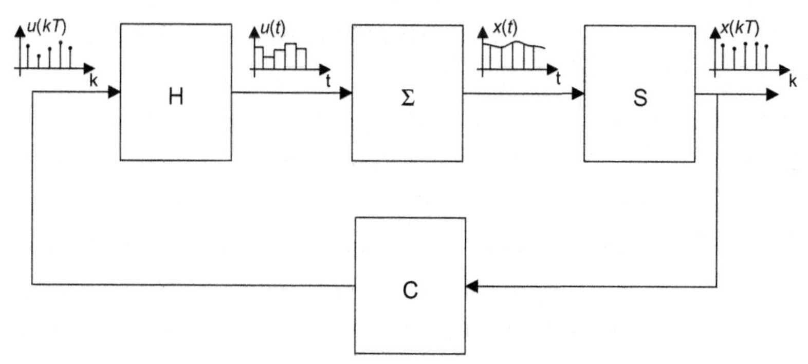

Figure 7.2 The discretization process.

transforms the continuous-time signal in a sequence of points separated by time intervals. Is it clear that the block represents the operation of the analog-to-digital converter.

- The block C represents the digital controller that calculates the control output $u(k)$ based on the samples $x(k)$ and on the specific control algorithm that uses the samples of the state. The control algorithm, usually defined in the continuous-time, must be discretized as well in order to be implemented in the digital controller.
- The block H represents the *hold*. It converts the digital control signal sent by block C into a continuous-time signal that can be used in the plant. The block holds the sample at time kT until the next sample is available. The result is a piecewise constant continuous-time signal $u(t)$, as described in Fig. 7.2. This represents the mathematical representation of the digital-to-analog converter.
- The block Σ represents the plant, which maps the control input $u(t)$ to the state $x(t)$.

The calculation of the discrete-time system is different between the linear time-invariant (LTI) system and the more generic nonlinear system [1].

For the first case, the discretization follows a well-defined procedure that is based on the exact calculation of the linear differential equation. The LTI plant is described by the following equation:

$$\dot{x} = f(x, u) = Ax + Bu \tag{7.3}$$

The differential Eq. (7.3) is solved with the initial condition $x(kT)$ at time $t(0) = kT$ and the input constant signal $u(kT)$.

The solution of Eq. (7.3) results in:

$$x(t) = e^{A(t-t_0)}x(t_0) + \int_{t_0}^{t} e^{A(t-\tau)}Bu(\tau)\mathrm{d}\tau \tag{7.4}$$

The continuous-time variables are substituted with their equivalent in the discrete form:

$$\begin{cases} x(t) = x((k+1)T) \\ x(t_0) = x(kT) \\ t = (k+1)T \\ t_0 = kT \\ u(\tau) = u(kT) \end{cases} \tag{7.5}$$

The discrete form of the LTI plant is obtained as:

$$\begin{aligned} x((k+1)T) &= e^{AT}x(kT) \\ &+ \int_{kT}^{(k+1)T} e^{A((k+1)T-\tau)}\mathrm{d}\tau Bu(kT) \\ &= e^{AT}x(kT) + \int_0^T e^{A\tau}\mathrm{d}\tau Bu(kT) \end{aligned} \tag{7.6}$$

Therefore, the Eq. (7.6) allows the adoption of the discrete-time system description:

$$x(k+1) = Fx(k) + Gu(k) \tag{7.7}$$

The procedure for obtaining the discrete system can be performed in Matlab by means of the command *c2d* [2].

In the case of nonlinear system Σ, the exact discrete model is usually impossible to find, given that the nonlinear differential equation is almost impossible to solve. Therefore, the discrete nonlinear model can be obtained only by applying an approximation. Among the different methods, which have different degree of approximation, the simpler and more popular method is the Euler approximation [1]. This is based on the fact that the time derivative of a variable x can be written as:

$$\dot{x} = \frac{dx}{dt} = \lim_{\Delta T \to 0} \frac{x(t+\Delta T) - x(t)}{\Delta T} \approx \frac{x(t+T) - x(t)}{T} \tag{7.8}$$

The application of (7.8) in the differential Eq. (7.1) results in:

$$x(k+1) \approx x(k) + T \cdot f(x(k), u(k)) \tag{7.9}$$

The Euler approximation is a general method that can be applied also in the case of a linear system. A general rule is that the time constant T must be small enough not to bring the system to instability.

The Euler approximation can be applied to the nonlinear controllers previously described, which means applying (7.9) to the control output calculated in continuous time.

In the case of using the LQR or the LQG control strategies, the standard control theory describes the procedure for obtaining the equivalent discrete form of the two controllers, based on the calculation of the discrete form of the Riccati Equation [3]. Matlab integrates these functionalities with the functions *lqrd* [4] and *kalmd* [5] and *lqg* [6], if the system is already in the discrete form.

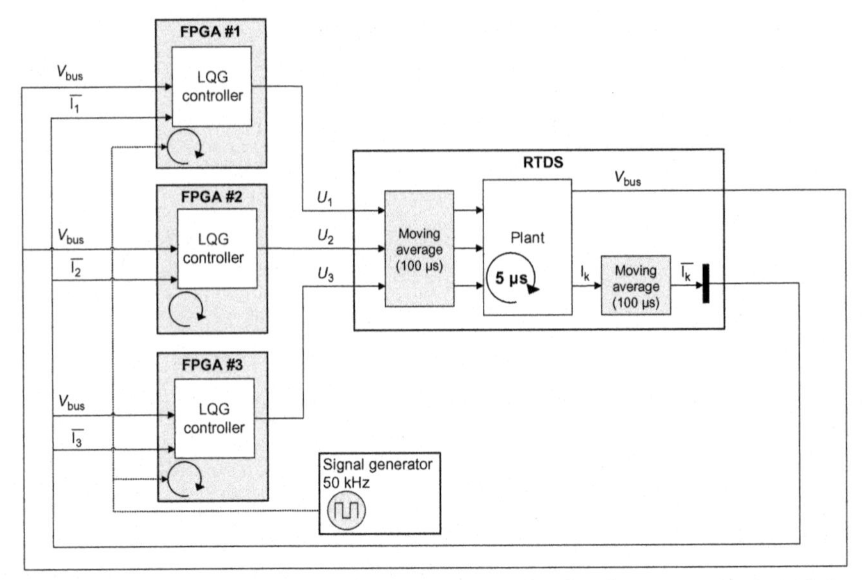

Figure 7.3 The implementation of the LQG controller for the HiL simulation of the MVDC ISPS.

For the LSF algorithm, the desired closed-loop poles are also discretized using the exponential matrix. Using the *place* command in Matlab the discrete state-feedback gains are calculated by means of the Ackermann Pole-Placement Technique using the discrete open-loop system equations and desired closed-loop discrete poles.

As an example of a possible application of one of the proposed control algorithms in a digital controller, the setup and implementation of the LQG controller is described in Fig. 7.3.

The controllers implemented in the FPGAs are synchronized by means of a signal generator that generates a square signal of 50 kHz. The measurement signal of the voltage is firstly filtered with a notch filter and used to calculate the estimate augmented vector. The nominal set-point generator and the LQR control make use of result of the Kalman filter calculation to calculate the control output, as described in Fig. 7.4.

As described in Fig. 7.3, the PWM modulation is performed in RTDS, in order to calculate the sole control strategy inside the physical hardware. The input into the PWM is averaged over 100 μs.

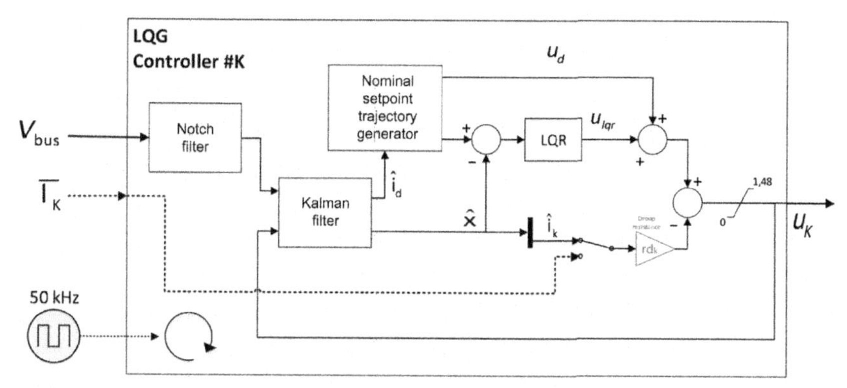

Figure 7.4 The implementation of the LQG controller in the FPGA.

Table 7.1 System Parameters—Even Power Sharing

	P_n (MW)	V_{in} (kV)	V_{out} (kV)	f_s (kHz)	R_f (mΩ)	L_f (mH)	C_f (μF)	C_{if} (μF)	P_{Linear} (MW)
$LRC_{1,2,3}$	20	8.9	5.85	1	99.7	2.1	659.72	–	
POL	50	5.85	3	1	33.2	1.1	2000	3608	1,56

7.4 SIMULATION CASE SHIPBOARD POWER SYSTEM

In this section, the results of the HiL implementation of LSF and LQG control and the validation of its effectiveness in voltage stabilization under large nonlinear load and generation parameters variations are shown. In all depicted figures the sampled values are colored blue and an averaging performed over ten samples is colored in red. The test cases comprise two scenarios, namely, even and uneven power sharing between the LRC. In each scenario an increase in load or a reduction in the installed generation capacity is performed to test the controllers' performance.

For testing the capability of the control algorithm to maintain a stiff DC bus voltage in the event of a sudden load increase, the following test case is performed. The base load condition is 20 MW, and then a load step of 25 MW is added to the system, which corresponds to a load factor increment from 0.33 to 0.75.

In this scenario the load is shared evenly between the LRC and the system data is given in Table 7.1.

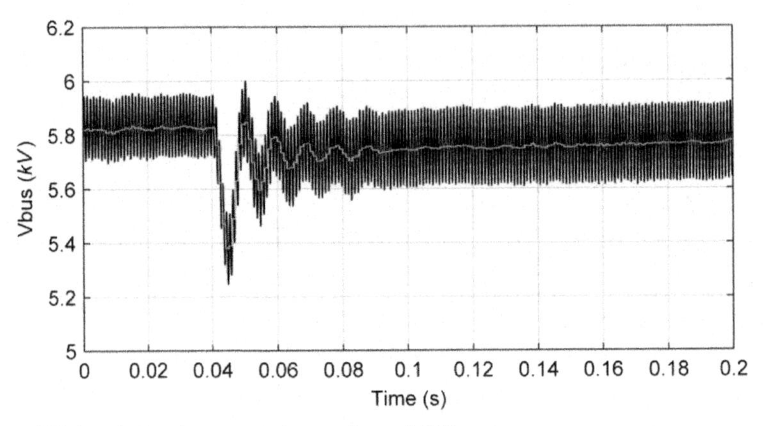

Figure 7.5 Load step increase—bus voltage (LQG).

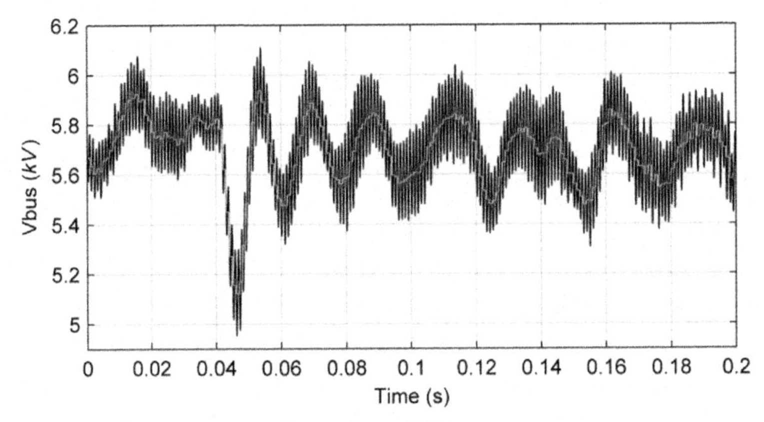

Figure 7.6 Load step increase—bus voltage (LSF).

The transient of the bus voltage under LQG control is shown in Fig. 7.5, whereas it is displayed in Fig. 7.6 for the LSF control. The control is able in both cases to stabilize the DC bus voltage. One important remark is the bus voltage level, which decreased from 6 to 5.85 kV, due to the droop resistance introduced for power sharing reasons. In Fig. 7.5 the LQG controller response shows a steady-state error, no integral action was implemented given the FPGA size constraints. The ripple in the bus voltage is 0.021 p.u.

In Fig. 7.6 for the LSF control a base oscillation can be observed similar to the cascaded system case with a POL of 3 kHz, although this time those oscillations are caused by the ripple induced by the other two LRC on the bus and therefore only allowing a partial compensation of the

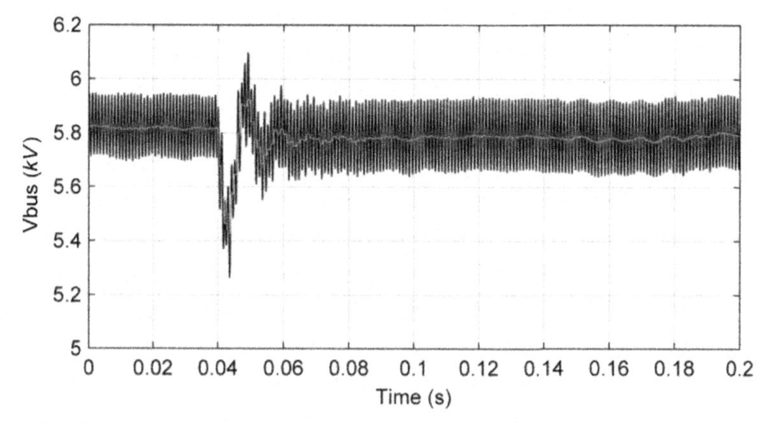

Figure 7.7 Converter disconnection—bus voltage (LQG).

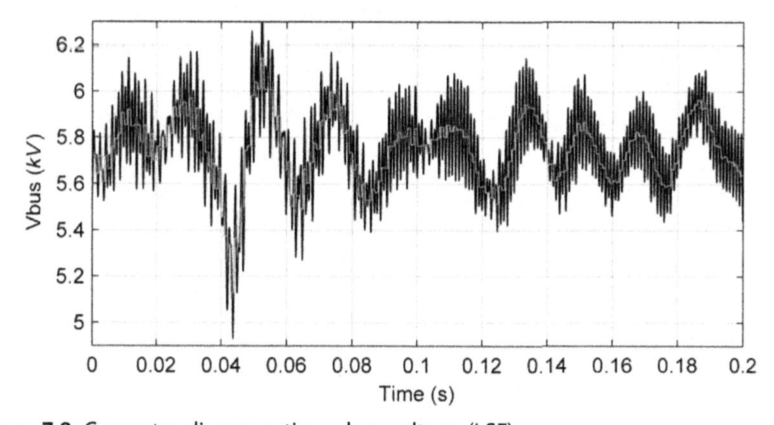

Figure 7.8 Converter disconnection—bus voltage (LSF).

nonlinearity. The oscillation which can be observed in Fig. 7.6 remains in steady state but is bounded to ± 0.1425 kV which is equal to 2.5% of the steady-state averaged value. The system under the control of the LSF is consequently bounded-input, bounded-output (BIBO) stable.

The second test case investigates the capability of the control algorithm to maintain a stiff DC bus voltage in the event of a sudden loss of generation capability. The test case starts from a base load of 25 MW while 60 MW of installed generation is present, at the instant of $t = 0.04$ s LRC_3 is disconnected, which reduces the available generation power to 40 MW.

The transient of the bus voltage under LQG control is shown in Fig. 7.7, whereas for the LSF control it is displayed in Fig. 7.8 during the disconnection of LRC_3. In Fig. 7.7 it is seen that the response has a faster settling time and the peak to peak oscillation is larger than in the load connection case. The reason for the faster settling time lies in the load factor which is 0.5, smaller compared to 0.66 load factor in the step increase test case.

The LSF exhibits, in Fig. 7.8, a transient behavior which is less smooth than during the step increase. The reason is that the compensation and linearization of the nonlinearity was split over three converters but only two converters need to fulfill the task without an update of their compensation coefficients. This shows a remarkable result in term of robustness that such a drastic change from the model assumption can be withstood by the LSF. Similar to Fig. 7.6, also in Fig. 7.8 an oscillation can be detected that remains in steady state but is bounded to ± 0.2280 kV which is equal to 3.986% of the averaged steady-state value. The system under the control of the LSF is consequently BIBO stable.

The experimental results can be summarized in the following way: The corresponding transient response parameters are presented and summarized in Table 7.2 considering that:

- Settling time, over-, and undershoot values are based on switched quantities.
- Overshoot percentages are calculated based on mean values of steady-state voltages.
- Each controller has a different base voltage, which is due to the involved droop sharing difference.

Table 7.2 Summary of Shipboard Power System Scenario

	Scenario 1			
Power sharing (%)	33/33/33			
Generation power (MW)	60			
Base load (MW)	20			
	LQG		**LSF**	
Mean (kV)	5.85		5.72	
Load step (MW)	+25	–	+25	–
Generation loss (MW)	–	−20	–	−20
Settling time (ms) (5%)	15.9	5.6	30.9	36.3
Undershoot (%)	9.87	9.58	13.4	13.7
Overshoot (%)	3.10	4.77	6.85	11.8

REFERENCES

[1] H.J. Marquez, Nonlinear Control Systems: Analysis and Design, John Wiley, Hoboken, 2003.
[2] MathWorks, 2017, Convert model from continuous to discrete time, accessed 17 October 2017 available at https://de.mathworks.com/help/control/ref/c2d.html.
[3] R.F. Stengel, Optimal control and estimation. Courier Corporation, 2012.
[4] MathWorks, 2017, Design discrete linear-quadratic (LQ) regulator for continuous plant, accessed 17 October 2017 available at https://de.mathworks.com/help/control/ref/lqrd.html.
[5] MathWorks, 2017, Design discrete Kalman estimator for continuous plant, accessed 17 October 2017 available at https://de.mathworks.com/help/control/ref/kalmd.html.
[6] MathWorks, 2017, Linear-Quadratic-Gaussian (LQG) design, accessed 17 October 2017 available at https://de.mathworks.com/help/control/ref/lqg.html.

FURTHER READING

Kazantzis and Kravaris, 1999 N. Kazantzis, C. Kravaris, Time-discretization of nonlinear control systems via Taylor methods, Comp. Chem. Eng. 23 (6) (1999) 763–784.

INDEX

Note: Page numbers followed by "*f*" and "*t*" refer to figures and tables, respectively.

K

L

M

R

S

W

X

Printed in the United States
By Bookmasters